Sensors and Systems for Indoor Positioning

Sensors and Systems for Indoor Positioning

Editors

Riccardo Carotenuto
Massimo Merenda
Demetrio Iero

MDPI • Basel • Beijing • Wuhan • Barcelona • Belgrade • Manchester • Tokyo • Cluj • Tianjin

Editors

Riccardo Carotenuto
DIIES
Universitá Mediterranea di
Reggio Calabria
Reggio Calabria
Italy

Massimo Merenda
Competence Unit
Cooperative Digital
Technologies
Austrian Institute of
Technology
Wien
Austria

Demetrio Iero
DIIES
Universitá Mediterranea di
Reggio Calabria
Reggio Calabria
Italy

Editorial Office
MDPI
St. Alban-Anlage 66
4052 Basel, Switzerland

This is a reprint of articles from the Special Issue published online in the open access journal *Sensors* (ISSN 1424-8220) (available at: www.mdpi.com/journal/sensors/special_issues/sensors_systems_indoor_positioning).

For citation purposes, cite each article independently as indicated on the article page online and as indicated below:

LastName, A.A.; LastName, B.B.; LastName, C.C. Article Title. *Journal Name* **Year**, *Volume Number*, Page Range.

ISBN 978-3-0365-4370-3 (Hbk)
ISBN 978-3-0365-4369-7 (PDF)

Contents

About the Editors

Riccardo Carotenuto

Riccardo Carotenuto was born in Rome, Italy. He received his Dr. Sc. degree in Electronic Engineering and PhD from the University "La Sapienza"of Rome, Rome, Italy. He has been an Associate Professor of Electronics with the University "Mediterranea"of Reggio Calabria, Reggio Calabria Italy, since 2002. His main interests include power conversion, energy harvesting, indoor localization, ultrasound imaging, ultrasound actuators, neural networks theory and applications. He authored or co-authored more than 120 papers published in international journals and conferences proceedings.

Massimo Merenda

Massimo Merenda received his Bachelor's, Master's and PhD degrees in electronic engineering from the "Mediterranea"University of Reggio Calabria (UNIRC), Italy, in 2002, 2005 and 2009, respectively. From 2003 to 2005, he was a fellow at the Institute of Microelectronics and Microsystems of the National Research Council IMM-CNR, Naples, Italy. From 2011 to 2018, he was a PostDoc researcher at UNIRC. He was a Researcher at the DIIES Department of UNIRC, and CNIT, from 2018 to 2021. He is currently a Senior Scientist at the Cooperative Digital Technologies Competence Unit of the Austrian Institute of Technology (AIT), in Vienna. He is a member of the AIOTI alliance and of the COST action CA20120 INTERACT. In the past, he explored the design of CMOS integrated circuits, RFID, silicon sensors, embedded systems, energy harvesting and the Internet of Things (IoT). He is currently working in the research field of edge computing for applications of the Internet of Conscious Things and beyond.

Demetrio Iero

Demetrio Iero was born in Reggio Calabria, Italy, in 1982. He received his master's degree in Electronic Engineering from the University "Mediterranea"of Reggio Calabria, Reggio Calabria, Italy in 2010, and PhD in 2014. He is now a temporary Researcher with the DIIES department of the University "Mediterranea"of Reggio Calabria. His main research activities include power electronics and switching power loss measurement, microcontrollers, IoT, and RFID platforms.

Preface to "Sensors and Systems for Indoor Positioning"

There is an increasing interest in indoor positioning, which is an emerging technology with a wide range of applications. Accurate and real-time positioning enables augmented and mixed-reality applications, human–machine and home automation gestural interfaces, and navigation in shopping centers. Relevant applications include robotics, acquiring the position of flexible arms, the navigation of unmanned automatic vehicles, security, the virtual fencing of sensitive locations, safety, and preventing accidents through the recognition of dangerous postures and positions in workers. Further fields of application include medicine, such as monitoring elderly people's movements or rehabilitative exercises; logistics, such as the positioning of goods in warehouses; and sport, such as monitoring body and limb position during training exercises and in game consoles.

This reprint contains the articles that appeared in {Sensors'} (MDPI) Special Issue on "Sensors and Systems for Indoor Positioning". The published original contributions focused on systems and technologies to enable indoor applications.

Riccardo Carotenuto, Massimo Merenda, and Demetrio Iero

Editors

sensors

Editorial

Advanced Sensors and Systems Technologies for Indoor Positioning

Riccardo Carotenuto [1,*], **Demetrio Iero** [1] **and Massimo Merenda** [2]

1. Department of Information Engineering Infrastructures and Sustainable Energy (DIIES), "Mediterranea" University, 89122 Reggio Calabria, Italy; demetrio.iero@unirc.it
2. Center of Digital Safety & Security, AIT Austrian Institute of Technology GmbH, 1210 Vienna, Austria; massimo.merenda@ait.ac.at
* Correspondence: r.carotenuto@unirc.it

1. Introduction

There is an increasing interest about indoor positioning, which is an emerging technology with a wide range of applications. Accurate and real-time positioning enables augmented and mixed reality applications, human–machine and home automation gestural interfaces, and navigation in shopping centers. Relevant applications include robotics, acquiring the position of flexible arms, navigation of unmanned automatic vehicles, security, virtual fencing of sensitive locations, safety, and preventing accidents through the recognition of dangerous postures and positions in workers. Further fields of application include medicine, such as monitoring elderly people's movements or rehabilitative exercises; logistics, such as the positioning of goods in warehouses; sport, such as monitoring body and limb position during training exercises and in game consoles.

Currently, research effort needs to be directed to new algorithms, architectures, sensor technologies, coverage, power consumption, size, and increased spatial and temporal resolution of indoor positioning systems based on the physical and economic constraints of various applications. In this framework, we are glad to edit this Special Issue on "Sensors and Systems for Indoor Positioning". Original contributions focused on systems and technologies to enable the indoor applications listed above are welcome.

There are many challenges in this area that need to be solved or improved. Research effort needs to be directed to new algorithms, architectures, sensor technologies, coverage, power consumption, size, and increased spatial and temporal resolution of indoor positioning systems, based on the physical and economic constraints of the various applications. In this outline, the Special Issue on "Sensors and Systems for Indoor Positioning" of the *Sensors* journal seeks to explore original contribution on systems and technologies to enable the indoor applications listed above are welcome.

From several received manuscripts, eleven original and high-quality papers were selected to be included in this Special Issue, each one reviewed by multiple expert reviewers and passed through several rounds of peer review.

2. Relevant Contributions

In [1], a carrier phase technology in wireless orthogonal frequency division multiplex (OFDM) systems is applied to improve ranging and positioning accuracy. Carrier phase measurement is a ranging technique that uses the phase difference between the received signal and the transmitted signal. Compared with positioning systems using only time of arrival (TOA), carrier phase information has a higher resolution and is more accurate, providing indoor high-precision positioning. Carrier phase ranging is widely used in global navigation satellite systems (GNSS) systems but is not yet commonly used in OFDM systems. Applying this technology can significantly improve positioning accuracy. However, using the OFDM carrier phase has two problems that the authors intend to

Citation: Carotenuto, R.; Iero, D.; Merenda, M. Advanced Sensors and Systems Technologies for Indoor Positioning. *Sensors* **2022**, *22*, 3605. https://doi.org/10.3390/s22103605

Received: 5 May 2022
Accepted: 6 May 2022
Published: 10 May 2022

Publisher's Note: MDPI stays neutral with regard to jurisdictional claims in published maps and institutional affiliations.

solve: (1) phase measurements in multipath environments and non-line-of-sight (NLOS) propagation; (2) integer ambiguity resolution in real-time positioning applications.

The paper presents a ranging scheme based on the carrier phase in a multipath environment; it analyzes the effect of multipath propagation on phase measurement and reports a correlation profile-based carrier phase measurement method. An extended Kalman filter (EKF) algorithm is also presented to estimate the integer ambiguity by SD carrier and TDOA measurements. The algorithm also considers the effect of NLOS error and mitigation efforts. Simulation shows that the proposed algorithm quickly solves the integer ambiguity even when NLOS errors occur. The carrier phase measurements combined with the accurately valued integer ambiguity led to a positioning error below 30 cm for 90% of the terminals.

The effective implementation of a UHF-RFID Smart Gate, an identification point placed at warehouse key points for forklift monitoring, is presented in [2]. The system is part of the I-READ 4.0 project aimed at developing an integrated and autonomous Cyber-Physical System for the automatic management of large warehouses with a high-stock rotation. The management of assets and forklifts is possible thanks to a network of Radio Frequency Identification (RFID) readers operating in the Ultra-High-Frequency (UHF) band.

The UHF-RFID Smart Gate consists of a checkpoint infrastructure based on RFID technology to identify forklifts and their direction of transit. The authors present an implementation with a single reader antenna, thus reducing infrastructure complexity and cost. The action classification method exploits the signal phase backscattered by RFID tags placed on forklifts, allowing the classification of two movements (entering or leaving a certain area) that the forklift can perform with respect to the gate.

The proposed system does not require calibration procedures, it does not need long computational times, and it can be implemented with commercial-off-the-shelf (COTS) components. The performance and the method capabilities were demonstrated in a real warehouse, and in 100% of cases, the forklift was correctly detected, and a 98% classification accuracy was achieved when the forklift speed ranged between 0.5 m/s and 1.5 m/s.

In [3], a new technique was presented to measure the distance between an emitter and a receiver, which is based on the different attenuation levels that ultrasonic signals of different frequencies undergo when propagating in the air.

Distance measurements are usually performed by measuring the Time of Flight (TOF) of an ultrasonic signal traveling from an emitter to receiving sensors. However, this requires close synchronization between the emitter and the sensors. This synchronization is usually conducted using a radio or optical channel, which requires additional hardware complexity, while for many applications, low-cost small lightweight sensors are required.

Intending to reduce the complexity of the measurement process and of the sensors, the paper proposes an innovative technique that measures the distance between emitter and receiver from the amount of attenuation suffered by signals emitted at different frequencies, without the need for any synchronization between them.

Simulation results showed that, using a 0.5 mm diameter emitter aperture, a ranging error of less than 2.75 cm and a mean error of 1.25 cm can be achieved. The technique does not reach the level of accuracy of other techniques but works in the absence of synchronization without limits on the distance measurement rate, with an unlimited number of sensors using the same emitter and with reduced computational power and device dimensions.

In [4], the authors compare two methods for the acoustic indoor localization of persons based on the time difference of arrival of the first-order reflection to interpret the returned signals in a small office room. They draw the approach from bats which can perceive the incoming reflected wave's direction. The first method is Direct Intersection, which determines a coordinate point based on the intersection of spheroids defined by observed distances of high-intensity reverberations. The second method, Sonogram analysis, overlays all channels' room impulse responses to generate an intensity map for the observed environment.

The authors investigate the two algorithms and both approaches yield mean distance localization errors ranging between 0.3 and 0.9 m. Direct Intersection shows a higher precision, while the Sonogram Estimation method provides more accurate results. Moreover, the former method has a lower computational cost and performs faster with comparable precision and accuracy.

In [5], a deep learning solution involving a clustering processing scheme in a fingerprint indoor positioning system was developed. Wi-Fi fingerprint-based positioning systems have a simple layout and a low cost; however, the multipath propagation of signals caused by obstacles, interference of moving objects, and changes in Wi-Fi APs affect the positioning accuracy based on a received signal strength indicator (RSSI) with traditional dataset and a deep learning classifier. To overcome this issue, the authors propose a clustering-based noise elimination scheme (CNES) for RSSI-based datasets, in which the dataset is preprocessed and noise samples are removed.

Experiments carried out in a dynamic environment showed that applying CNES to the test database will increase the average positioning accuracy up to 22.4%, archiving a positioning accuracy of 90.4%, which is much higher than the accuracy of the dataset without pre-processing.

A smartphone-based navigation and information service for a University library employing Wi-Fi fingerprinting is developed in [6]. The motivation of this study is to help students, employees, and visitors of the TU Wien University to find the correct bookshelf. The authors carried out a study of the availability, performance, and usability of Wi-Fi in areas of the library using different smartphones in different modes, such as static, kinematic, and stop-and-go, evaluating positioning accuracies in the various modes. The investigations showed that Wi-Fi fingerprinting can be used to achieve positioning accuracies on the meter level. Accuracy can be increased by the installation of additional access points to provide better distribution and geometry for localization and also by deploying additional hardware based on low-cost Raspberry Pi units that broadcast and receive Wi-Fi signals.

In [7], a three-dimensional visible light positioning system with multiple photodiodes and reinforcement learning (RL) is demonstrated. The system can realize accurate 3D positioning without the need of data for offline training. The authors propose and compare experimentally three methods developed to improve the 3D positioning accuracy over a basic 3D positioning model based on the RSSI trilateration without RL.

The experimental results show that the three RL-based methods outperform the basic one, providing higher position accuracy. Among the three methods, the third, which is a combination of the first two, offers the highest positioning accuracy, with an average positioning error of 2.6 cm and at least 20% improvement compared to the basic model.

In [8], the authors propose an indoor localization system based on an infrared angle-of-arrival (AoA) sensor network for accurate and inexpensive real-time. The authors attempt to overcome the disadvantage of state-of-the-art indoor localization systems relying on complex NLOS signal propagation with multiangulation and multilateration methods that have high installation costs, computational demands, and energy requirements. The paper presents a novel sensor utilizing infrared (IR) signal in the line-of-sight (LOS) context using the AoA technique that avoids NLOS propagation issues by exploiting the concept of the wireless sensor network (WSN).

To demonstrate the proposal, a supermarket cart navigation system was realized as a proof-of-concept using an IR-AoA sensor prototype, server-side component, and an application for smartphones and smartwatches. The localization performance ranged from centimeter-level accuracy achieved in a static context to 1 m mean error in a mobile cart context. The implementation demonstrated that inexpensive and easily deployable wireless sensors nodes can be utilized to provide appropriate localization accuracy.

In [9] an adaptive residual weighted K-Nearest neighbor (WKNN) fingerprint positioning algorithm based on visible light is proposed. The WKNN algorithm is a commonly used fingerprint positioning algorithm for which its difficulty consists in the optimization of K to obtain the minimum positioning error. The authors propose an adaptive algorithm in which, initially, the target matches the fingerprints according to the RSSI, and K is a dynamic value according to the matched RSSI residual.

Simulation results show that the proposed algorithm presents a reduced average positioning error when compared with random forest (81.82%), extreme learning machine (83.93%), artificial neural network (86.06%), grid-independent least square (60.15%), self-adaptive WKNN (43.84%), WKNN (47.81%), and KNN (73.36%). Moreover, it achieves a significant reduction in positioning error while maintaining lower algorithm complexity.

In [10], the use of software Field II is proposed to simulate signal aberration and ranging error in ultrasonic indoor positioning applications. Ultrasonic systems have already been demonstrating their effectiveness in achieving high positioning accuracy and refresh rates, but attention must be paid to certain aspects of signal propagation. In this paper, Field II, an acoustical simulation software that is well-established in medical imaging, has been applied to the acoustic field in the air for the evaluation of ranging techniques.

In this study, it is shown how a typical chirp signal used in ultrasonic positioning systems undergoes a shape aberration depending on the shape and size of the transducer and on the angle under which the transducer is seen by the receiver. Such signal shape aberrations produce results affected by a much greater error than expected. The spatial distributions of the ranging error are provided, showing favorable low error regions. The work also demonstrates that particular attention must be paid to the design of the acoustic section of the ultrasonic positioning systems, considering both the shape and size of the ultrasonic emitters and the shape of the acoustic signal used.

In perspective, the advantages of the proposed approach are the possibility of examinations, while in the design phase, advantages include the acoustic field over time in the region of interest as a function of the aperture and the type of signal emitted and the capability to easily test several algorithms in different operating situations.

In [11], a study on a recursive algorithm for indoor positioning using pulse-echo ultrasonic signals was investigated. Ultrasounds are widely used for real-time applications in short-range communication systems and one of the parameters widely used is TOF, which can be evaluated by using different techniques. In the paper, a nonstandard cross-correlation method is investigated for TOF estimation, with a procedure based on the use of template signals to improve the accuracy of recursive TOF evaluations.

Experimental results were compared with both the standard threshold and cross-correlation techniques, showing an average improvement of 30% and 19% in terms of standard error, and an enhancement in repeatability of about 10%. However, an increase of 70% in computational load has been estimated in the evaluation of TOF.

Funding: This research received no external funding.

Acknowledgments: We would like to thank all authors for their valuable collaboration and contributions to this Special Issue and the reviewers for their hard work during the review process.

Conflicts of Interest: The authors declare no conflict of interest.

References

1. Zhang, Z.; Kang, S.; Zhang, X. Indoor Carrier Phase Positioning Technology Based on OFDM System. *Sensors* **2021**, *21*, 6731. [CrossRef] [PubMed]
2. Motroni, A.; Buffi, A.; Nepa, P.; Pesi, M.; Congi, A. An Action Classification Method for Forklift Monitoring in Industry 4.0 Scenarios. *Sensors* **2021**, *21*, 5183. [CrossRef] [PubMed]
3. Carotenuto, R.; Pezzimenti, F.; Della Corte, F.G.; Iero, D.; Merenda, M. Ranging with Frequency Dependent Ultrasound Air Attenuation. *Sensors* **2021**, *21*, 4963. [CrossRef] [PubMed]
4. Schott, D.J.; Saphala, A.; Fischer, G.; Xiong, W.; Gabbrielli, A.; Bordoy, J.; Höflinger, F.; Fischer, K.; Schindelhauer, C.; Rupitsch, S.J. Comparison of Direct Intersection and Sonogram Methods for Acoustic Indoor Localization of Persons. *Sensors* **2021**, *21*, 4465. [CrossRef] [PubMed]
5. Liu, S.; Sinha, R.S.; Hwang, S.-H. Clustering-Based Noise Elimination Scheme for Data Pre-Processing for Deep Learning Classifier in Fingerprint Indoor Positioning System. *Sensors* **2021**, *21*, 4349. [CrossRef] [PubMed]

6. Retscher, G.; Leb, A. Development of a Smartphone-Based University Library Navigation and Information Service Employing Wi-Fi Location Fingerprinting. *Sensors* **2021**, *21*, 432. [CrossRef]
7. Zhang, Z.; Chen, H.; Zeng, W.; Cao, X.; Hong, X.; Chen, J. Demonstration of Three-Dimensional Indoor Visible Light Positioning with Multiple Photodiodes and Reinforcement Learning. *Sensors* **2020**, *20*, 6470. [CrossRef]
8. Arbula, D.; Ljubic, S. Indoor Localization Based on Infrared Angle of Arrival Sensor Network. *Sensors* **2020**, *20*, 6278. [CrossRef] [PubMed]
9. Xu, S.; Chen, C.-C.; Wu, Y.; Wang, X.; Wei, F. Adaptive Residual Weighted K-Nearest Neighbor Fingerprint Positioning Algorithm Based on Visible Light Communication. *Sensors* **2020**, *20*, 4432. [CrossRef] [PubMed]
10. Carotenuto, R.; Merenda, M.; Iero, D.; Della Corte, F.G. Simulating Signal Aberration and Ranging Error for Ultrasonic Indoor Positioning. *Sensors* **2020**, *20*, 3548. [CrossRef] [PubMed]
11. Pullano, S.A.; Bianco, M.G.; Critello, D.C.; Menniti, M.; La Gatta, A.; Fiorillo, A.S. A Recursive Algorithm for Indoor Positioning Using Pulse-Echo Ultrasonic Signals. *Sensors* **2020**, *20*, 5042. [CrossRef] [PubMed]

Article

Indoor Carrier Phase Positioning Technology Based on OFDM System

Zhenyu Zhang [1,2,*], Shaoli Kang [2] and Xiang Zhang [2]

[1] School of Electronic and Information Engineering, Beihang University, Beijing 100191, China
[2] State Key Laboratory of Wireless Mobile Communications, China Academy of Telecommunications Technology, Beijing 100083, China; kangshaoli@catt.cn (S.K.); zhangxiang4@datangmobile.cn (X.Z.)
* Correspondence: zhangzhenyu@buaa.edu.cn

Abstract: Carrier phase measurement is a ranging technique that uses the receiver to determine the phase difference between the received signal and the transmitted signal. Carrier phase ranging has a high resolution; thus, it is an important research direction for high precision positioning. It is widely used in global navigation satellite systems (GNSS) systems but is not yet commonly used in wireless orthogonal frequency division multiplex (OFDM) systems. Applying carrier phase technology to OFDM systems can significantly improve positioning accuracy. Like GNSS carrier phase positioning, using the OFDM carrier phase for positioning has the following two problems. First, multipath and non-line-of-sight (NLOS) propagation have severe effects on carrier phase measurements. Secondly, ambiguity resolution is also a primary issue in the carrier phase positioning. This paper presents a ranging scheme based on the carrier phase in a multipath environment. Moreover, an algorithm based on the extended Kalman filter (EKF) is developed for fast integer ambiguity resolution and NLOS error mitigation. The simulation results show that the EKF algorithm proposed in this paper solves the integer ambiguity quickly. Further, the high-resolution carrier phase measurements combined with the accurately estimated integer ambiguity lead to less than 30-centimeter positioning error for 90% of the terminals. In conclusion, the presented methods gain excellent performance, even when NLOS error occur.

Keywords: extended Kalman filter; localization; time of arrival; carrier phase; ambiguity resolution

Citation: Zhang, Z.; Kang , S.; Zhang, X. Indoor Carrier Phase Positioning Technology Based on OFDM System. *Sensors* **2021**, *21*, 6731. https:// doi.org/10.3390/s21206731

Academic Editor: Riccardo Carotenuto

Received: 12 July 2021
Accepted: 6 October 2021
Published: 11 October 2021

Publisher's Note: MDPI stays neutral with regard to jurisdictional claims in published maps and institutional affiliations.

1. Introduction

With the rapid development of industries such as the Internet of Things and industrial control, high-precision indoor positioning technology has become an important issue to be solved. It is challenging to receive valid satellite navigation signals in the indoor environment, and other high-precision positioning technologies need to be studied. In recent years, positioning services based on wireless communications are rapidly developing. The mobile cellular network covers a wide area and is one option for dense urban areas and indoor positioning. Benefit from the advance of 5G technology, high-precision positioning using the wireless access network has become a hot research direction. In wireless networks, traditional ranging-based positioning methods include angle of arrival (AOA), received signal strength (RSS), and time-of-arrival (TOA). Among them, AOA determines location of the user by measuring the angle between the terminal and the base station (BS) [1]. Since measuring angle often requires a sufficient number of antennas at the receiver, the application range of AOA technology is limited. RSS technology needs to establish an accurate signal energy propagation model, making it challenging to achieve high measurement accuracy [2]. TOA-based positioning technology converts arrival time to a distance and then uses the distance information for positioning. TOA has been widely used due to its low requirements of positioning equipment [3]. This paper mainly studies high-precision TOA measurement and positioning algorithms based on orthogonal frequency division multiplex (OFDM) systems.

TOA estimation can be considered as a channel estimation problem. Many documents carry out channel impulse response (CIR) estimation from the time domain or frequency domain perspective in the OFDM multi-carrier system [4–9]. Typical schemes are cross-correlation algorithms based on the pseudo-randomness of the transmission sequence, including the maximum criterion algorithm and the threshold algorithm[4]. However, this positioning method is limited by the signal bandwidth and receiver resolution, and it is challenging to achieve sub-meter accuracy. Besides, TOA-based parameter estimation techniques have been widely studied, including multiple signal classification algorithm (MUSIC) [10], Signal Parameters via Rotational Invariant Techniques (ESPRIT) [11], and Space-Alternating Generalized Expectation maximization (SAGE) [12] algorithms. These algorithms are not limited by the system sampling rate but are determined by the search interval for time delay estimation. A smaller search interval will effectively improve the accuracy of TOA estimation but will significantly increase the computational complexity. The vast computational overhead makes it difficult to apply such algorithms to real-time user localization scenarios. Other studies have attempted to apply phase ranging techniques to indoor localization [13–15]. However, specific phase emission and measurement devices are challenging to reduce the cost of localization effectively. Phase measurement techniques based on OFDM systems can provide high accuracy positioning measurements while satisfying communication requirements, and therefore are a research direction for phase-based positioning. References [16–18] describe methods for distance measurement through the phase difference of subcarriers in OFDM systems. Still, such schemes are often only suitable for LOS propagation or multipath propagation when the Rice factor is high. Carrier phase localization techniques based on wireless cellular networks have been proposed in the literatures [19,20]. Furthermore, carrier phase measurement in multipath environments and the suppression of non-line-of-sight (NLOS) error require further research.

There are two modes for receivers in global navigation satellite systems (GNSS): pseudorandom code (C/A code or P-code) and the carrier phase [21,22]. The ranging principle of pseudorandom code mode is similar to TOA, and the measurement error of pseudorandom code mode is vast. The carrier phase ranging method uses the carrier phase of the measurement signal to extract the propagation distance information. Under line-of-sight (LOS) conditions, the carrier phase's measurement error is a small fraction of the carrier wavelength and can reach the centimeter range. However, the carrier phase measurement includes unknown integer ambiguity: the distance between the user and the BS in terms of carrier wavelength can be divided into an integral part of the wavelength plus a fractional function. During the initialization of positioning, the phase measurement is in the range $[0, 2\pi]$, so only fractional multiples of the distance can be measured, which causes the problem of integer ambiguity. Once this problem is solved, the carrier phase positioning can meet the requirements of high accuracy.

Inspired by reference [19], the positioning accuracy might be significantly improved if the carrier phase technology can be extended in the indoor location system. Compared with GNSS positioning, wireless networks can work in challenging scenarios and have more flexible carrier frequency configurations, fewer error sources, and more minor path losses. These characteristics constitute the advantages of supporting carrier phase technology in wireless networks. However, despite these advantages, there are many challenges while applying carrier phase positioning in wireless networks: Multipath and NLOS propagation in the indoor environment, fast resolution of integer ambiguities in wireless networks, etc.

In summary, this paper proposes a carrier phase-ranging scheme based on the OFDM system. Under the premise of high accuracy ranging, this paper focuses on two aspects: carrier phase measurement in a multipath environment and how to solve the integer ambiguity quickly and accurately. First, we model the carrier phase measurement in a multipath environment and analyze the integer ambiguity generation. Second, we propose an extended Kalman filter (EKF) for solving the integer ambiguity. The EKF-based algorithm can solve the position of the terminal while solving the integer ambiguity. Further, we describe how to utilize the EKF to mitigate the errors caused by NLOS propagation.

The overall structure of this study contains five chapters. Section 2 describes the model for studying TOA and phase estimation in OFDM systems. In Section 3, we propose an EKF that combines carrier phase and TOA measurements to enhance positioning accuracy and reduce the impact of NLOS error on mobile positioning. Numerical simulation results are presented in Section 4 to prove the effectiveness of the new methods. In Section 5, we summarize conclusions drawn from this paper.

In this paper, vectors and matrices are denoted by boldface lower-case letters and boldface upper-case letters. The superscript $[.]^T$ denotes the transpose. The superscript $[.]^{ij}$ and the subscript $[.]_{ab}$ represent, respectively, the single-difference (SD) between the transmitters and between the receivers.

2. Ranging System

Consider OFDM transmission with N subcarriers, subcarrier spacing Δf_{SCS} and sampling interval $T_S = 1/(N\Delta f_{SCS})$. OFDM transmission is block oriented. Assume N quadrature-amplitude modulation (QAM) symbols $X_k^m, k \in \{1, \ldots, N\}$ are grouped into a vector $\mathbf{X^m} = \left[X_1^m, \ldots, X_N^m\right]^T$ and transmitted in the m-th OFDM symbol in a slot. A unitary inverse discrete-time Fourier transform (IDFT) on $\mathbf{X^m}$ gives a continuous time representation of the complex envelope of an OFDM symbol of duration $T = NT_s = 1/\Delta f_{SCS}$ (note: here T does not include cyclic prefix).

$$x^m(t) = \frac{1}{\sqrt{N}} \sum_{k=1}^{N} X_k^m e^{j2\pi(k-1)t/T}; 0 \leq t \leq T, \tag{1}$$

the time-domain signal $x^m(t)$ is up-converted to the carrier frequency f_c for transmission.

$$\begin{aligned} s^m(t) &= x^m(t)e^{j2\pi f_c t} \\ &= \frac{1}{\sqrt{N}} \sum_{k=1}^{N} X_k^m e^{j2\pi((k-1)/T + f_c)t}; \quad 0 \leq t \leq T. \end{aligned} \tag{2}$$

Assume the channel is the quasi-static channel, i.e., the channel does not change during the transmission of one OFDM symbol, the quasi-static channel can then be described by a time discrete CIR $\mathbf{h} = \left[h_0(t), h_1(t), \ldots, h_{L_p-1}(t)\right]^T$, multipath channel model can be expressed as:

$$h(t, \tau) = \sum_{l=1}^{L_p} h_l(t)\delta(t - \tau_l(t)) + h_d(t, \tau), \tag{3}$$

L_p is the total number of paths which include one LOS path and $L_p - 1$ NOLS paths, $h_l(t)$ is the gain for the l-th path, $\delta(t - \tau_l(t))$ is the Dirac delta function, $\tau_l(t)$ is the TOA of the l-th path, $h_d(t, \tau)$ are the diffuse multipath components (DMC) [23], which represent the non-discrete part of the channel. The received signal after passing through the multipath channel can be expressed as:

$$y^m(t) = s^m(t) \otimes h(t, \tau) + w_n^m = \int_{-\infty}^{\infty} s^m(\xi)h(t - \xi, \tau)d\xi + w^m, \tag{4}$$

w^m is the color noise consisting of w_n^m and $s^m(t) \otimes h_d(t, \tau)$, where $w_n^m \sim N(0, \sigma^2)$ is the complex additive noise with zero mean and σ^2 variance. If the received signal contains color noise, it is necessary to consider the use of whitening filters to convert the color noise to white noise [24]. Furthermore, the received m-th OFDM symbol can be expressed by:

$$
\begin{aligned}
y^m(t) &= \frac{1}{\sqrt{N}} \sum_{k=1}^{N} X_k^m \left[\sum_{l=1}^{L_p} h_l(t) e^{-j2\pi\left(f_c+\frac{k-1}{T}\right)\tau_l(t)} \right] e^{j2\pi\left(f_c+\frac{k-1}{T}\right)t} + w^m \\
&= \frac{1}{\sqrt{N}} \sum_{k=1}^{N} X_k^m \left[\sum_{l=1}^{L_p} h_l(t) e^{-j2\pi\left(f_c+\frac{k-1}{T}\right)\tau_l(t)} + w_k^m \right] e^{j2\pi\left(f_c+\frac{k-1}{T}\right)t}
\end{aligned}
\tag{5}
$$

After down-conversion and removal of the samples of the received signal which belong to the cyclic prefix, the received signal $y^m(t)$ is converted into a discrete time domain signal $y^m[n]$:

$$
y^m[n] = \frac{1}{\sqrt{N}} \sum_{k=1}^{N} X_k^m \left[\sum_{l=1}^{L_p} h_l[nT_s] e^{-j2\pi\left(f_c+\frac{k-1}{T}\right)\tau_l[nT_s]} + w_k^m \right] e^{j2\pi\frac{n(k-1)}{N}}.
\tag{6}
$$

2.1. Conventional Cross-Correlation TOA Estimator

Signal arrival time needs to be obtained from reference signals. In Long Term Evolution (LTE) Release 9, positioning reference signals (PRS) were used to improve TOA-based positioning. PRS are pseudo-random sequences with good autocorrelation. With the help of the autocorrelation characteristics of the PRS sequence, it is easier to find the direct path in the environment of multipath transmission. The cross-correlation expression is:

$$
\begin{aligned}
R_{xy} &= \sum_{n=1}^{N} \overline{x^m[n-\tau]} y^m[n] \\
&= h_1[nT_s] e^{-j2\pi f_c \tau_1[nT_s]} R_m^{xx}[\tau - \tau_1[nT_s]] + \sum_{i=2}^{L} h_i[nT_s] e^{-j2\pi f_c \tau_i[nT_s]} R_m^{xx}[\tau - \tau_i[nT_s]] \\
&+ R_m^{xw}[\tau], \tau \in [1,...N];
\end{aligned}
\tag{7}
$$

where $\overline{(.)}$ denotes the complex conjugate function, $x[n]$ is the replica of the transmitted PRS. Furthermore, based on the autocorrelation of the PRS series, we have:

$$
\begin{aligned}
R_m^{xx}[\tau] &= \sum_{n=1}^{N} \overline{x^m[n-\tau]} x^m[n] = \delta[\tau] \\
R_m^{xw}[\tau] &= \sum_{n=1}^{N} \overline{x^m[n-\tau]} w^m[n] \approx 0
\end{aligned}
\tag{8}
$$

where $w^m[n]$ is the downsampled additive noise. From Equation (7), it can be seen that the magnitude of the correlation function is affected by the carrier phase $2\pi f_c \tau_i[nT_s]$. To exclude this effect, we use $|R_{xy}[\tau]|$ instead of $R_{xy}[\tau]$ for TOA estimation. Taking the threshold method as an example, TOA is determined by estimating the time delay of the first (earliest) peak in the magnitude of the normalized cross-correlation function above a certain threshold [4].

$$
\hat{\tau} = \arg \min_{\tau} \left\{ \frac{|R_{xy}[\tau]|}{\max\{|R_{xy}|\}} \geq \zeta \right\},
\tag{9}
$$

here, ζ is the preset threshold. Correlation profile-based methods can estimate the propagation delay of the first path in a multipath environment. Still, due to the limited sampling rate of the system, the measurement accuracy of this method is low. Rewriting $\hat{\tau}$ to T_r^i and introducing terminal r and the BS i, then the estimated TOA can be modeled as [25]:

$$
T_r^i = (d_r^i + w_{r,T}^i)/c.
\tag{10}
$$

- T_r^i (known) is the TOA measurement from terminal r to BS i (unit:s).
- c is the speed of radio waves in vacuum, 299,792,458 (unit: m/s).
- $d_r^i = \sqrt{(x_i - x)^2 + (y_i - y)^2}$ (unknown) is the geometric distance between the antennas of transmitter i and receiver r (unit: m).
- (x_i, y_i) (known) is the two-dimensional vector giving the coordinates of BS i.
- (x, y) (unknown) is the location of the terminal to be solved.
- $w_{r,T}^i \sim \mathcal{N}\left(0, \sigma_{r,i}^2\right)$ (unknown) is a random Gaussian variable accounting for the residual estimation error (unit: m).

2.2. High-Precision TOA Estimation Scheme Based on Carrier Phase

From Equation (5), combine the known PRS signal, the frequency domain channel response is written as:

$$H^m(k) = \sum_{l=1}^{L_p} h_l[nT_s] e^{-j2\pi \frac{k-1}{N} \dot{\tau}_l - j\phi_l} + w_k^m, \tag{11}$$

where $\dot{\tau}_l = N \Delta f_{SCS} \tau_l[nT_s]$ is the transmission delay in units of sampling interval. $\phi_l = 2\pi f_c \tau_l[nT_s]$ is the phase shift caused by free-space propagation.

As can be seen from Equation (11), the distance between the BS and the terminal is reflected in each subcarrier phase. However, due to the signal aliasing of multiple transmission paths, it is difficult to directly estimate phase information of the first path from the unprocessed subcarrier phase. Therefore, we convert the frequency domain channel response to the time domain for further analysis. Furthermore, when the distance (in units of sampling interval) is not an integer multiple of the sampling interval, the time domain channel response is subject to energy leakage [26]:

$$h_n^m = \frac{\sin(\pi \dot{\tau}_l)}{\sqrt{N} \sin\left(\frac{\pi}{N}(\dot{\tau}_l - n)\right)} \sum_{L_p} h_l[nT_s] e^{-j\frac{\pi}{N}(n+(N-1)\dot{\tau}_l) - j\phi_l}. \tag{12}$$

Much of the literature [27,28] describes using Equation (7) or (9) to find the integer multiple sampling points closest to the transmission delay, denoted as $[\dot{\tau}_1]$, and $[.]$ is a rounding function. The time domain signal is processed to eliminate the effects of multipath effects. The window can be expressed as: for $n \in [[\dot{\tau}_1] - \frac{W}{2}, [\dot{\tau}_1] + \frac{W}{2}]$, $\tilde{h}_n^m = h_n^m$, else $\tilde{h}_n^m = 0$. Furthermore, W is the length of the window.

We use the tapped delay line model to characterize the frequency-selective channel, and each tap represents a different channel delay in units of the sampling interval of the receiver. Figure 1 shows a schematic of the window at $W = 0$. Based on the correlation of the transmit sequence, the terminal can determine the arrival delay of the direct path. The manipulation of the power delay profile further eliminates the effects of multipath. Additionally, it is worth noting that actual distance to the BS in this example is 12.4 sampling intervals. Due to the limitation of the system sampling rate, the TOA obtained by the cross-correlation algorithm is 12 sampling intervals, which generates a significant measurement error. Furthermore, as shown in the figure, the distance is a non-integer multiple of the sampling interval resulting in leakage between taps.

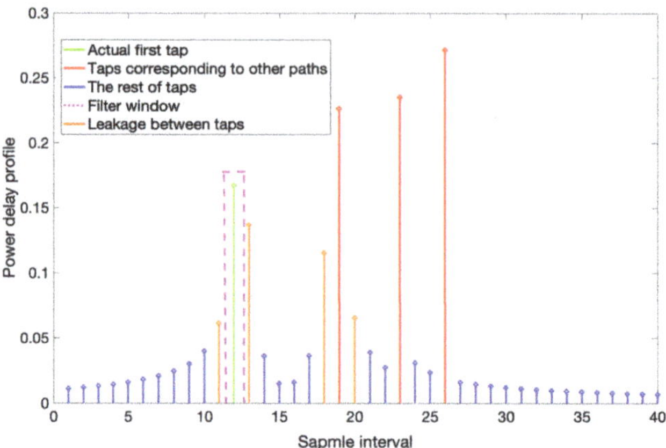

Figure 1. Power delay profile.

For the convenience of analysis, we consider $W = 0$ and the window can be expressed as:

$$\begin{cases} \tilde{h}_n^m = \dfrac{h_1[nT_s]\sin(\pi\dot{\tau}_1)}{\sqrt{N}\sin\left(\frac{\pi}{N}(\dot{\tau}_1-n)\right)}e^{-j\frac{\pi}{N}(n+(N-1)\dot{\tau}_1)-j\phi_1}, n = [\dot{\tau}_1] \\ \tilde{h}_n^m = 0, n \in \mathcal{N}_{|\dot{\tau}_1} \end{cases}, \tag{13}$$

here, $\mathcal{N}_{|\dot{\tau}_1} = \{1, 2, [\dot{\tau}_1]-1, [\dot{\tau}_1]+1, \cdots, N\}$. The frequency domain channel response after the window function can be written as:

$$H_k^m = \frac{h_1[nT_s]\sin(\pi\dot{\tau}_1)e^{-j\frac{\pi}{N}([\dot{\tau}_1]+(N-1)\dot{\tau}_1)-j\phi_1-j2\pi\frac{[\dot{\tau}_1]}{N}k}}{\sqrt{N}\sin\left(\frac{\pi}{N}(\dot{\tau}_1 - [\dot{\tau}_1])\right)}. \tag{14}$$

The time-domain window eliminates the effect of multipath on the carrier phase, but introduces additional problems at the same time.

(1) The phase difference between sub-carriers $e^{-j2\pi\frac{[\dot{\tau}_1]}{N}}$ can no longer accurately reflect the distance. The system resolution also limits the time delay measured through the subcarrier phase due to the effect of the time domain window.

(2) The time-domain window processing introduces some phase noise. For example, $e^{-j\frac{\pi}{N}([\dot{\tau}_1]+(N-1)\dot{\tau}_1)}$. Furthermore, the sign of $\dfrac{\sin(\pi\dot{\tau}_1)}{\sin\left(\frac{\pi}{N}(\dot{\tau}_1-[\dot{\tau}_1])\right)}$ will also affect the phase of sub-carriers.

It can be proved that the subcarrier phase of $k = N/2$ can effectively reflect the distance information from the terminal to the BS. Furthermore, the channel frequency response of subcarrier $k = N/2$ can be approximated as:

$$H_{k=N/2}^m = \frac{h_1[nT_s]\sin(\pi\dot{\tau}_1)}{\sqrt{N}\sin\left(\frac{\pi}{N}(\dot{\tau}_1 - n)\right)}e^{-j\pi\dot{\tau}_1-j\phi_1} + w_{N/2}^m, \tag{15}$$

the proof is given in the Appendix A. Therefore the phase at $k = N/2$ can be written as:

$$\hat{\phi} = -angle\left(e^{-j\pi\dot{\tau}-j\phi_1-\hat{w}_P}\right); 0 \le \hat{\phi} \le 2\pi, \tag{16}$$

here, $\hat{w}_P$ is the phase noise caused by $w_{N/2}^m$. Due to the trigonometric function properties, the part beyond 2π cannot be found when solving for the phase; thus, the integer ambiguity arises. Considering the phase shifts experienced in the channel, e.g., phase noise, base on $\dot{\tau}_1 = N\Delta f_{SCS}\tau_1[nT_s]$ and $\phi_1 = 2\pi f_c\tau_1[nT_s]$, the phase-based ranging can be written as:

$$\hat{\phi} + 2\pi N_I = \pi N\Delta f_{SCS}\tau_1[nT_s] + 2\pi f_c\tau_1[nT_s] + \hat{w}_P, \tag{17}$$

here, N_I is the unknown integer ambiguity. Divide Equation (17) by 2π and simplify the equation:

$$\phi = \frac{N}{2}\Delta f_{SCS}\tau_1[nT_s] + f_c\tau_1[nT_s] - N_I + w_P$$
$$= \frac{d}{\lambda} - N_I + w_P,$$
(18)

here, $\phi = \frac{\tilde{\phi}}{2\pi}$ is the normalized phase measurement, $w_P = \frac{\tilde{w}_P}{2\pi}$ is the normalized phase noise, $d = c\tau_1[nT_s]$ is the geometric distance between the antennas of transmitter and receiver, $\lambda = \frac{c}{f_c + \frac{N}{2}\Delta f_{SCS}}$ is the equivalent wavelength. Further, introducing terminal r and the BS i, we have:

$$\phi_r^i = \frac{d_r^i}{\lambda} - N_r^i + w_{r,P}^i.$$
(19)

- ϕ_r^i (known) is the carrier phase measurement (unit: carrier circle).
- λ (known) is the wavelength calculated from c, f_c, N, and Δf_{SCS}, (unit: m).
- N_r^i (unknown) is the integer ambiguity (unit: carrier circle).
- $w_{r,P}^i \sim \mathcal{N}\left(0, \tilde{\sigma}_{r,i}^2\right)$ (unknown) is the residual estimation error (unit: carrier circle).

The time delay from the user to the BS can be deduced by measuring the phase of the $N/2$ subcarrier. Equation (16) shows that the system sampling rate does not limit the carrier phase measurement, and thus the accuracy of the carrier phase-based ranging technique is high. Furthermore, we use the phase-lock-loop (PLL) [29] to measure the carrier phase. At the initial locking moment of the PLL, the carrier phase measurement is between $[0, 1]$. After that, the change of user position will be reflected in the measured phase (continuous phase tracking allows carrier phase more than 1 or less than 0), thus ensuring that the integer ambiguity is constant during the user positioning. However, since the integer ambiguity is unknown, the carrier phase measurements are challenging to be used directly for user location solutions. Therefore, we propose a location algorithm combining carrier phase and TOA measurements in the following.

3. Positioning Algorithm

The ambiguity resolution is one of the primary problems in carrier phase measurement. In this section, we propose an EFK algorithm based on TOA and carrier phase measurements. This algorithm can estimate the position while estimating the integer ambiguity.

3.1. TOA and Carrier Phase Measurements

According to Equations (10) and (19), further considering the non-ideal factors such as clock error and NLOS error, the TOA measurements and carrier phase between the i-th BS and user equipment (UE) r at a specific epoch can be written as:

$$T_r^i = (d_r^i + m_r^i + w_{r,T}^i)/c + \delta t^i - \delta t_r$$
$$\phi_r^i = \frac{d_r^i + c(\delta t^i - \delta t_r) + m_r^i}{\lambda} - N_r^i + w_{r,P}^i$$
(20)

- δt^i (unknown) is the clock error of the transmitter i (unit: s).
- δt_r (unknown) is the clock error of the receiver r (unit: s).
- m_r^i (unknown) represents the channel bias introduced by NLOS reflections (unit: m).

The SD of the TOA and carrier phase measurements from the receiver r by measuring the signals from two transmitters i and j can be expressed as:

$$T_r^{ij} = (d_r^{ij} + m_r^{ij} + w_{r,T}^{ij})/c + \delta t^{ij}$$
$$\phi_r^{ij} = \frac{d_r^{ij} + c\delta t^{ij} + m_r^{ij}}{\lambda} - N_r^{ij} + w_{r,P}^{ij}$$
(21)

where the double superscript "ij" indicates the differential operation between transmitters i and j, i.e., $s_r^{ij} = s_r^i - s_r^j; s \in \{T, \phi, d, \delta t, N, m, w\}$. According to Equations (10) and (19), the measurement noise $w_{r,T}^{ij}$ and $w_{r,P}^{ij}$ are still independent Gaussian noise with following distributions, i.e.,

$$E\left[w_{r,T}^{ij}, w_{r,T}^{kj}\right] = \begin{cases} \sigma_{r,i}^2 + \sigma_{r,j}^2; & i = k \\ \sigma_{r,j}^2; & i \neq k \end{cases}. \tag{22}$$

The SD operation of Equation (21) removes the measurement errors common to the receiver, e.g., the receiver clock offset δt_r. Furthermore, the double-difference (DD) TOA and carrier phase measurements from two transmitters i and j, and two receivers r and u can be expressed as:

$$\begin{aligned} T_{ru}^{ij} &= (d_{ru}^{ij} + m_{ru}^{ij} + w_{ru,T}^{ij})/c \\ \phi_{ru}^{ij} &= \frac{d_{ru}^{ij} + m_{ru}^{ij}}{\lambda} - N_{ru}^{ij} + w_{ru,P}^{ij} \end{aligned}, \tag{23}$$

where the double superscript "ij" indicates the differential operation between transmitters i and j, and double subscript "ru" indicates the differential operation between receivers r and u, $s_{ru}^{ij} = s_r^{ij} - s_u^{ij} = \left(s_r^i - s_r^j\right) - \left(s_u^i - s_u^j\right); s \in \{T, \phi, d, \delta t, N, m, w\}$. DD measurement noise $w_{ru,T}^{ij}$ and $w_{ru,P}^{ij}$ are no longer independent Gaussian noise. Assume the transmitter j is selected as the reference, we have:

$$E\left[w_{ru,T}^{ij}, w_{ru,T}^{kj}\right] = \begin{cases} \sigma_{r,j}^2 + \sigma_{u,j}^2; & i \neq k \\ \sigma_{r,i}^2 + \sigma_{u,i}^2 + \sigma_{r,j}^2 + \sigma_{u,j}^2; & i = k \end{cases}. \tag{24}$$

DD operation removes the measurement biases related to the transmitters and the receivers, such as the transmitter clock offsets and receivers clock offsets. We introduce the concept of reference device, where it is assumed that u is the reference device and that the location of terminal u is known. It can be seen by Equation (23) that the introduction of the reference device helps to eliminate the clock error. We can construct the SD measurements from the DD measurements, which are not impacted by the receiver and the transmitter clock biases. Given that d_u^{ij} can be obtained from the known locations of the reference device u and the BSs, we can construct the SD measurements $\hat{T}_r^{ij}$ and Φ_r^{ij}:

$$\begin{aligned} \hat{T}_r^{ij} &\triangleq cT_{ru}^{ij} + d_u^{ij} = d_r^{ij} + m_{ru}^{ij} + w_{ru,T}^{ij} \\ \Phi_r^{ij} &\triangleq \phi_{ru}^{ij} + \frac{d_u^{ij}}{\lambda} = \frac{d_r^{ij} + m_{ru}^{ij}}{\lambda} - N_{ru}^{ij} + w_{ru,P}^{ij} \end{aligned}, \tag{25}$$

Equation (25) shows the $\hat{T}_r^{ij}$ and Φ_r^{ij} are not impacted by the receiver and the transmitter clock biases. It is worth noting that the reference device can be either a UE with a known exact location or a BS. For some positioning scenarios, the deployment of additional hardware can cause a significant overhead; therefore, 3GPP has agreed on selecting the reference device, i.e., the device with the known location can be a UE and/or a BS (also known as evolved gNB) [30].

3.2. Extended Kalman Filter

For an EKF design, one needs first to define the unknown EKF states. An EKF for carrier phase positioning may include the following EKF states:

- UE position. EKF for positioning needs to include the states associated with the unknown UE position. The EKF may use the 2D (or 3D) UE position coordinates directly as the EKF states. For example, in the following discussion of the EKF design, we assume the EKF states include a 2D position.
- UE velocity. With the consideration of UE mobility, the EKF states may also include the UE velocity. The number of states for UE velocity is generally the same as the number of states for UE position.

- Integer ambiguities. The premise of using the carrier phase for location is to solve integer ambiguities. According to Equation (25), it is necessary to solve the DD integer ambiguities while solving the user position.

Let the position of the UE at epoch k be $\mathbf{s}(k)$. In the absence of other information, assume that the velocity of UE keeps constant, the position at the next epoch can be expressed as $\mathbf{s}(k+1) = \mathbf{s}(k) + \mathbf{v}T$. Furthermore, in the case of no cycle slip, the ambiguities remain consistent in each epoch. Assume the system states include 2D position, 2D velocity, and the DD integer ambiguities are obtained from m cells, and the system can be represented as:

$$\begin{cases} x(k+1) = x(k) + v_x(k)T \\ v_x(k+1) = v_x(k) \\ y(k+1) = y(k) + v_y(k)T \\ v_y(k+1) = v_y(k) \\ N_{ru}^{ij}(k+1) = N_{ru}^{ij}(k) \end{cases} \tag{26}$$

Assume the j-th cell is selected as the reference cell. The EKF state vector x can be expressed as follows:

$$\begin{aligned} \mathbf{x} &\triangleq [\mathbf{s}, \mathbf{v}, \mathbf{N}]^T \\ &= \left[x, y, v_x, v_y, N_{ru}^{1j}, \ldots, N_{ru}^{(j-1)j}, N_{ru}^{(j+1)j}, \ldots, N_{ru}^{mj} \right]^T , \end{aligned} \tag{27}$$

where $\mathbf{s} = (x, y)$ models the UE position; $\mathbf{v} = (v_x, v_y)$ is the UE velocity, and $\mathbf{N} = [N_{ru}^{1j}, \ldots, N_{ru}^{(j-1)j}, N_{ru}^{(j+1)j}, \ldots, N_{ru}^{mj}]$ includes the DD integer ambiguities. Based on the selected EKF states, the state transition equation of the discrete EKF for carrier phase positioning can be written as:

$$\mathbf{x}(k+1) = \mathbf{F}(k)\mathbf{x}(k) + \mathbf{W}_\mathbf{x}(k). \tag{28}$$

The one-step state transition matrix is as follows:

$$\mathbf{F} = \begin{bmatrix} \mathbf{I}(2 \times 2) & \mathbf{F}_{12} & \mathbf{0} \\ \mathbf{0} & \mathbf{I}(2 \times 2) & \mathbf{0} \\ \mathbf{0} & \mathbf{0} & \mathbf{I}(m-1 \times m-1) \end{bmatrix}, \tag{29}$$

where $F_{12} = \begin{bmatrix} \Delta T & 0 \\ 0 & \Delta T \end{bmatrix}$, $E[\mathbf{W_x}] = 0$, and $\mathbf{Q} = E[\mathbf{W_x W_x^T}] = \text{diag}(\mathbf{Q}_r; \mathbf{Q}_v; \mathbf{0}(m-1 \times m-1))$, $\mathbf{Q}_r = \text{diag}\{\sigma_x^2, \sigma_y^2\}$, $\mathbf{Q}_v = \text{diag}\{\sigma_{v_x}^2, \sigma_{v_y}^2\}$, $\mathbf{I}$ represents an identity matrix, and $\mathbf{0}$ represents a zero matrix. ΔT is the time interval of the state transition of the Kalman filter. σ_x^2, σ_y^2 and $\sigma_{v_x}^2, \sigma_{v_y}^2$ represent the uncertainty in the prediction of the UE position and velocity.

The measurement equations of the discrete EKF as:

$$\mathbf{Z}(k+1) = \mathbf{h}(\mathbf{x}(k+1)) + \mathbf{W_Z}(k+1) \tag{30}$$

$$\mathbf{Z}(k+1) = \begin{bmatrix} \mathbf{T} \\ \mathbf{\Theta} \end{bmatrix}, \mathbf{T}(k+1) = \begin{bmatrix} \hat{T}_r^{1j}(k+1) \\ \vdots \\ \hat{T}_r^{(j-1)j}(k+1) \\ \hat{T}_r^{(j+1)j}(k+1) \\ \vdots \\ \hat{T}_r^{mj}(k+1) \end{bmatrix} \tag{31}$$

$$\mathbf{\Theta}(k+1) = \begin{bmatrix} \Phi_r^{1j}(k+1) \\ \vdots \\ \Phi_r^{(j-1)j}(k+1) \\ \Phi_r^{(j+1)j}(k+1) \\ \vdots \\ \Phi_r^{mj}(k+1) \end{bmatrix} \tag{32}$$

$$\mathbf{W}_z(k+1) = \begin{bmatrix} \mathbf{W_T}(k+1) \\ \mathbf{W_P}(k+1) \end{bmatrix}, E[\mathbf{W}_z] = 0$$

$$\mathbf{R} = E[\mathbf{W_z W_z^T}] = \begin{bmatrix} \mathbf{R_T} & 0 \\ 0 & \mathbf{R_P} \end{bmatrix} \tag{33}$$

$Z(k+1)$ is the SD measurement vector, $\mathbf{W_Z}(k+1)$ is the measurement noises, $\mathbf{R_T}$ and $\mathbf{R_P}$ represent, respectively, the convince matrix of the measurement noises $\mathbf{W_T}$ and $\mathbf{W_P}$. $\mathbf{R_T}$ is non-diagonal matrixes due to DD operation on the measurements as shown in Equation (24). $\mathbf{R_P}$ can be obtained similarly.

$h(x(k+1))$ is a nonlinear function that describes the relationship between the state vector and the measurement vector:

$$\mathbf{h}(\mathbf{x}(k+1)) = \begin{bmatrix} \mathbf{h}(\mathbf{x}(k+1))_T \\ \mathbf{h}(\mathbf{x}(k+1))_P \end{bmatrix}$$

$$\mathbf{h}(\mathbf{x}(k+1))_T = \begin{bmatrix} h_T^{1j} \\ \vdots \\ h_T^{(j-1)j} \\ h_T^{(j+1)j} \\ \vdots \\ h_T^{mj} \end{bmatrix}, \mathbf{h}(\mathbf{x}(k+1))_P = \begin{bmatrix} h_P^{1j} \\ \vdots \\ h_P^{(j-1)j} \\ h_P^{(j+1)j} \\ \vdots \\ h_P^{mj} \end{bmatrix} \tag{34}$$

$$h_T^{ij} = h_T^i - h_T^j; (i = 1, \ldots, m; i \neq j)$$
$$h_P^{ij} = h_P^i - h_P^j; (i = 1, \ldots, m; i \neq j)$$
$$h_T^i = \sqrt{(x(k+1|k) - x_i)^2 + (y(k+1|k) - y_i)^2}; (i = 1, \ldots, m)$$
$$h_P^i = \frac{\sqrt{(x(k+1|k)-x_i)^2+(y(k+1|k)-y_i)^2}}{\lambda} - N_{ru}^i; (i = 1, \ldots, m)$$

There is a need to linearize the measurement Equation (30) around the estimated UE location to use the EKF algorithm. The Jacobian matrix $\mathbf{H}$ can be obtained as:

$$\mathbf{H}(\mathbf{x}(k+1|k)) = \frac{\partial \mathbf{h}}{\partial \mathbf{x}}\Big|_{\mathbf{x}(k+1|k)} = \begin{bmatrix} \frac{\partial \mathbf{h}_T}{\partial \mathbf{x}}\big|_{\mathbf{x}(k+1|k)} \\ \frac{\partial \mathbf{h}_P}{\partial \mathbf{x}}\big|_{\mathbf{x}(k+1|k)} \end{bmatrix} \tag{35}$$

$$\frac{\partial \mathbf{h}_T}{\partial \mathbf{x}}\Big|_{\mathbf{x}(k+1|k)} = \begin{bmatrix} \frac{\partial h_T^{1j}}{\partial x}\big|_{x(k+1|k)} & \frac{\partial h_T^{1j}}{\partial y}\big|_{y(k+1|k)} \\ \vdots & \vdots \\ \frac{\partial h_T^{(j-1)j}}{\partial x}\big|_{x(k+1|k)} & \frac{\partial h_T^{(j-1)j}}{\partial y}\big|_{y(k+1|k)} & 0(m-1 \times 2) & 0(m-1 \times m-1) \\ \frac{\partial h_T^{(j+1)j}}{\partial x}\big|_{x(k+1|k)} & \frac{\partial h_T^{(j+1)j}}{\partial y}\big|_{y(k+1|k)} \\ \vdots & \vdots \\ \frac{\partial h_T^{mj}}{\partial x}\big|_{x(k+1|k)} & \frac{\partial h_T^{mj}}{\partial y}\big|_{y(k+1|k)} \end{bmatrix} \tag{36}$$

$$\frac{\partial \mathbf{h}_P}{\partial \mathbf{x}}\Big|_{\mathbf{x}(k+1|k)} = \begin{bmatrix} \frac{\partial h_P^{1j}}{\partial x}\big|_{x(k+1|k)} & \frac{\partial h_P^{1j}}{\partial y}\big|_{y(k+1|k)} & & \\ \vdots & \vdots & & \\ \frac{\partial h_P^{(j-1)j}}{\partial x}\big|_{x(k+1|k)} & \frac{\partial h_P^{(j-1)j}}{\partial y}\big|_{y(k+1|k)} & \mathbf{0}(m-1 \times 2) & -\mathbf{I}(m-1 \times m-1) \\ \frac{\partial h_P^{(j+1)j}}{\partial x}\big|_{x(k+1|k)} & \frac{\partial h_P^{(j+1)j}}{\partial y}\big|_{y(k+1|k)} & & \\ \vdots & \vdots & & \\ \frac{\partial h_P^{mj}}{\partial x}\big|_{x(k+1|k)} & \frac{\partial h_P^{mj}}{\partial y}\big|_{y(k+1|k)} & & \end{bmatrix} \tag{37}$$

$$
\begin{aligned}
\frac{\partial h_T^{ij}}{\partial x}\Big|_{x(k+1|k)} &= \frac{\partial h_T^{i}}{\partial x}\Big|_{x(k+1|k)} - \frac{\partial h_T^{j}}{\partial x}\Big|_{x(k+1|k)}; (i = 1,\ldots,m; i \neq j) \\
\frac{\partial h_P^{ij}}{\partial x}\Big|_{x(k+1|k)} &= \frac{\partial h_P^{i}}{\partial x}\Big|_{x(k+1|k)} - \frac{\partial h_P^{j}}{\partial x}\Big|_{x(k+1|k)}; (i = 1,\ldots,m; i \neq j) \\
\frac{\partial h_T^{i}}{\partial x}\Big|_{x(k+1|k)} &= \lambda \frac{\partial h_P^{i}}{\partial x}\Big|_{x(k+1|k)} = \frac{x(k+1|k)-x_i}{\sqrt{(x(k+1|k)-x_i)^2+(y(k+1|k)-y_i)^2}} \\
\frac{\partial h_T^{i}}{\partial y}\Big|_{y(k+1|k)} &= \lambda \frac{\partial h_P^{i}}{\partial y}\Big|_{y(k+1|k)} = \frac{y(k+1|k)-y_i}{\sqrt{(x(k+1|k)-x_i)^2+(y(k+1|k)-y_i)^2}} \\
\frac{\partial h_P^{ij}}{\partial N_{ru}^{ij}} &= -1
\end{aligned}
\tag{38}
$$

Given the state and measurement equations in previous sections, the EKF algorithm can be applied to calculate the estimate of $\mathbf{x}(k+1)$ based on the SD measurements. EKF algorithm [31,32] includes the following time-update and measurement update equations. Furthermore, the time-update equation are:

$$
\begin{aligned}
\mathbf{x}(k+1|k) &= \mathbf{F}(k)\mathbf{x}(k|k); \\
\mathbf{P}(k+1|k) &= \mathbf{F}(k)\mathbf{P}(k|k)\mathbf{F}^T(k) + \mathbf{Q}(k);
\end{aligned}
\tag{39}
$$

where $\mathbf{x}(k|k)$ and $\mathbf{P}(k|k)$ are, respectively, the estimated state vector and its covariance matrix at the epoch $t = t_k$. $\mathbf{x}(k+1|k)$ and $\mathbf{P}(k+1|k)$ represent, respectively, the predicted state vector and its covariance matrix at the epoch $t = t_{k+1}$, based on $\mathbf{x}(k|k)$ and $\mathbf{P}(k|k)$. The matrixes $\mathbf{F}(k)$ and $\mathbf{Q}(k)$ are defined in Equation (29). Furthermore, the measurement update equation are:

$$
\begin{aligned}
\mathbf{K}(k+1) &= \mathbf{P}(k+1|k)\mathbf{H}(\mathbf{x}(k+1|k))\left[\mathbf{H}(\mathbf{x}(k+1|k))\mathbf{P}(k+1|k)\mathbf{H}^T(\mathbf{x}(k+1|k)) + \mathbf{R}(k)\right]^{-1}; \\
\mathbf{x}(k+1|k+1) &= \mathbf{x}(k+1|k) + \mathbf{K}(k+1)[\mathbf{Z}(k+1) - \mathbf{h}(\mathbf{x}(k+1|k))]; \\
\mathbf{P}(k+1|k+1) &= [\mathbf{I} - \mathbf{K}(k+1)\mathbf{H}(\mathbf{x}(k+1|k))]\mathbf{P}(k+1|k);
\end{aligned}
\tag{40}
$$

$\mathbf{H}(\mathbf{x}(k+1|k))$ is the Jacobian matrix given by Equation (35), the measurement equation $\mathbf{h}(\mathbf{x}(k+1|k))$ is defined in Equation (34), and calculated based on the predicted position $(x(k+1|k), y(k+1|k))$ at time $t = t_{k+1}$.

3.2.1. NLOS Error Recognition and Elimination Based on EKF

Equation (25) shows that DD operation may not cancel out the impact of the NLOS. Furthermore, we propose an EKF-based scheme for NLOS error identification and elimination.

$$
\begin{aligned}
\hat{T}_r^{ij} &\triangleq cT_{ru}^{ij} + d_u^{ij} = d_r^{ij} + m_{ru}^{ij} + w_{ru,T}^{ij} \\
\Phi_r^{ij} &\triangleq \phi_{ru}^{ij} + \frac{d_u^{ij}}{\lambda} = \frac{d_r^{ij}+m_{ru}^{ij}}{\lambda} - N_{ru}^{ij} + w_{ru,P}^{ij}
\end{aligned}
\tag{41}
$$

According to the state and measurement equations at $t = t_k$, EKF can predict the SD measurements at $t = t_{k+1}$. Because the NLOS error reaches several meters, if there is NLOS propagation at $t = t_{k+1}$, the SD measurements will deviate greatly from the predicted value of EKF. NLOS error can be identified and corrected according to the deviation:

$$\text{if } \left| \hat{T}_r^{ij} - h_T^{ij} \right| > \Lambda$$
$$\text{then } \hat{T}_r^{ij} = h_T^{ij}, \Phi_r^{ij} = h_P^{ij} \quad . \tag{42}$$

The threshold setting depends on the maximum DD measurement noise. For the deviation greater than Λ, the NLOS error needs to be updated. The predicted measurements of EKF are used to improve the positioning accuracy.

3.2.2. EKF Initialization

$$\mathbf{x}(0) = \left[x(0), y(0), v_x(0), v_y(0), N_{ru}^{1j}(0), \dots, N_{ru}^{mj}(0) \right]^T. \tag{43}$$

For the first step of the EKF ($t = 0$), the estimated initial UE position $(x(0), y(0))$ is obtained from the time difference of arrival (TDOA) or other approaches [33]. The initial estimates of $(v_x(0), v_y(0))$ can be set to $\mathbf{0}$. The initial ambiguities $N_{ru}^{1j}(0), N_{ru}^{2j}(0), \dots, N_r^{mj}(0)$ can be simply determined based on the initial UE position and known positions of cell, i.e.,

$$N_r^i(0) = \frac{\sqrt{(x(0)-x_i)^2+(y(0)-y_i)^2}}{\lambda} - \phi_r^i(0);$$
$$N_r^{ij} = N_r^i - N_r^j;$$
$$N_{ru}^{ij} = N_r^{ij} - N_u^{ij}; \tag{44}$$

here, N_u^{ij} is the SD integer ambiguity of the reference device. The initial covariance matrix $\mathbf{P}_0$ can be set as the diagonal matrix as follows:

$$\mathbf{P}(0) = \text{diag}\left\{ P_x(0), P_y(0), P_{v_x}(0), P_{v_y}(0), \mathbf{P_N}(0) \right\};$$
$$\mathbf{P_N}(0) = \left\{ P_{N_{1j}}(0), \dots, P_{N_{(j-1)j}}(0), P_{N_{(j+1)j}}(0), \dots, P_{N_{mj}}(0) \right\}; \tag{45}$$

where $P_x(0), P_y(0)$ can be set based on the assumed maximum positioning error of the TDOA. $P_{v_x}(0), P_{v_y}(0)$ can be set based on the expected maximum velocity of the UE; $P_{N_{1j}}(0), \dots, P_{N_{mj}}(0)$ are set based on the maximum assumed DD measurement error.

3.2.3. Interaction with the Ambiguity Resolution Block

The EKF estimated float DD carrier-phase ambiguities would be sent to ambiguity resolution block to get integer DD carrier-phase ambiguities to improve positioning accuracy. For this purpose, after each EKF step k, the float solution of the DD carrier-phase ambiguities $\hat{\mathbf{N}}(k \mid k)$ and the corresponding to covariance matrix $\mathbf{P_N}(k)$ are provided to the ambiguity resolution block for searching the DD integer ambiguities $\overline{\mathbf{N}}(k \mid k)$. To fix integer ambiguities, we use MLAMBDA, a modified LAMBDA method for integer least squares ambiguity determination [34,35].

DD integer ambiguities $\overline{\mathbf{N}}(k \mid k)$ can be used to update $\hat{\mathbf{N}}(k \mid k)$. However, the EKF performance may be degraded if unreliable $\overline{\mathbf{N}}(k \mid k)$ is used to update $\hat{\mathbf{N}}(k \mid k)$. Thus, before using $\overline{\mathbf{N}}(k \mid k)$ to update $\hat{\mathbf{N}}(k \mid k)$, there is a need to test the reliability of the DD integer ambiguities $\overline{\mathbf{N}}(k \mid k)$.

The following approach is used to test the reliability of DD integer ambiguities $\overline{\mathbf{N}}(k \mid k)$.

- Initialization: Set a predefined threshold for the ratio test: $\epsilon > 0$, e.g., $\epsilon = 0.5$. Set a predefined maximum count n_{max}, e.g., $n_{max} = 5$. Set counter $n(0) = 0$.
- Step 1: For each epoch k, requesting the MLAMBDA to output two sets of the DD integer ambiguities. With the request, MLAMDA will return one group of the best

estimates and one group of the second-best of the DD integer ambiguities and together with the corresponding residuals, say $r_1(k)$ and $r_2(k)$.

- Step 2: Calculate the ratio of $r_1(k)/r_2(k)$, and compared it with a predefined threshold ϵ. The smaller $r_1(k)/r_2(k)$ indicates that the best DD integer ambiguities estimates and the second-best estimates are close. If $r_1(k)/r_2(k) < \epsilon$, the counter $n(k)$ is increased by 1, i.e., $n(k) = n(k-1) + 1$. Otherwise, set $n(k) = 0$.
- Step 3: If $n(k) > n_{max}$, declared that the $\overline{\mathbf{N}}(k \mid k)$ is reliable DD integer ambiguity resolution. Once reliable DD integer ambiguity resolution $\overline{\mathbf{N}}(k \mid k)$ is obtained, it can be used to update the EKF.

3.2.4. Interaction with the Pre-Processing Measurement Block

Before each EKF operation, the EKF needs to adjust the state variables and covariance matrices based on TOA and carrier phase measurements.

If at time $t = t_k$, it is detected that there is a cycle slip for the phase measurements from i-th cell, the corresponding state, and covariance of the cell need to be reset. N_{ru}^{ij} can be reset based on the TDOA measurement $\hat{T}_r^{ij}$ and the SD carrier phase measurement Φ_r^{ij}. The diagonal element $\mathbf{P}(k|k)$ corresponding to N_{ru}^{ij} will be set based on the maximum assumed integer ambiguities measurement error.

Suppose at time $t = t_k$, the measurements associated with an existing i-th cell are no longer available. In that case, the corresponding state N_{ru}^{ij} needs to be removed from EKF, and so the elements of covariance matrix $\mathbf{P}(k|k)$ related the N_{ru}^{ij}. The dimension of the EKF will be reduced correspondingly.

If $t = t_k$, the measurements associated with a new cell are available, the EKF will add a new state of integer ambiguity for that cell. The corresponding state of the cell is calculated based on the TDOA measurement $\hat{T}_r^{ij}$ and the SD carrier phase measurement Φ_r^{ij}. The diagonal element of $\mathbf{P}(k|k)$ corresponding to N_{ru}^{ij} will be set based on the maximum assumed integer ambiguities measurement error.

Figure 2 shows the signal processing diagram for the real-time kinematic positioning based on TOA and carrier phase measurements.

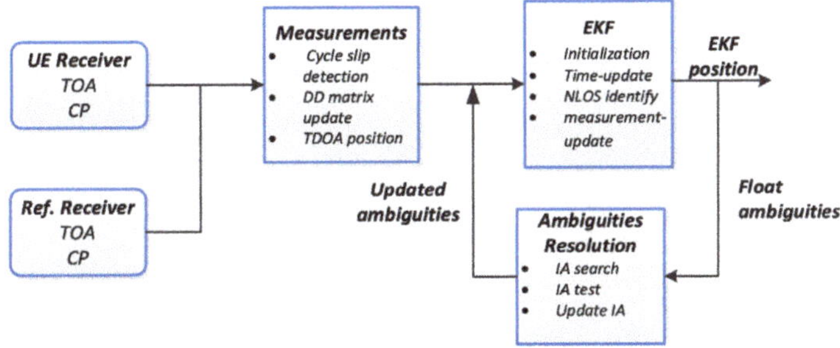

Figure 2. Flowchart of TOA/Carrier Phase combined Real-Time Kinematic.

4. Numerical Results

In this paper, MATLAB is used to verify the algorithm. Furthermore, one PRS subframe is used in each PRS positioning occasion. Perfect muting is assumed in the simulation. The positioning scene is shown in Figure 3, where six BSs are regularly distributed in the building, and the reference device is located in the center of the scene. In the simulation, it is assumed that there is a synchronization error in Network. Therefore, there are time-varying synchronization errors at the BS side and the terminal side. The detailed simulation parameters are listed in Table 1. For other parameters including the number of multipath in the indoor scenarios, the criteria for generating LOS/NLOS, and the path loss, please refer to [36].

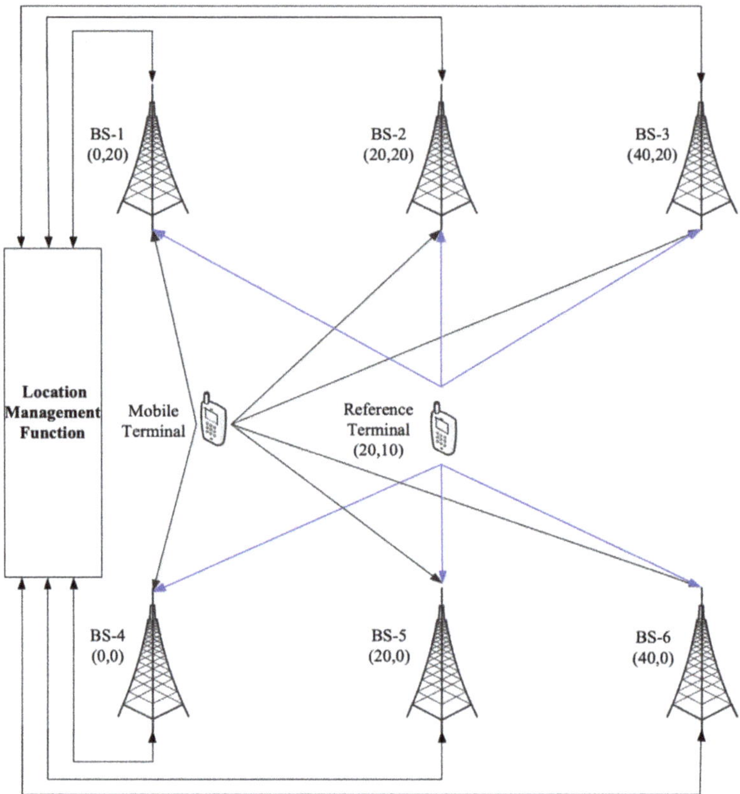

Figure 3. Terminal positioning environment.

In the simulation, moving speed of the terminal is 1 m/s, and the position solution interval is 0.1 s. The terminal moves according to a specific track in which the length is 60 m, and the number of sampling epoch is 600. High accuracy localization using the carrier phase requires a fast and accurate solution of the integer ambiguity. Therefore, we use the following perspectives to evaluate the effectiveness of the algorithm:

- The accuracy and convergence speed of the integer ambiguity.
- The terminal that can judge whether the solved integer ambiguity is reliable or not.
- The real-time positioning accuracy of the terminal position.
- The cumulative distribution curve of positioning error.

Equation (25) shows that the SD carrier phase measurement Φ_r^{ij} contains the DD integer ambiguity N_{ru}^{ij}. Therefore, we use the DD integer ambiguity for performance comparison. Define the integer ambiguity estimation error as:

$$e_N = |N_{ru,ture}^{ij} - N_{ru}^{ij}| \tag{46}$$

Figure 4 illustrates the four DD integer ambiguity estimation errors. When $e_N = 0$, it represents that the estimated integer ambiguity is the same as the actual ambiguity. Furthermore, the first BS is used as the reference BS in our experiment. All integer ambiguity errors were significant at epoch 0 due to the sizeable initial position estimation error. In the 92*nd* epoch, BS21, BS41, and BS51 all estimate the integer ambiguity correctly and remain unchanged in the subsequent epochs; BS31 always has an error of 1 circle during the experiment. The mistake of BS31 did not affect the localization accuracy because the other integer ambiguities were correctly estimated.

Table 1. System Parameters.

Parameters	Values
Channel model	5G New Radio (NR) channel model (Indoor-Mixed office[36]).
Carrier frequency	3.5 GHz
Carrier wavelength	0.085 m
Inter-site distance	20 m
Room size	40 m × 20 m
Subcarrier spacing	15 KHz
Reference signal	New Radio PRS Structure from [37].
Reference Signal Transmission Bandwidth	50 MHz
Number of BSs	6
UE-antennas	4
Number of subcarriers	3240
FFT Length	4096 for 50 MHz
Sampling rate	61.44 MHz for 50 MHz
Number of occasions used per positioning estimate	1
Interference modelling	Perfect muting
Clock error between BSs	Gaussian distribution with a mean of 25 ns and a variance of 10 ns.
Clock error of the terminal	Gaussian distribution with a mean of 50 ns and a variance of 15 ns.
Delay spread	Exponential distribution with a mean of 22 ns.
Total transmission power	24 dBm
Maximum directional gain of an antenna element	5 dBi
UE speed	1 m/s
Position solution interval	0.1 s
NLOS error identification threshold	$\Lambda = 3$
Ratio test threshold	$\epsilon = 0.5$
LOS generation probability	Table 7.4.2-1 in the literature [36].
Fading model	Large scale fading: Table 7.4.1-1 in the literature [36]; Fast fading: Section 7.5 of [36].
Channel independence	The channel model of the reference device and the channel model of the user terminal are independent of each other.

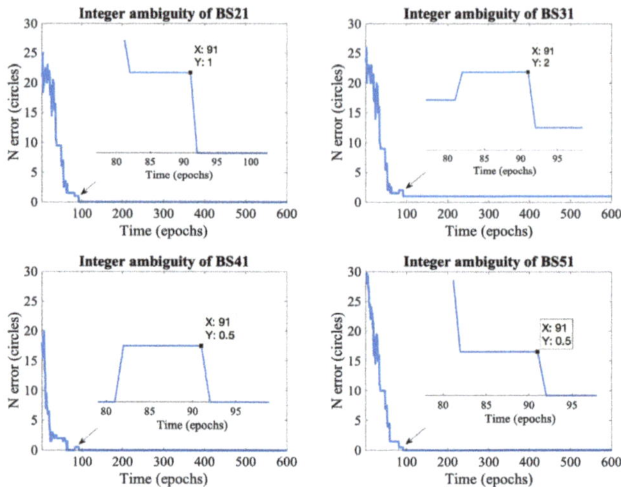

Figure 4. Schematic diagram of DD integer ambiguity convergence.

Figure 5 shows the ratio test was used to check whether the DD integer ambiguities output by the EKF are reliable at the current epoch. The dashed line represents the preset threshold $\epsilon = 0.5$. After the 113th epoch, the reliability rates are all below the threshold. Therefore the algorithm determines that the obtained integer ambiguities are reliable after the 113*th* epoch. Combined with Figure 4, it can be seen that it is valid to use the ratio test to determine whether the DD integer ambiguities converge to the actual value.

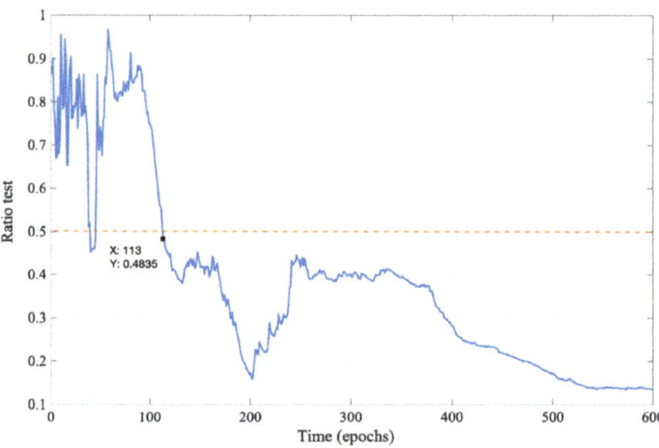

Figure 5. Test of the reliability of DD integer ambiguities.

We evaluate the performance of the 'TOA+CP EKF' based differential positioning method as in Figure 6. In addition, we also list two other cases for comparison, wherein 'GMM EKF' [31] is a method to perform positioning solution by TDOA measurement, which eliminates the NLOS error by model NLOS propagation as Gaussian mixture model; 'EKF' [32] represents a commonly used EKF location algorithm based on TDOA measurement. To eliminate the effect of clock errors, all three algorithms mentioned above use TDOA obtained from Equation (25) instead of TOA for positioning, and the algorithm proposed in this paper also requires SD carrier phase measurements.

Figures 6 and 7 show the performance of the three algorithms during mobile localization. Since all three algorithms use differential measurements for position solving, it can be seen from the results that the time-varying synchronization errors do not affect the positioning accuracy. At the initial epoch, the accurate integer ambiguity has not been solved, so the carrier phase measurement is difficult to determine the initial position. Therefore, initial positions of all three algorithms are calculated from the Chan algorithm [33] using TDOA measurements. It can be seen from Figures 6 and 7 that in the first few epochs, the positioning error of 'TOA+CP EKF' is significant, which is caused by the inaccurate integer ambiguity. In subsequent periods, as the algorithm correctly fix the integer ambiguity, the positioning error gradually decreases. Furthermore, the carrier phase measurement is not limited by the system sampling rate, which, combined with the correct integer ambiguity, makes the carrier phase algorithm suitable for scenarios with high accuracy requirements. Comparatively, both 'GMM EKF' and 'EKF' use only TDOA for user position tracking, which leads to lower positioning accuracy.

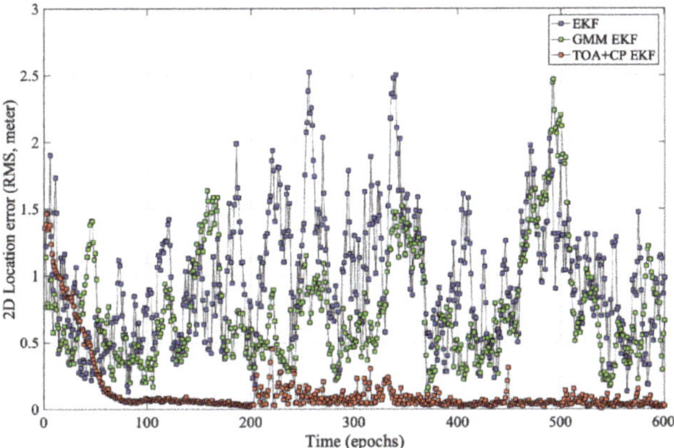

Figure 6. Statistic of mobile positioning error.

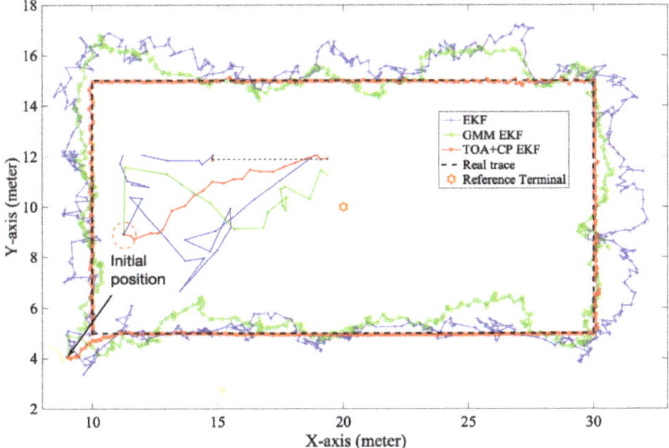

Figure 7. Localization performance of EKF.

The cumulative density function (CDF) curves of horizontal positioning errors are used as performance metrics in positioning evaluations. Define the positioning error as:

$$e_{pos} = \sqrt{(\hat{x} - x_{\text{ture}})^2 + (\hat{y} - y_{\text{ture}})^2} \tag{47}$$

The CDFs for the localization error from both methods are shown in Figure 8. The 'TOA+CP EKF' method has the best performance, with 90% of the horizontal positioning errors within 0.27 m. Therefore, the carrier phase-based localization technique can meet the high accuracy localization requirements. The 'GMM EKF' method has the middle performance due to the algorithm using TDOA for user location tracking and NLOS elimination. Since the system sampling rate limits the TDOA measurement resolution, the positioning accuracy is low. 'EKF' method has the worst performance because it only uses the TDOA and has limited effectiveness in eliminating NLOS error.

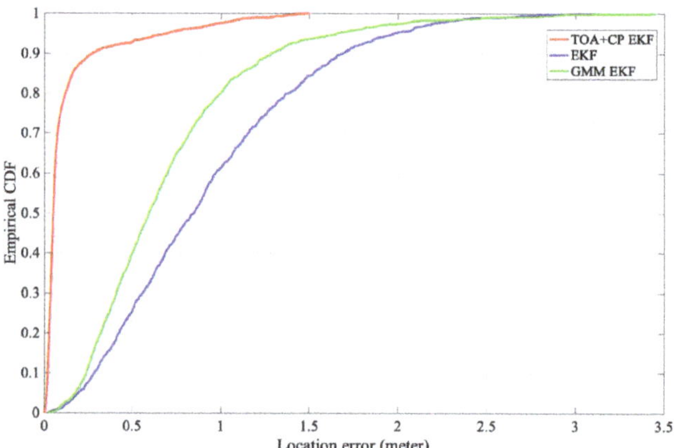

Figure 8. The CDF of horizontal localization error.

We simulated the localization accuracy of this algorithm with the different number of BSs. In our experiments, as shown in Figure 9, we set the length of the indoor scenario to 100 m and the width to 20 m. Furthermore, the coordinates of the six BSs are $[0,0]$, $[40,0]$, $[100,0]$, $[0,20]$, $[40,20]$, $[100,20]$, respectively. The coordinate of the reference UE is $[50,10]$. The actual distance between the user and the BS determines the probability of LOS. Thus, the expansion of the simulation environment decreases the LOS probability and equivalently simulates the case of increasing obstacles.

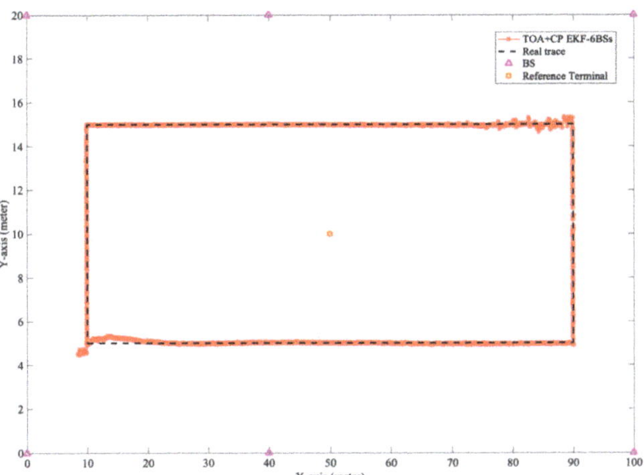

Figure 9. Layout of Indoor - Mixed office scenario.

When five BSs are used in the experiment, the BS located at $[40, 0]$ is removed. When four BSs are involved in localization, the two BSs located at $[40, 0], [40, 20]$ are removed. From Figure 10, it can be seen that the localization accuracy of the algorithm proposed in this paper decreases as the number of BSs decreases. The decrease in the number of available BSs leads to a more extended solution period for the integer ambiguity and thus decreases the localization accuracy. In addition, compared with Figure 8, the decrease in LOS probability does not cause severe degradation of the localization accuracy, so the NLOS error suppression scheme proposed in this paper is effective.

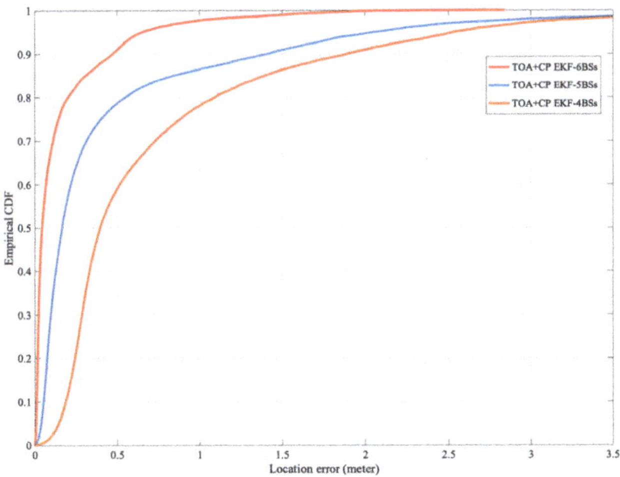

Figure 10. The CDF of horizontal localization error with differfent number of BSs.

5. Conclusions

The main research direction of this paper is to apply carrier phase technology in OFDM systems to improve ranging and positioning accuracy. Compared with single-point positioning using only TOA measurement, carrier phase information is more accurate than TOA measurement, and it is a possible choice for indoor high-precision positioning.

This paper intends to solve two problems of indoor carrier phase positioning: 1. Phase measurements in a multipath environment. 2. Fast and precise integer ambiguity resolution in real-time positioning scenarios. First, this paper analyzed the effect of multipath propagation on phase measurement in detail, and proposed a correlation profile-based carrier phase measurement method. Second, this paper presents an EKF algorithm to estimate the integer ambiguity by the SD carrier and TDOA measurements. In addition to the integer ambiguity estimation, the algorithm also considers the effect brought by NLOS error. Experiments show that the algorithm proposed in this paper can quickly find the integer ambiguity and virtually eliminate the NLOS error, thus improving the positioning accuracy.

Author Contributions: Methodology, Z.Z. and X.Z.; software, Z.Z. and X.Z.; validation, Z.Z. and X.Z.; investigation, Z.Z. and S.K.; resources, Z.Z.; data curation, Z.Z. and X.Z.; writing—original draft preparation, Z.Z. and X.Z.; writing—review and editing, S.K.; visualization, Z.Z.; supervision, S.K.; project administration, S.K.; All authors have read and agreed to the published version of the manuscript.

Funding: This research received no external funding.

Institutional Review Board Statement: Not applicable.

Informed Consent Statement: Not applicable.

Data Availability Statement: The data presented in this study are available on request from the corresponding author. The data are not publicly available due to privacy reasons.

Acknowledgments: We appreciate the guidance of Ren Da and Bin Ren, State Key Laboratory of Wireless Mobile Communications, China Academy of Telecommunications Technology. The authors appreciate the support from Datang Mobile Communications Equipment Co., Ltd. for providing the channel modeling for one of the experiments.

Conflicts of Interest: The authors declare no conflict of interest.

Appendix A

For interference $e^{-j\frac{\pi}{N}([\dot{t}_1]+(N-1)\dot{t}_1)}$ in Equation (14), the following approximation can be made. Since $[\dot{t}_1] \approx \dot{t}_1 \pm 0.5$, and $\frac{\pm 0.5}{N} \ll 1$, So we have $e^{-j\frac{\pi}{N}([\dot{t}_1]+(N-1)\dot{t}_1)} \approx e^{-j\pi\dot{t}_1}$. Thus, the phase of H_k^m can be expressed as:

$$
\begin{cases}
-\pi\dot{t}_1 - \phi_1 - 2\pi k\frac{[\dot{t}_1]}{N}, & \text{if } \frac{\sin(\pi\dot{t}_1)}{\sin\left(\frac{\pi}{N}(\dot{t}_1-[\dot{t}_1])\right)} \geq 0 \\
-\pi\dot{t}_1 - \phi_1 - 2\pi k\frac{[\dot{t}_1]}{N} - \pi, & \text{if } \frac{\sin(\pi\dot{t}_1)}{\sin\left(\frac{\pi}{N}(\dot{t}_1-[\dot{t}_1])\right)} < 0
\end{cases}
\tag{A1}
$$

Take $[\dot{t}] = 2\tilde{n}$ as an example, and $\tilde{n}$ is an arbitrary natural number. For specific analysis, it can be divided into two cases, $2\tilde{n} - 0.5 \leq \dot{t} \leq 2\tilde{n}$, and $2\tilde{n} \leq \dot{t} \leq 2\tilde{n} + 0.5$.

(1) For $2\tilde{n} - 0.5 \leq \dot{t} \leq 2\tilde{n}$, we have:

$\sin(\pi\dot{t}) < 0$, and $\dot{t} - [\dot{t}] < 0$, $\sin\left(\frac{\pi}{N}(\dot{t}_1 - [\dot{t}_1])\right) \leq 0$, so $\frac{\sin(\pi\dot{t}_1)}{\sin\left(\frac{\pi}{N}(\dot{t}_1-[\dot{t}_1])\right)} \geq 0$.

(2) For $2\tilde{n} \leq \dot{t} \leq 2\tilde{n} + 0.5$, we have:

$\sin(\pi\dot{t}) > 0$, and $\dot{t} - [\dot{t}] > 0$, $\sin\left(\frac{\pi}{N}(\dot{t}_1 - [\dot{t}_1])\right) \geq 0$ so $\frac{\sin(\pi\dot{t}_1)}{\sin\left(\frac{\pi}{N}(\dot{t}_1-[\dot{t}_1])\right)} \geq 0$.

We know that $\frac{\sin(\pi\dot{t}_1)}{\sin\left(\frac{\pi}{N}(\dot{t}_1-[\dot{t}_1])\right)} \geq 0$ when $[\dot{t}] = 2\tilde{n}$. Similarly, we can conclude that when $[\dot{t}] = 2\tilde{n} + 1$, $\frac{\sin(\pi\dot{t}_1)}{\sin\left(\frac{\pi}{N}(\dot{t}_1-[\dot{t}_1])\right)} \leq 0$. So (A1) can be simplified to:

$$
\begin{cases}
-\pi\dot{t}_1 - \phi_1 - 2\pi k\frac{[\dot{t}_1]}{N}, & \text{if } [\dot{t}_1] = 2\tilde{n} \\
-\pi\dot{t}_1 - \phi_1 - 2\pi k\frac{[\dot{t}_1]}{N} - \pi, & \text{if } [\dot{t}_1] = 2\tilde{n} + 1
\end{cases}
\tag{A2}
$$

when $k = N/2$, (A2) can be written as:

$$\begin{cases} -\pi\dot{\tau}_1 - \phi_1 - \pi[\dot{\tau}_1], & \text{if } [\dot{\tau}_1] = 2\tilde{n} \\ -\pi\dot{\tau}_1 - \phi_1 - \pi[\dot{\tau}_1] - \pi, & \text{if } [\dot{\tau}_1] = 2\tilde{n} + 1 \end{cases} \quad (A3)$$

Thus, at $k = N/2$, the phase processed by the window function is $-\pi\dot{\tau}_1 - \phi_1$, regardless of whether $[\dot{\tau}_1]$ is odd or even. The proof is completed.

References

1. Garcia, N.; Wymeersch, H.; Larsson, E.; Haimovich, A.; Coulon, M. Direct Localization for Massive MIMO. *IEEE Trans. Signal Process.* **2010**, *65*, 2475–2487. [CrossRef]
2. Rappaport, T.; Xing, Y.; MacCartney, G.; Molisch, A.; Mellios, E.; Zhang, J. Overview of Millimeter Wave Communications for Fifth Generation (5G) Wireless Networks With a Focus on Propagation Models. *IEEE Trans. Antennas Propag.* **2017**, *65*, 6213–6230. [CrossRef]
3. Zhang, P.; Chen, H. A Survey of Positioning Technology for 5G. *J. Beijing Univ. Posts Telecommun.* **2018**, *41*, 1–12.
4. Ryden, H.; Zaidi, A.A.; Razavi, S.M.; Gunnarsson, F.; Siomina, I. Enhanced time of arrival estimation and quantization for positioning in LTE networks. In Proceedings of the IEEE 27th Annual International Symposium on Personal, Indoor, and Mobile Radio Communications (PIMRC), Valencia, Spain, 4–7 September 2016; pp. 1–6.
5. Bialer, O.; Raphaeli, D.; Weiss, A.J. Robust time-of-arrival estimation in multipath channels with OFDM signals. In Proceedings of the 25th European Signal Processing Conference (EUSIPCO), Kos, Greece, 28 August–2 September 2017; pp. 2724–2728.
6. Wang, D.; Fattouche, M. OFDM transmission for timebased range estimation. *IEEE Signal Process Lett.* **2010**, *17*, 571–574. [CrossRef]
7. Wang, T.; Shen, Y.; Mazuelas, S. Bounds for OFDM ranging accuracy in multipath channels. In Proceedings of the IEEE International Conference on Ultra-Wideband (ICUWB), Bologna, Italy, 14–16 September 2011; pp. 450–454.
8. Hu, S.; Berg, A.; Li, X.; Rusek, F. Improving the performance of OTDOA based positioning in NB-IoT systems. In Proceedings of the IEEE Global Communications Conference, Singapore, 4–8 December 2017; pp. 1–7.
9. He, Z.; Ma,Y.; Tafazolli, R. Improved high resolution TOA estimation for OFDM-WLAN based Indoor ranging. *IEEE Wirel. Commun. Lett.* **2013**, *2*, 163–166. [CrossRef]
10. Li, X.; Pahlavan, K. Super-resolution TOA estimation with diversity for Indoor geolocation. *IEEE Trans. Wireless Commun.* **2004**, *3*, 224–234. [CrossRef]
11. Driusso, M.; Marshall, C.; Sabathy, M.; Knutti, F.; Mathis, H.; Babich, F. Vehicular Position Tracking Using LTE Signals. *IEEE Trans. Veh. Technol.* **2017**, *66*, 3376–3391. [CrossRef]
12. Chong, C.; Laurenson, D.; Tan, C.; McLaughlin, S.; Beach, M.; Nix, A. Joint detection-estimation of directional channel parameters using the 2-D frequency domain SAGE algorithm with serial interference cancellation. In Proceedings of the IEEE International Conference on Communications, ICC 2002 (Cat. No.02CH37333), New York, NY, USA, 28 April–2 May 2002; Volume 2, pp. 906–910.
13. Ma, Y.; Pahlavan, K.; Geng, Y. Comparison of POA and TOA based ranging behavior for RFID application. In Proceedings of the IEEE 25th Annual International Symposium on Personal, Indoor, and Mobile Radio Communication (PIMRC), Washington, DC, USA, 2–5 September 2014.
14. DiGiampaolo, E.; Martinelli, F. Mobile Robot Localization Using the Phase of Passive UHF RFID Signals. *IEEE Trans. Ind. Electron.* **2014**, *61*, 365–376. [CrossRef]
15. Ma, Y.; Zhou, L.; Liu, K.; Wang, J. Iterative Phase Reconstruction and Weighted Localization Algorithm for Indoor RFID-Based Localization in NLOS Environment. *IEEE Sens. J.* **2014**, *14*, 597–611. [CrossRef]
16. Noschese, M.; Babich, F.; Comisso, M.; Marshall, C.; Driusso, M. A low-complexity approach for time of arrival estimation in OFDM systems. In Proceedings of the International Symposium on Wireless Communication Systems (ISWCS), Bologna, Italy, 28–31 August 2017.
17. Makki, A.; Siddig, A.; Saad, M.; Bleakley, C.; Cavallaro, J. High-resolution time of arrival estimation for OFDM-based transceivers. *Electron. Lett.* **2015**, *51*, 294–296. [CrossRef]
18. Babich, F.; Noschese, M.; Marshall, C.; Driusso, M. A simple method for TOA estimation in OFDM systems. In Proceedings of the European Navigation Conference (ENC), Lausanne, Switzerland, 9–12 May 2017; pp. 305–310.
19. Shamaei, K.; Kassas, Z. Sub-meter accurate UAV navigation and cycle slip detection with LTE carrier phase measurements. In Proceedings of the ION Global Navigation Satellite Systems Conference, Miami, FL, USA, 16–20 September 2019.
20. Abdallah, A.A.; Shamaei, K.M.; Kassas, Z. Performance Characterization of an Indoor Localization System with LTE Code and Carrier Phase Measurements and an IMU. In Proceedings of the 2019 International Conference on Indoor Positioning and Indoor Navigation (IPIN), Pisa, Italy, 3–30 September 2019.
21. Takasu, T.; Yasuda, A. A Kalman-filter-based integer ambiguity resolution strategy for long-baseline RTK with ionosphere and troposphere estimation. In Proceedings of the ION GNSS, Portland, OR, USA, 21–24 September 2010.
22. Zhao, S.; Cui, X.; Guan, F.; Liu, M. A Kalman filter-based short baseline RTK algorithm for single-frequency combination of GPS and BDS. *Sensors* **2014**, *14*, 15415–15433. [CrossRef] [PubMed]

23. Richter, A. On the Estimation of Radio Channel Parameters: Models and Algorithms. Ph.D. Thesis, Ilmenau University of Technology, Thuringia, Germany, 2015.
24. Kay, S. Fundamentals of Statistical Signal Processing. *PTR Prentice Hall.* August, 1994. Available online: https://www.sciencedirect.com/science/article/pii/0967066194901953 (accessed on 1 July 2021).
25. Geng, C.; Yuan, X.; Huang, H. Exploiting Channel Correlations for NLOS ToA Localization with Multivariate Gaussian Mixture Models. *IEEE Wirel. Commun. Lett* **2020**, *9*, 70–73. [CrossRef]
26. van de Beek, J.; Edfors, O.; Sandell, M.; Wilson, S.; Borjesson, P. On channel estimation in OFDM systems. In Proceedings of the IEEE 45th Vehicular Technology Conference, Countdown to the Wireless Twenty-First Century, Chicago, IL, USA, 25–28 July 1995.
27. Xu, W.; Huang, M.; Zhu, C.; Dammann, A. Maximum likelihood TOA and OTDOA estimation with first arriving path detection for 3GPP LTE system. *Trans. Emerg. Telecommun. Technol.* **2016**, *27*, 339–356. [CrossRef]
28. Zhang, Z.; Kang, S. Time of Arrival Estimation Based on Clustering for Positioning in OFDM System. *IET Commun.* **2020**, *14*, 2584–2591. [CrossRef]
29. Kuang, L.; Ni, Z.; Lu, J; Zheng, J. A time-frequency decision-feedback loop for carrier frequency offset tracking in OFDM systems. *IEEE Trans. Wireless Commun.* **2005**, *4*, 367–373. [CrossRef]
30. 3GPP, FL Summary for Accuracy Improvements by Mitigating UE Rx/Tx and/or gNB Rx/Tx Timing Delays. R1-2103781, Rel.17, 12–20 April 2021. Available online: https://www.3gpp.org/ftp/tsg_ran/WG1_RL1/TSGR1_104b-e/Docs (accessed on 1 July 2021).
31. Cui, W; Li, B.; Zhang, L.; Meng, W. Robust Mobile Location Estimation in NLOS Environment Using GMM, IMM, and EKF. *IEEE Syst. J.* **2019**, *13*, 3490–3500. [CrossRef]
32. Chen, B.; Yang, C.; Liao, F.; Liao, J. Mobile Location Estimator in a Rough Wireless Environment Using Extended Kalman-Based IMM and Data Fusion. *IEEE Trans. Veh. Technol.* **2009**, *58*, 1157–1169. [CrossRef]
33. Chan, Y.; Ho, K. A Simple and Efficient Estimator for Hyperbolic Location. *IEEE Trans. Signal Process.* **1994**, *42*, 1905–1915. [CrossRef]
34. Chang, X.; Yang, X.; Zhou, T. MLAMBDA: A modified LAMBDA method for integer least-squares estimation. *J. Geod.* **2005**, *79*, 552–565. [CrossRef]
35. Al Borno, M.; Chang, X.; Xie, X. On "decorrelation" in solving integer least-squares problems for ambiguity determination. *Surv. Rev.* **2014**, *46*, 37–49. [CrossRef]
36. 3GPP, Study on Channel Model for Frequencies from 0.5 to 100 GHz. TR 38.901, Rel.15, 11 January 2020. Available online: https://www.3gpp.org/ftp/Specs/archive/38_series/38.901 (accessed on 1 July 2021).
37. 3GPP, Physical Channels and Modulation. TS 38.211, Rel.16, 28, September, 2021. Available online: https://www.3gpp.org/ftp/Specs/archive/38_series/38.211 (accessed on 1 July 2021).

sensors

MDPI

Article

An Action Classification Method for Forklift Monitoring in Industry 4.0 Scenarios

Andrea Motroni [1,*], Alice Buffi [2], Paolo Nepa [1,3], Mario Pesi [4] and Antonio Congi [4]

1 Department of Information Engineering, University of Pisa, Via G. Caruso, 56122 Pisa, Italy; paolo.nepa@unipi.it
2 Department of Energy, Systems, Territory and Constructions Engineering, University of Pisa, 56122 Pisa, Italy; alice.buffi@unipi.it
3 Institute of Electronics, Computer and Telecommunication Engineering (IEIIT), Italian National Research Council (CNR), 10129 Turin, Italy
4 Sofidel SpA, 55016 Porcari, Italy; mario.pesi@sofidel.com (M.P.); antonio.congi@sofidel.com (A.C.)
* Correspondence: andrea.motroni@ing.unipi.it

Abstract: The I-READ 4.0 project is aimed at developing an integrated and autonomous Cyber-Physical System for automatic management of very large warehouses with a high-stock rotation index. Thanks to a network of Radio Frequency Identification (RFID) readers operating in the Ultra-High-Frequency (UHF) band, both fixed and mobile, it is possible to implement an efficient management of assets and forklifts operating in an indoor scenario. A key component to accomplish this goal is the UHF-RFID Smart Gate, which consists of a checkpoint infrastructure based on RFID technology to identify forklifts and their direction of transit. This paper presents the implementation of a UHF-RFID Smart Gate with a single reader antenna with asymmetrical deployment, thus allowing the correct action classification with reduced infrastructure complexity and cost. The action classification method exploits the signal phase backscattered by RFID tags placed on the forklifts. The performance and the method capabilities are demonstrated through an on-site demonstrator in a real warehouse.

Keywords: cyber-physical system; Industry 4.0; internet-of-reader; IREAD 4.0; radio frequency identification; RFID classification method; smart gate; smart forklift; smart warehouse

Citation: Motroni, A.; Buffi, A.; Nepa, P.; Pesi, M.; Congi, A. An Action Classification Method for Forklift Monitoring in Industry 4.0 Scenarios. *Sensors* **2021**, *21*, 5183. https://doi.org/10.3390/s21155183

Academic Editors: Riccardo Carotenuto, Massimo Merenda, Demetrio Iero and Roberto Teti

Received: 26 May 2021
Accepted: 27 July 2021
Published: 30 July 2021

Publisher's Note: MDPI stays neutral with regard to jurisdictional claims in published maps and institutional affiliations.

1. Introduction

The term "Industry 4.0" was born in 2013 when the German government promoted the "High-Tech Strategy 2020 Action Plan" for a planned "4th industrial revolution" [1]. Since then, notable efforts have been carried out toward the implementation of Smart Factories [2] and Smart Warehouses [3]. The underlying concept concerns the integration of industrial technologies with information and communication technologies, which leads to the implementation of a Cyber-Physical-System (CPS) [4]. Each part of the system becomes able to autonomously exchange information, trigger actions and control each other [5]. In other words, a CPS allows the implementation of a digital and intelligent factory in order to promote manufacturing to become more digital, information-led, customized, and green [6]. Furthermore, several enabling technologies have been developed for the Industry 4.0 paradigm, e.g., Internet of Things (IoT) [7], Near-Field Communication (NFC) [8], Radio Frequency Identification (RFID) [9], Wireless Sensor Network (WSN) [10], and Block Chain (BC) [11], to name but a few.

The last few years have seen more widespread diffusion of solutions and systems put into practice for the fourth industrial revolution. The aim is to implement an interconnection between production facilities, storage systems, and factory machinery in such a way to allow a real-time interaction between workers, devices and items in the whole supply chain. Consequently, both factory and warehouse facilities may become *smart*.

In such a framework, the implementation of a smart warehouse concerns two different aspects. From one side, the possibility of a real-time inventory of items within the warehouse allows the definition of a proper company production-plan based on the market demand, by avoiding excesses of production and warehouse congestion. On the other hand, the development of a location-based system makes sure of not only the awareness of the item presence but also of its position within the warehouse, together with the position of the vehicles employed for procurement operations. It follows the development of lots of additional functionalities such as the optimization of item placement and of the vehicle paths during the loading/unloading operations with a consequent improvement of operator work-quality and safety.

The I-READ 4.0 project, funded by Regione Toscana, Italy, fits into this context. In particular, it concerns the implementation of an integrated and autonomous CPS for the automatic management of very large warehouses. The system consists of a network of RFID readers in the Ultra-High-Frequency (UHF) operating band, which are able to automatically collect data from the warehouse pallets equipped with UHF-RFID tags and stored within the tissue-paper warehouse of the Sofidel Italian Company in Porcari, Lucca. Firstly presented in [12], the I-READ 4.0 system consists of two main technological elements: UHF-RFID Smart Gate and UHF-RFID Smart Forklift. The Smart Gates use fixed readers able to detect forklifts/pallets entering or exiting from areas of interest. The Smart Forklifts are equipped with UHF-RFID readers able to auto-localize themselves by exploiting data from UHF-RFID reference tags in the scenario and then localize the tagged pallets in the indoor warehouse. The system is low-cost, reconfigurable, flexible and scalable regardless of several factors, e.g. warehouse sizes, good typology and spatial resolution required for item localization.

In this paper, the main idea of the I-READ 4.0 system is a detailed description with particular focus on the UHF-RFID Smart Gate implementation for the forklift action classification. In particular, with the term "action", we refer to two particular movements that the forklift can do with respect to a UHF-RFID Gate. The *IN* action represents the forklift entering a certain area by crossing the gate. The *OUT* action, instead, refers to a forklift leaving a certain area by crossing the gate. The UHF-RFID Smart Gate proposed here is based on an asymmetrical deployment of the reader antenna to allow for a correct forklift discrimination with no additional sensors. The proposed system does not require calibration procedures, and it can be implemented with commercial-off-the-shelf (COTS) hardware. The designed classification method also presents a low computational burden. The Smart Gate implementation is described together with the performance evaluation of an on-site demonstrator. The paper is organized as follows: in Section 2, a state-of-the-art analysis of RFID Gates for good crossing identification is reported; Section 3 describes the I-READ 4.0 architecture, the UHF-RFID Smart Gate and the proposed phase-based action classification method; Section 4 shows the performance of the UHF-RFID Smart Gate, and finally, Section 5 sets conclusions and discusses some future developments.

2. RFID Gates

A UHF-RFID gate is usually composed of a UHF-RFID reader connected to one or more antennas and possibly with other optional devices. Typically its main task is the identification of crossing tagged assets, being goods, people, or vehicles, such as forklifts or pallet trucks. However, an RFID gate able to provide the direction of transit of the identified object/person, can allow a complete awareness of the asset locations in plants or warehouses.

Typically, two main problems occur when deploying an RFID gate in an industrial environment. First, due to the large beamwidth of standard reader antennas and the multipath effects typical of an indoor scenario with metallic objects and surfaces, the target assets crossing the gate are identified together with other static or moving tagged items nearby the gate, so stray read events may occur [13]. Second, the tag reading rate can be slowed due to the presence of the other tags demanding the communication channel resources, thus introducing a non-null probability that the tag on the target asset does not respond to any interrogation query during the crossing action [14]. The multipath effect could also affect the correct detection of the target RFID tags due to the fading effect of the communication channel [15]. To mitigate these issues, solutions relying on shielded reading zones using tunnel gates [16] were proposed. However, such solutions are required for a strong modification of the work environment and are not always suitable or easy to deploy. To avoid shielding structures, other solutions were proposed in [17–19]. In [17], a localization technique is combined with the gate functionality to solve the problem of discriminating among moving and static tags. Keller et al. [18] suggested using various aggregated attributes based on the low-level reader data, e.g., Electronic Product Code (EPC), Received Signal Strength Indicator (RSSI), timestamp, and reading antenna, to perform a classification algorithm in forklift truck applications, getting an overall accuracy of 95.5%. To improve the performance, the same authors extended the method by using an advanced reader antenna setup [19]. By employing a portal configuration with two readers and eight antennas, an overall accuracy of around 99% is obtained at the expense of a relatively high infrastructure cost.

To determine the crossing direction of the assets, additional devices such as light or ultrasound motion sensors [20] can be used, despite the high complexity and cost of the system. Moreover, light or ultrasound motion sensors are prone to false-positives or interruptions as unexpected entities obstruct the sensors. Other systems may employ Computer Vision (CV) and RFID systems as the concept presented in [21], but CV may give rise to privacy issues and also suffers from the outage problem if the light conditions of the environment are not adequate.

To limit cost and complexity of the system, solutions based only on RFID technology have been proposed. The first systems employed more than one antenna to estimate the crossing direction of assets by processing the detection information and the RSSI measurements. In [22], a method was proposed that uses the difference in the crossing time of two antennas aligned along the gate crossing direction without additional external sensors. In [23], a similar method was proposed relying on active RFID tags and based on creating different interrogation zones for each antenna. In [24], a double antenna scheme to control the access of children at a school door was proposed. The antennas are placed on the school door, one facing the inside, the other the outside.

Phase-based solutions [25] can be useful as the backscattered signal phase varies significantly with the motion of tagged assets, and can be profitably used to allow the usage of a single antenna, thus reducing the infrastructure cost. An example of an RFID phase-based access control system exploiting a single antenna was presented in [26] for tagged people crossing-direction discrimination. It is noteworthy that phase-based techniques can also allow to discriminate tags carried out by a forklift [27] or moving along a conveyor belt [28] with respect to static tags in the warehouse/plant scenario. The concept of phase measurements applied to conveyor belts was also explored in [29], where a two-antenna architecture was proposed for measuring the Direction of Arrival (DoA) of moving RFID tags for localization purposes. The Doppler Effect can be indeed profitably exploited for the tag localization on conveyor belts, as demonstrated by [30].

More recently, machine learning techniques were investigated in RFID systems both for localization purposes [31,32] and RFID Smart Gate implementation [33–35]. In [33], a single antenna architecture was proposed to determine the direction of people crossing an indoor RFID gate based on an Artificial Neural Network (ANN). Consecutive RSSI data are aggregated within frames, and the mean RSSI for each time frame is fed as an input feature for the neural network. The obtained accuracy is higher than 99%. Machine Learning solutions were also employed to solve the issue of stray reads [34], where a 97.5% classification accuracy among actual RFID tags crossing the gate and static or other tags moving close to the gate without crossing it was achieved with a single antenna architecture. However, such a system does not allow the crossing direction estimation. In [35], both the RSSI and the phase are processed through different machine learning techniques to discriminate among moving and static RFID tags. In fact, when the relative distance between the reader antenna and the tag changes, both the received power and phase change significantly.

A concept for asset tracking was proposed in the patent [36] as a device-free user localization scheme. Basically, a set of antennas is attached at the ceiling facing the floor, whereas a set of tags is placed on the floor. A moving object can shadow the tags and create a signature of the motion of the object itself. The same scheme is applicable for RFID gates, as proposed in [37,38] to solve the problem of pallet trucks crossing a key point (e.g., to monitor the charging of goods on a truck). In both solutions, an antenna is placed at the ceiling facing downwards, and a regular grid of 24 tags is placed on the floor. When a metallic cart crosses the target area, the tags are shadowed. Such information is given in input to a Long-Short Term Memory (LSTM) [39] Recurrent Neural Network (RNN) [37] or a convolutional neural network [38]. In both cases, a classification accuracy of 100% is obtained. Despite the robustness of these solutions, the deployment of the tags on the floor is unfortunately not always possible in warehouse scenarios, as the tags cannot stand high pressures caused by the weight and encumbrance of industrial vehicles such as forklifts.

3. Materials and Methods

3.1. The I-READ4.0 System Architecture

The I-READ 4.0 system was conceived by considering large-area warehouses with a high pallet-handling per day. The demonstrator was designed to operate in the *Tassignano* warehouse of the Sofidel paper industry with headquarters in Porcari, Lucca (https://www.sofidel.com/, accessed on 26 May 2021). It has an area of around 20,000 m^2 (Figure 1) with an average handling of 2000–3000 pallets per day. Figure 2 illustrates the I-READ 4.0 framework, which comprises two main technological elements: the UHF-RFID Smart Gate (studied in this paper) and the UHF-RFID Smart Forklift.

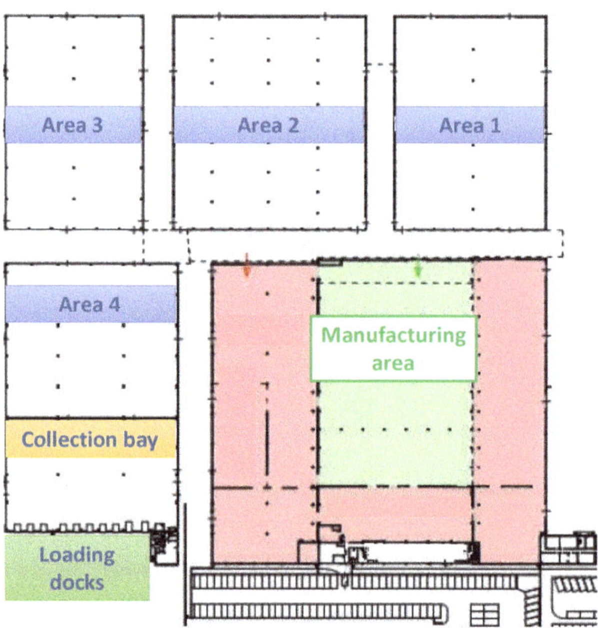

Figure 1. *Tassignano* warehouse plan.

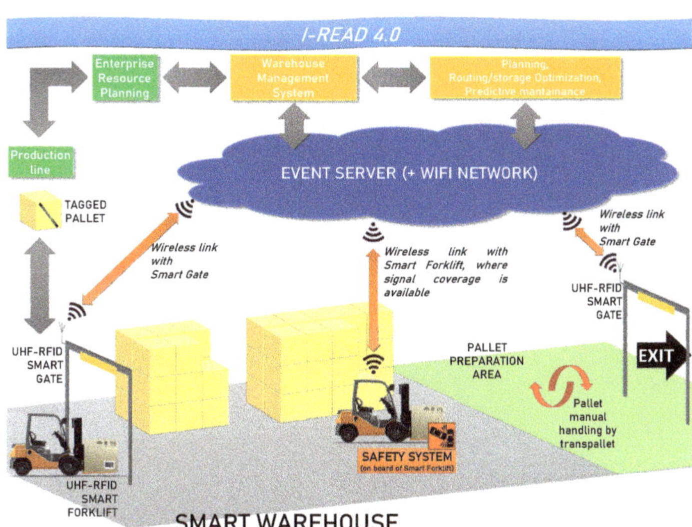

Figure 2. The I-READ 4.0 framework.

The general architecture of the system is briefly described here. The proposed solution uses the passive UHF-RFID technology and particularly an integrated network of RFID readers, some fixed (UHF-RFID Smart Gates) and other mobile (UHF-RFID Smart Forklifts), capable of identifying individual pallets, their status (loaded by a forklift or unloaded), and their location (Figure 2). The tagged objects are pallets containing the final product, e.g., tissue paper. The pallets exit from the end of the production line and are brought into the storage warehouse carried by forklifts. Again, when the product must be shipped, a forklift lifts the pallet and brings it to the loading area (pallet preparation area). Then, each single pallet is loaded onto the truck manually handled by a pallet truck. For the correct management of the warehouse, it is essential to trace all these steps. Both the warehouse entrance at the end of the production line and the exit are equipped with a UHF-RFID Smart Gate, described later in this manuscript, which is capable of monitoring all the access/departure of products and forklifts to/from the warehouse. When the UHF-RFID Smart Forklift moves inside the warehouse, it is localized with a tracking system to allow the real-time pallet localization. In fact, the pallet location is associated to the forklift location at the time of the unloading event. In this context, the presence of the UHF-RFID Smart Gates can be fruitfully exploited to set the initial position of the forklift when developing tracking systems. Through the Wi-Fi network, the Smart Gate and the Smart Forklift send the data regarding position and status of each pallet to the warehouse central server. The knowledge of the position of pallets and forklifts allows to produce a real-time map of warehouse occupation and therefore enables to implement an optimization algorithm to improve the management of good flows and the occupation of warehouse areas. Furthermore, the information on the forklift position, combined with the data of the collision detection system installed on each forklift, allows to carry out a statistical analysis about the areas with the highest risk of collision. The detection of these potential collisions (near miss) will be shown to the forklift drivers and the Warehouse Management System (WMS) through the Event Server. For the aim of this paper, the design, development and testing of the UHF-RFID Smart Gate are relevant. That is, we are going to focus on that component of the global system architecture.

Items coming out from the production lines are assembled in pallets. Each pallet is around 80×120 cm wide, and it has to be equipped with an identification label of size 148×105 mm according to the Global Standard GS1 (Figure 3). The label is printed at the end of the production line and shows the Serial Shipping Container Code (SSCC). Behind the label, there is a UH101 tag by LAB-ID measuring 95 × 88 mm and equipped with the NXP UCODE 7 chip (http://www.lab-id.com/wordpress/wp-content/uploads/2017/06/UH101.pdf, accessed on 26 May 2021) with −21 dBm sensitivity. The tag on the *smart label* is initialized through the CAEN RFID Proton R4320P reader (https://www.caenrfid.com/en/products/proton-r4320p/, accessed on 26 May 2021) connected to the CAEN RFID ANT-024 SPIN antenna. A picture of the end of the production line along with the RFID hardware to write the tag EPC is shown in Figure 4. In particular, the EPC is properly derived from the translation of the SSCC code according to the GS1 standard (https://www.gs1.org/sites/default/files/docs/epc/EPC-RTIPalletTagging-ImpGuide-i2.pdf, accessed on 26 May 2021).

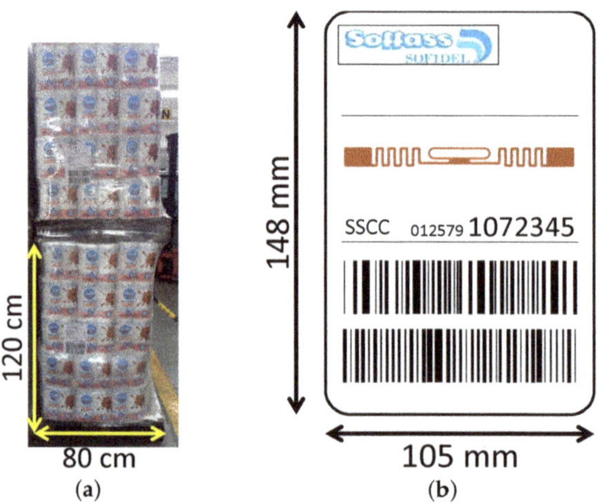

Figure 3. (**a**) Column composed by two tagged pallets and (**b**) sketch of the tagged label applied on the pallet (the tag is on the label rear side).

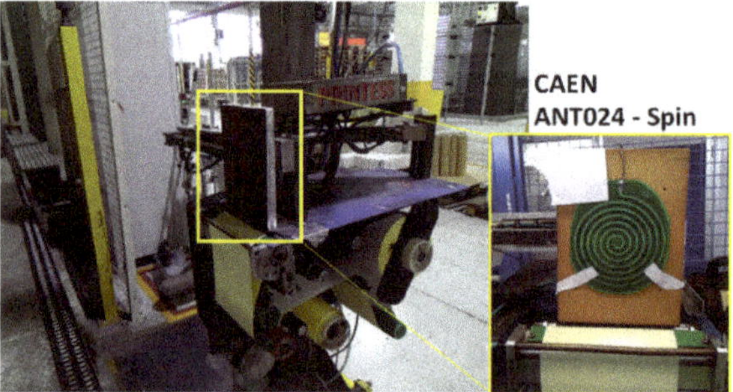

Figure 4. RFID label printer at the end of the production line.

Pallets are taken from the manufacturing area by Laser Guided Vehicles (LGVs) and carried at the entrance of the warehouse, which is composed of four storage areas (Figure 1). Here, workers handle them through the RFID Smart Forklift and bring them to a specific warehouse area passing through the RFID Smart Gate.

3.2. The UHF-RFID Smart Gate

Within the I-READ 4.0 system, the goal of the UHF-RFID Smart Gate is to monitor the crossing of goods at a point of interest within the warehouse. The gate must be able to completely identify the loaded pallets carried out by the forklift, to identify the forklift, and to understand its crossing direction. In fact, most gates can be crossed in both directions, and it is important to correctly determine if a product enters or leaves a certain warehouse area. That is, the UHF-RFID Smart Gate must implement an action classification method to understand whether the forklift is entering (*IN*) or leaving (*OUT*) a certain zone. The UHF-RFID Smart Gate is composed of the following hardware:

- An Impinj Speedway Revolution R420 UHF-RFID reader;
- A circularly-polarized (CP) CAEN WANTENNAX019 antenna;
- A linearly-polarized (LP) CAEN WANTENNAX007 antenna;
- Two coaxial cables;
- An ethernet cable to connect the reader to the Event Server;
- A power supply for the reader.

The circularly-polarized (CP) antenna CAEN WANTENNAX019 (https://www.caenrfid.com/it/products/wantennax019/, accessed on on 26 May 2021) was installed in the upper part of the gate, at a height of about 4.5 m. It has a gain of 8.5 dBc and a half power beam width of 65° on both planes ($HPBW_H = HPBW_V = 65°$). It was installed with a tilt angle of about 30° with respect to the horizontal plane to create an asymmetrical radiation-pattern footprint with respect to the gate. Thanks to this particular configuration characterized by an asymmetrical antenna deployment, the forklift crossing direction can be determined by using only a single antenna, as described later. Such CP antenna is mainly used to identify the forklift tags and to perform the action classification method.

With the intention of increasing the reliability of the gate when detecting all the carried pallets, a second antenna was installed at the gate side. Since the RFID labels on the pallets are always applied at the same position and parallel to the ground, a linearly-polarized (LP) antenna was chosen to maximize the power radiated to the tag. The chosen model is the CAEN WANTENNAX007 (https://pdf.directindustry.com/pdf/caen-rfid/wantennax007/113435-366469.html, accessed on on 26 May 2021) with gain equal to 8.0 ± 0.5 dBi, and half power beam width equal to 65° on the horizontal plane ($HPBW_H = 65°$) and 68° on the vertical plane ($HPBW_V = 68°$). The antenna was fixed to the wall at a height of about 3 m from the ground and tilted to about 45° with respect to the horizontal plane. In Figure 5, two of the UHF-RFID Smart Gates installed at the warehouse entrance are shown. It must be highlighted that the gate infrastructure does not include additional invasive metallic structures as typical for tunnel gates [16].

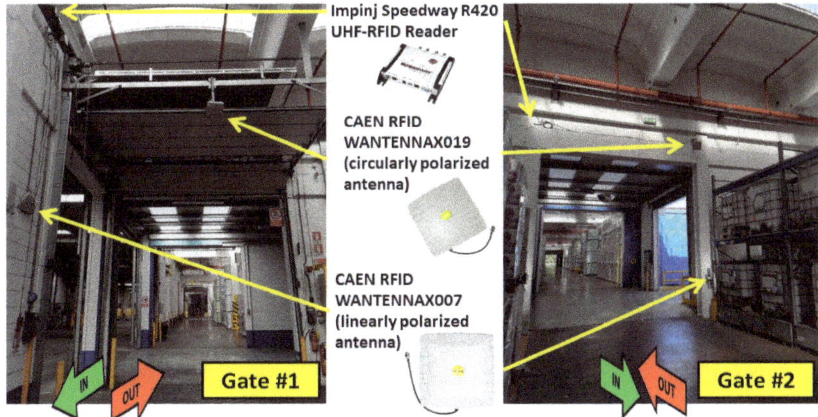

Figure 5. RFID Smart Gates installed at two entrances of the *Tassignano* warehouse.

The forklifts are equipped with two OMNI-ID EXO 2000 on-metal RFID tags (https://omni-id.com/datasheet/1373, accessed on 26 May 2021) to be identified by the gate. One tag is placed on the forklift upright at a height of 2.6 m (Figure 6a), while the second tag is placed on the forklift roof at a height of 2.2 m (Figure 6b) for redundancy purposes.

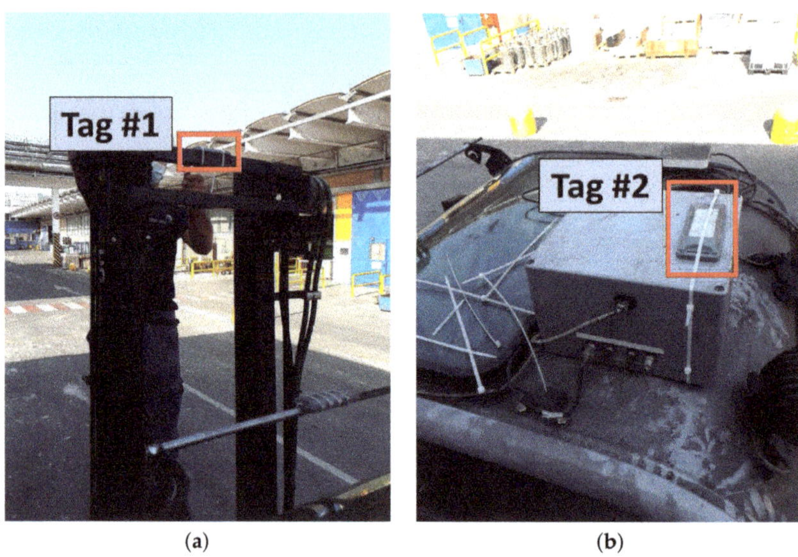

(a)

(b)

Figure 6. RFID tags placed on the forklift. (**a**) Tag placed on the upright, and (**b**) tag placed on the forklift top.

Two photocell barrier sensors SICK WTT12L-B2561 are placed in proximity of the gates to evaluate the performance of the phase-based action classification method and to get an estimate of the forklift speed v. A picture of the photocells is in Figure 7.

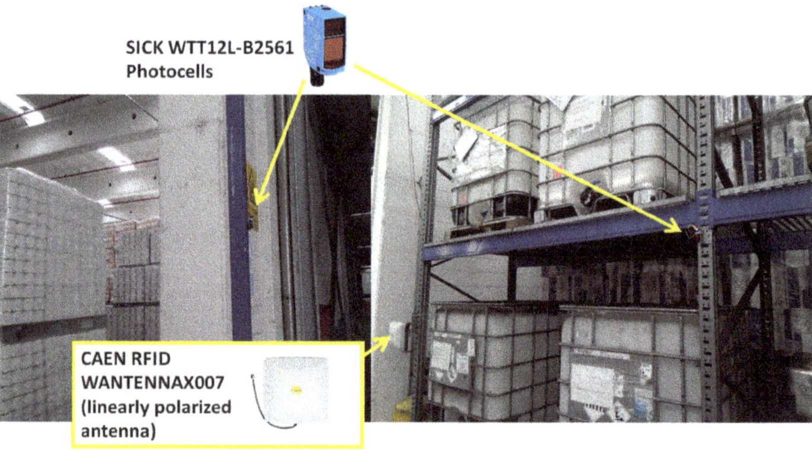

Figure 7. Photocells installed at one of the UHF-RFID Smart Gates.

3.3. Signal Model

The phase-based action classification method proposed here enables a smart-gate operation with a single antenna to determine the moving direction of the forklift crossing the gate. When the reader interrogates a tag, the latter backscatters the impinging signal, thus enabling the reader to measure a phase proportional to the distance between the reader and the tag. When the tagged forklift crosses the gate, the reader antenna performs several queries of the moving tag and measures the phase of the signal at different time steps t_n, being $n \in \{0, ..., N_R - 1\}$ and N_R the number of successful queries of the tag during the vehicle crossing. To be more precise, the phase of the signal measured by the reader at time t_n can be resumed as:

$$\phi(t_n) = \mathrm{mod}\left(\frac{-4\pi d(t_n)}{\lambda} + \phi_0(\theta_R, \psi_R, \theta_T, \psi_T, t_n) + \phi_m(t_n)\right)_{2\pi} \tag{1}$$

where $d(t_n)$ is the distance between the tag and the reader at time t_n, λ is the carrier wavelength in free-space, $\phi_0(\theta_R, \psi_R, \theta_T, \psi_T, t_n)$ is the phase bias caused by reader and tag antennas and by the electrical circuitry, where θ_R and ψ_R are the elevation and azimuth angle at time t_n, respectively, at the reader antenna side, and θ_T and ψ_T are the elevation and azimuth angle at time t_n, respectively, at the tag antenna side. $\phi_m(t_n)$ is the contribution to the phase caused by multipath phenomena at time t_n. The distance $d(t_n)$ is defined as:

$$d(t_n) = \|\mathbf{p_{ant}} - \mathbf{p_{tag}}(t_n)\| \tag{2}$$

where $\mathbf{p_{ant}}$ is the vector $[x_{ant}, y_{ant}, z_{ant}]^T \in \mathbb{R}^3$ of the reader antenna location, and $\mathbf{p_{tag}}(t_n)$ is the vector $[x_{tag}(t_n), y_{tag}(t_n), z_{tag}(t_n)]^T \in \mathbb{R}^3$ of the tag trajectory sample at time t_n.

The value of $\phi_0(\theta_R, \psi_R, \theta_T, \psi_T, t_n)$ is defined as:

$$\phi_0(\theta_R, \psi_R, \theta_T, \psi_T, t_n) = \phi_{TX}(\theta_R, \psi_R, t_n) + \phi_{RX}(\theta_R, \psi_R, t_n) + \phi_{tag}(\theta_T, \psi_T, t_n) \tag{3}$$

where ϕ_{TX} and ϕ_{RX} are the phase offsets caused by the transmitting and receiving circuitry of the reader, and ϕ_{tag} is a phase offset that depends on the tag itself and may be different even among tags of the same model. The $\phi_0(\theta_R, \psi_R, \theta_T, \psi_T, t_n)$ term is almost constant over consecutive tag query responses within the reader antenna's main beam, and it will be indicated in the rest of the paper as ϕ_0.

To overcome the problem of the phase $2\pi-$ambiguity, we can perform phase unwrapping [40]:

$$\phi^u(t_n) = \frac{-4\pi d(t_n)}{\lambda} + \phi_0 + \phi_m(t_n) \tag{4}$$

To correctly execute the phase unwrapping, consecutive phase samples must not differ more than π. If we consider the value of $\phi_m(t_n) - \phi_m(t_{n-1}) \approx 0$, meaning that the phase difference caused by the multipath between consecutive time steps is negligible, only the condition $d(t_n) - d(t_{n-1}) < \lambda/4$ must be satisfied. This fact is a direct consequence of the Nyquist–Shannon Sampling Theorem, which states the condition for which a signal is sampled without aliasing. Further considerations on the topic applied to the RFID field can be found in [41,42]. As it will be discussed later, a relatively high forklift speed or a poor RFID reader sampling rate may both lead to errors during the phase unwrapping process and, therefore, to classification errors.

Now, for the sake of simplicity, the value of $\phi^u(t_n)$ is normalized by the first sample acquired at $n = 0$. We represent the normalized unwrapped phase with $\phi^n(t_n)$:

$$\phi^n(t_n) = \frac{-4\pi\Delta d(t_n)}{\lambda} + \Delta\phi_m(t_n) \tag{5}$$

where $\Delta d(t_n) = d(t_n) - d(t_0)$, and $\Delta\phi_m(t_n) = \phi_m(t_n) - \phi_m(t_0)$.

3.4. RFID Gate with Antenna in Symmetrical Configuration

By referring to Figure 8, we consider a bi-dimensional scenario in which the forklift moves mainly along the x-axis with a constant speed v; such a hypothesis is plausible in a few-second interval, when considering the forklift weight and inertia. When the forklift performs an *IN* action, it moves towards the positive direction of the x-axis with positive speed, whereas when performing an *OUT* action, it moves towards the negative direction with a negative speed. The tag is placed on the forklift top, at a height h_{tag}. The gate antenna is placed in $[x_{ant}, y_{ant}, z_{ant}]^T = [0, 0, h_{ant}]^T$, and it is facing the floor in such a way that its coverage area is symmetrical in the xy-plane with respect to the z-axis. The coverage area on the tag plane is determined by the antenna $HPBW$ through the following equation:

$$l = \Delta h \arctan(HPBW/2) \tag{6}$$

where Δh is the height difference between the antenna and the tag: $\Delta h = h_{ant} - h_{tag}$. This means that the tag is detectable when the forklift is inside the region $|x| < l$.

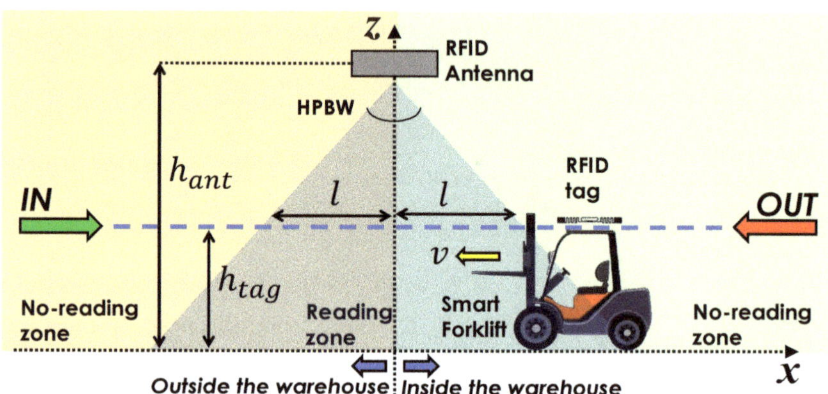

Figure 8. Sketch of the symmetrical configuration of the RFID Smart Gate.

Let us suppose the forklift is performing an *IN* action. The time variation of the x-coordinate is:

$$x(t) = -l + vt \tag{7}$$

being $t \geq 0$. By considering a constant sampling time T, the acquisition time steps t_n can be written as $t_n = nT$. By denoting $x[n] = x(t_n) = x(nT)$, we can also derive the normalized unwrapped phase sequence $\phi^n[n]$ with (4) as follows:

$$\phi^n[n] = \frac{-4\pi}{\lambda}\left(\sqrt{(-l + vnT)^2 + \Delta h^2} - \sqrt{(-l)^2 + \Delta h^2}\right) \tag{8}$$

where we neglected the effect of the multipath for simplicity. Let us consider an RFID gate operating at the frequency $f = 865.7$ MHz. The unwrapped normalized phase $\phi^n[n]$, is depicted in Figure 9 when $l = 3$ m, $v = 2$ m/s, $\Delta h = 2.5$ m, and $T = 50$ ms, for both *IN* and *OUT* actions. As expected, during an *IN* action, the normalized unwrapped phase decreases when the forklift (tag) is approaching the antenna in the region $x \leq 0$, while it increases once the forklift (tag) has crossed the gate and gets further from the antenna in the region $x > 0$. For the *OUT* action, instead, the normalized unwrapped phase decreases when the forklift (tag) is approaching the antenna in the region $x \geq 0$, and increases once the forklift (tag) has crossed the gate and gets further from the antenna in the region $x \leq 0$. It appears straightforward that the time behavior of $\phi^n[n]$ is the same for both *IN* and *OUT* actions, as the antenna coverage area is symmetrical. Therefore, it is not possible to discriminate between the two actions by using this gate configuration.

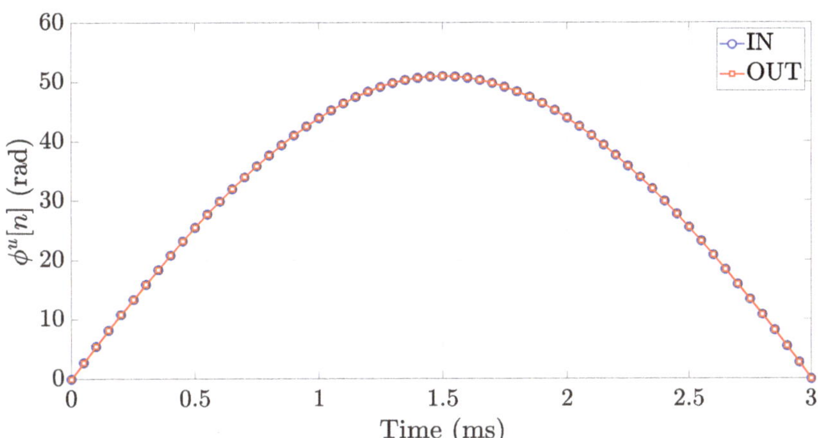

Figure 9. Time behavior of the unwrapped normalized phase in the symmetrical configuration of the RFID Smart Gate for the following system parameters: $f = 865.7$ MHz, $l = 3$ m, $v = 2$ m/s, $\Delta h = 2.5$ m, and $T = 50$ ms.

3.5. RFID Gate with Asymmetrical Antenna Deployment and Action Classification Method

To make the $\phi^n[n]$ time behavior different between the two actions, *IN* and *OUT*, and to allow correct action discrimination, the reader antenna is tilted of an angle θ with respect to the vertical axis (z-axis) to make it point towards the inside of the warehouse in such a way that the reader cannot detect tags outside the room, as shown in Figure 10. Let us suppose that the antenna is pointed in such a way that it can only detect tags within the region $l_1 \leq x \leq l_2$, with l_1 and l_2 real positive values and $l_1 < l_2$. When the forklift performs an *IN* action, the tag will be detected only when it is getting further from the antenna, so the $\phi^n[n]$ will be a decreasing function. On the other hand, when the forklift performs an *OUT* action, the tag will be detected only when it is getting closer to the antenna, so the $\phi^n[n]$ will be an increasing function. The time behavior of $\phi^n[n]$ for *IN* (blue circular markers) and *OUT* (red squared markers) actions is depicted in Figure 11 when $l_1 = 1$ m, $l_2 = 4$ m, $v = 2$ m/s, $\Delta h = 2.5$ m, and $T = 50$ ms. These results confirm that the asymmetrical configuration of the gate antenna guarantees the capability of recognizing the *IN* and *OUT* actions, without requiring additional antennas or sensors.

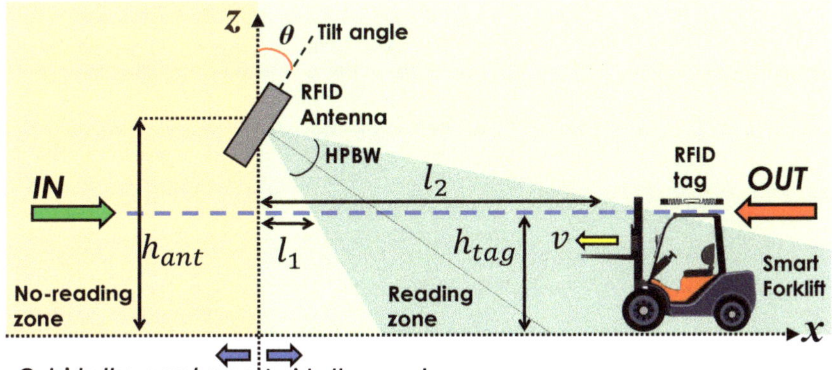

Figure 10. Sketch of the asymmetrical gate.

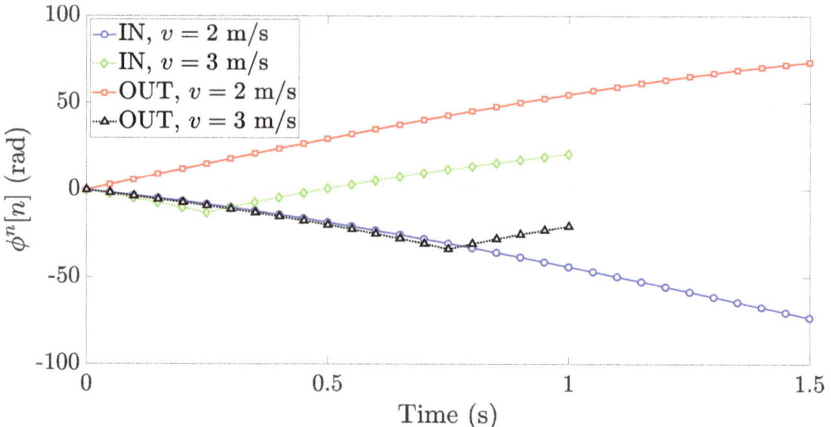

Figure 11. Time behavior of the normalized unwrapped phase in the asymmetrical antenna deployment for *IN* and *OUT* actions by varying the speed v when the parameters are the following: $f = 865.7\,\text{MHz}$, $l_1 = 1\,\text{m}$, $l_2 = 4\,\text{m}$, $\Delta h = 2.5\,\text{m}$, and $T = 50\,\text{ms}$.

The classification algorithm is straightforward. If the measured $\phi^n[n]$ is a decreasing function, the estimated action is *IN*; otherwise, the estimated action is *OUT*. To do that, we first interpolate the measured curve with a first-order polynomial function. Then, we calculate the slope coefficient m and execute the following decision criterion:

$$\begin{cases} \text{Classified action: IN} & \text{if } m \leq 0, \\ \text{Classified action: OUT} & \text{if } m > 0 \end{cases} \tag{9}$$

As already said, to operate correctly, this algorithm must rely on a correct phase unwrapping of the measured phase. When the forklift speed increases, the average spatial sampling may be greater than $\lambda/4$. This effect makes the Nyquist sampling condition not satisfied, and the slope of the normalized unwrapped phase may change at some points. By leaving all the other parameters unchanged, Figure 11 also shows the normalized unwrapped phase for the forklift speed $v = 3\,\text{m/s}$, instead of $v = 2\,\text{m/s}$. The aforementioned slope change is strongly evident for both the *IN* (green diamond markers) and *OUT* (black triangle markers) actions. This means that, on the basis of the forklift speed, the estimation of the curve slope m could fail by leading to a possible classification error. As a consequence, the reader queries have to be sent with a time interval able to guarantee the Nyquist sampling condition by knowing the maximum allowed speed for the forklift.

As we will see in the next section, the influence of the environment can also introduce errors in the classification method.

Moreover, static tagged forklifts or pallet tags nearby the gates can be filtered out from the classification method, as their measured phase is almost constant. An advantage of this algorithm is the low-effort computational burden which allows the method implementation on low-power computers, as it will be shown in the next section. Alternatively, the method can be directly executed on an RFID reader dedicated PC if this is present. Another solution is to transmit the data on an external PC that controls all the RFID Smart Gates of the warehouse, as was done in this proof of concept.

4. Experimental Analysis

4.1. Experimental Results

Figure 12a,b shows an example of a successful and unsuccessful *IN* classification, respectively. As apparent in Figure 12b, the unwrapping fails by causing a wrong sign estimation of the slope coefficient m. Similarly, Figure 12c,d shows an example of a successful and unsuccessful *OUT* classification. In such a case, the slope coefficient m of Figure 12d is wrongly estimated as negative. Table 1 resumes the principal features of the showed curves in terms of the number of samples N_R, time duration of the crossing action T_d, forklift speed v, average sampling time Δ_T, average spatial sampling Δ_S and measured slope m.

There are multiple causes of unsuccessful classification, mainly related to a failed phase unwrapping in industrial scenarios. First of all, the multipath phenomena can introduce strong and unpredictable contributions on the phase variation $\phi^n[n]$. Second, the presence of multiple tags close to the gate demanding for the communication channel may slow the forklift-tag reading rate. Finally, the speed of the forklifts, which can move up to 3.5 m/s, may cause a poor sampling of the phase curve and consequently a wrong phase unwrapping.

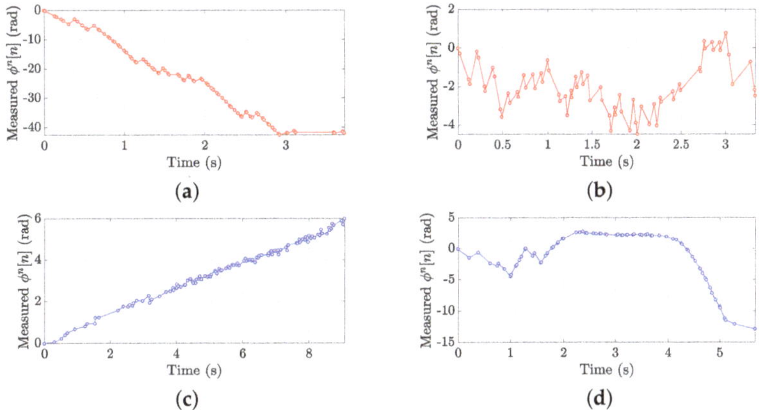

Figure 12. Examples of measured normalized unwrapped phase $\phi^n[n]$. (**a**) Successful *IN* classification, (**b**) wrong *IN* classification, (**c**) successful *OUT* classification, (**d**) wrong *OUT* classification.

Table 1. Main features of the trials represented in Figure 12.

Trial	N_R	T_d (s)	v (m/s)	Δ_T (ms)	Δ_S (cm)	m (rad/s)
Successful *IN*	85	3.7	1.57	44	7	−13.28
Wrong *IN*	72	3.3	0.98	46	4.5	0.1063
Successful *OUT*	114	9.1	0.49	80	3.3	0.62
Wrong *OUT*	76	5.7	0.77	74	5.7	−0.97

To better understand the effect of the forklift speed v and evaluate the classification accuracy, an experimental campaign was conducted. We analyzed a total of $N_T = 264$ trials acquired from the gate placed at the production line end during the regular forklift operations. The number of recorded *IN* and *OUT* actions is $N_{IN} = 164$ and $N_{OUT} = 100$, respectively. The reason for such difference is due to the exclusion from the experimental analysis of all the cases where the optical barrier sensors failed, so it was not possible to determine the forklift speed and recognize the ground truth of the forklift passage. In 100% of the cases, at least one of the two tags placed on the forklift was detected by the CP antenna at least once.

The classification accuracy computed for different ranges of the forklift speed v is shown in Figure 13. The overall action classification accuracy of the method is 92% but reaches a maximum value of 97–98% when the forklift travels at a speed between 0.5 m/s. and 1.5 m/s. It is apparent that, when the forklift overpasses the speed of 1.5 m/s, the accuracy of the action classification method decreases as the phase unwrapping fails. On the other hand, a low forklift speed can be detrimental too, since the phase slope could be too close to zero, making the action classification less reliable. This effect is apparent in Figure 13 for $v < 0.5$ m/s.

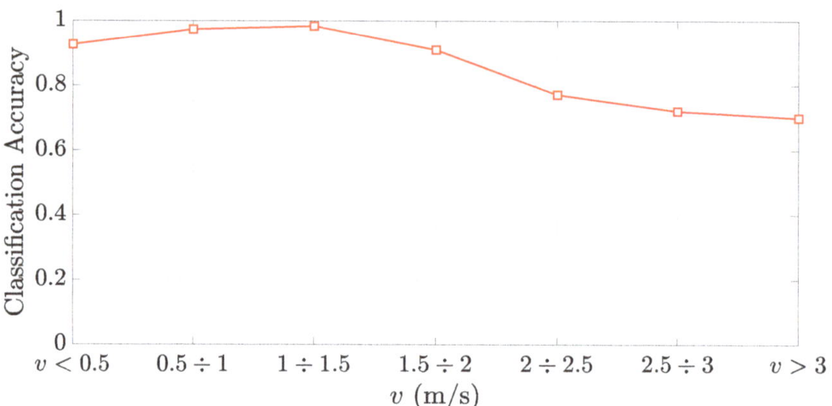

Figure 13. Classification accuracy vs. forklift speed v.

The number of tag readings is a crucial parameter for the success of the classification algorithm. Therefore, the average number of samples with respect to the forklift speed is also reported (Figure 14). As expected, the number of available readings decreases with the increase of the forklift speed. However, thanks to the proper reader configuration, the average number of readings never goes under 45 for $v < 3$ m/s. When this cannot be guaranteed, ad-hoc interpolation techniques could be adopted.

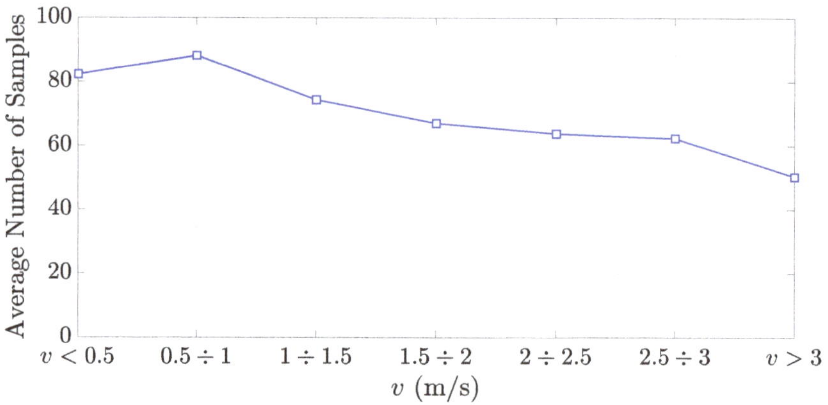

Figure 14. Average number of samples vs. forklift speed v.

Finally, to demonstrate the low computational burden of the proposed method, the elaboration time of the $N_T = 264$ trials has been depicted in Figure 15. The analysis was conducted on a laptop with an Intel(R) Core(TM) i7-7700HQ CPU @ 2.80 GHz and 16 GB RAM, showing a mean elaboration time of 0.13 ms with a standard deviation of 0.05 ms. The case totality required less than 1 ms to be processed. Such a time is negligible with respect to the acquisition time, which depends on the forklift speed and can be in the order of 1–2 s. Therefore, we can conclude that the computational burden of the algorithm is not an issue at all.

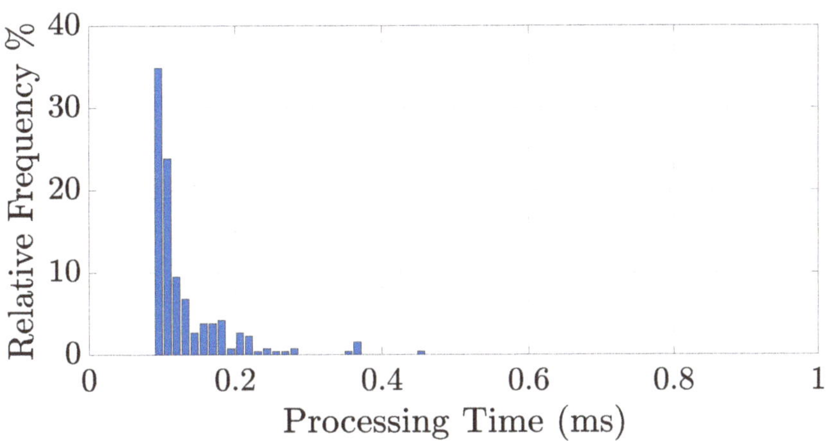

Figure 15. Histogram of the processing time (ms) for the $N_T = 264$ analyzed trials.

4.2. Discussion

A discussion on possible alternatives to this algorithm must be conducted. As reported in [25], it is possible to measure the radial speed v_r of a tag with respect to the reader antenna through the acquisition of the Doppler frequency shift. Indeed, the tag radial speed measurements in the asymmetrical antenna deployment can be profitably used for the forklift action classification similarly to (9). To obtain reliable Doppler frequency shift data, the reader manufacturer suggests to configure the Impinj Speedway R420 reader to low reading-rate modes [43]. In such way, the duration of the RFID signal packets is longer; therefore, the Doppler frequency shift is easier to be measured. However, such a condition does not fit with our need to have fast readings to ensure both the forklift and the goods detection and to satisfy the Nyquist–Shannon Sampling Theorem. Therefore, during the tests, we had to configure the reader to a fast reading-rate mode, so the Doppler frequency shift measurements were affected by severe detrimental noise. Consequently, the here proposed signal processing Equation (9) resulted in a more robust, reliable and accurate action classification method. Additionally, the fast-rate reader configuration allows minimizing the number of cases where the Sampling Theorem is not met and phase unwrapping fails. Another aspect that must be considered is the Doppler shift $\Delta f = 2fv_r/c$, when the forklift travels at high speed, e.g., $v = 3$ m/s, $\Delta f < 17.31$ Hz. Given that the bandwidth for a single RFID channel in the ETSI European lower band is 200 kHz [44], such Δf can be considered negligible and difficult to measure. Finally, the proposed method does not require any preliminary system calibrations, and can be implemented with COTS devices.

4.3. Comparison with the State-of-the-Art

Each state-of-the-art solution presented in Section 2 requires a different and custom architecture, so it is difficult to make a fair comparison by evaluating the classification performance of other pre-existing solutions directly on-site with the same antenna configuration and dataset. In any case, we can compare the proposed system with the others analyzed in Section 2 in terms of cost, encumbrance, and scalability. The cost of a COTS RFID system at the UHF-RFID band is mainly determined by the RFID reader, which may reach more than 1000$ (USD). Each RFID antenna costs around 100–200$ (USD) and, therefore, can be a significant cost for solutions requiring multiple antennas. The cost of a passive RFID tag can be considered negligible for small volumes of goods, as RFID inlay labels usually cost less than 0.1$ (USD). Some passive RFID tags designed for metallic surfaces can cost around 10–20$ (USD) each, but there are many models that can be bought for less than 5$. Battery-Assisted Passive (BAP), active, or sensor-equipped tags can reach a cost of 30$, but they are usually not necessary. Metallic supports or shields shall be included in the total cost of the system, and therefore, it turns out that shielded gates [16,18] are quite expensive solutions due to the large infrastructure required. The encumbrance is relative to the global volume occupied by the hardware needed to implement the gate, which could be very significant in the case of shielded gates. Cost and encumbrance together usually impact the scalability of the solution since a high cost, or alternatively, a high encumbrance, makes the solution less replicable inside the plant, factory, or warehouse. The scalability of a solution is also determined by the time required for the installation process. For instance, mounting several antennas at the ceiling, mounting several shielded gates, or installing several photocells or ultrasound barriers in addition to the RFID hardware could be a time-consuming operation, which must be considered as a significant cost. Finally, solutions based on a machine learning classification algorithm could require a supervised training process, which can be difficult to achieve in a short time, and huge amounts of data have to be collected in several operating conditions.

As summarized in Table 2, the solutions based on shielded gates [16,18] have been considered as "High Cost", "High Encumbrance" and "Low Scalability" due to the cost of the metal shields, their volume, and the installation complexity. On the other hand, shielded gates are the best options to filter out false positive readings.

By referring to [19], we opted for "Medium–High Cost", "Medium–High Encumbrance", and "Low–Medium scalability". Indeed, the proposed solution requires antennas aggregated in panels. The cost and the encumbrance are lower than the shielded gates, but the cost of the antenna panel is not negligible and must be considered when taking into account the system scalability. The systems proposed in [22–24] have been evaluated as "Low–Medium Cost", "Low–Medium Encumbrance" and "Medium–High Scalability". Indeed, the three systems require two antennas, which increase the cost with respect to solutions with a single antenna, and the encumbrance cannot be considered as "low", too, as it is required to find enough space for two antennas. On the other hand, the installation of two antennas is indeed a fast process, and therefore, the scalability of the solutions is good. The solutions in [26,28] are based on phase processing, such as the one presented in this paper, and also require a single antenna. Therefore, they are classified as "Low Cost", "Low Encumbrance", and "High Scalability" [33] as they still rely on a single antenna, but the scalability is considered "Medium" as the proposed solution is based on a neural network classifier, which requires a time-consuming training stage. Following the same reasoning, the two solutions exposed in [37,38], both based on neural networks classifiers, are considered "Medium Scalability" solutions. In this case, however, the presence of the reference RFID tags on the ground makes the encumbrance of the solution higher with respect to solutions that do not require reference tags. Finally, the solution proposed in this paper is considered "Low Cost", "Low Encumbrance" and "High Scalability", as it needs a single antenna and does not require any calibration stages at the installation time. In comparison with the solutions of the same category in terms of cost, encumbrance, and scalability, e.g., [28], the proposed solution is designed to work in more complex environments with respect to the conveyor belt, where the speed of the RFID tags is known in advance, and the tag motion is constrained along assigned paths. Reference [26] is indeed a solution with low cost, low encumbrance and high scalability, but the proposed method has been evaluated only in a laboratory/office environment, whereas the solution proposed in this paper has been verified in a real industrial environment during regular work activities.

Table 2. Comparison of the proposed solution with the state-of-the-art.

Reference	Cost	Encumbrance	Scalability	Architecture
[16]	High	High	Low	Shielded Gate
[18]	High	High	Low	Shielded Gate
[19]	Medium–High	Medium–High	Low–Medium	Antenna Panels
[22]	Low–Medium	Low–Medium	Medium–High	One reader and two antennas
[23]	Low–Medium	Low–Medium	Medium–High	One reader and two antennas
[24]	Low–Medium	Low–Medium	Medium–High	One reader and two antennas
[26]	Low	Low	High	One reader and one antenna
[28]	Low	Low	High	One reader and one antenna
[33]	Low	Low	Medium	One reader and one antenna
[37]	Low	Low–Medium	Medium	One reader, one antenna, reference tags
[38]	Low	Low–Medium	Medium	One reader, one antenna, reference tags
This paper	Low	Low	High	One reader and one antenna

5. Conclusions

This paper presented an effective implementation of a UHF-RFID Smart Gate, a fixed identification point placed at warehouse key points for forklift monitoring. Each Smart Gate implements an action classification method that exploits the phase of the backscattering RFID signal to determine the gate crossing direction of the forklifts with respect to the gate. Thanks to an asymmetrical deployment of the reader antenna and the phase acquisition of the signal exchanged by the fixed reader antenna and tags on the forklifts, a scalable and low-cost solution exploiting only one antenna can be used for each gate, with no additional sensors. Performance and method capabilities were investigated through an experimental demonstrator installed in a real warehouse. Data were gathered during the regular operations of the workers. In 100% of cases, the forklift was detected by the RFID gate, and a 98% classification accuracy was achieved when the forklift speed ranged between 0.5 m/s and 1.5 m/s. The accuracy decreases for higher speeds. The proposed method requires short computational time and is therefore suitable for the real-time monitoring of the forklift crossings. For future developments, artificial intelligence techniques will be designed and evaluated to improve classification accuracy even when forklifts are moving at higher speeds.

Author Contributions: Conceptualization, A.M., A.B., P.N.; software, A.M.; methodology, A.M., A.B., P.N., validation, A.M., A.B., M.P., resources, P.N., A.C., writing, A.M., A.B., P.N., review and editing, all authors. All authors have read and agreed to the published version of the manuscript.

Funding: This work was supported in part by Regione Toscana (POR FESR 2014-2020—Line 1 Research and Development Strategic Projects) through the Project IREAD4.0 under Grant CUP 7165.24052017.112000028, and in part by the Italian Ministry of Education and Research (MIUR) in the framework of the CrossLab project (Departments of Excellence). Info: Paolo Nepa (e-mail: paolo.nepa@unipi.it).

Acknowledgments: The authors wish to thank the company Sofidel S.p.a. for their precious support.

Conflicts of Interest: The authors declare no conflict of interest.

Abbreviations

BC	Block Chain
COTS	commercial-off-the-shelf
CP	Circular Polarization
CPS	Cyber-Physical System
CV	Computer Vision
EPC	Electronic Product Code
HPBW	Half-Power Beamwidth
IoT	Internet of Things
LGV	Laser-Guided Vehicle
LP	Linear Polarization
LSTM	Long Short-Term Memory
NFC	Near-Field Communication
NN	neural networks
RFID	Radio Frequency IDentification
RNN	Recurrent Neural Network
RSSI	Received Signal Strength Indicator
SSCC	Serial Shipping Container Code
UHF	Ultra-High Frequency
WMS	Warehouse Management System
WSN	Wireless Sensor Networks

References

1. Xu, H.; Yu, W.; Griffith, D.; Golmie, N. A Survey on Industrial Internet of Things: A Cyber-Physical Systems Perspective. *IEEE Access* **2018**, *6*, 78238–78259. [CrossRef]
2. Kim, B.H.; Cho, J.H. A Study on Modular Smart Plant Factory Using Morphological Image Processing. *Electronics* **2020**, *9*, 1661. [CrossRef]
3. Piardi, L.; Kalempa, V.C.; Limeira, M.; de Oliveira, A.S.; Leitão, P. ARENA—Augmented Reality to Enhanced Experimentation in Smart Warehouses. *Sensors* **2019**, *19*, 4308. [CrossRef] [PubMed]
4. Liu, X.; Cao, J.; Yang, Y.; Jiang, S. CPS-Based Smart Warehouse for Industry 4.0: A Survey of the Underlying Technologies. *Computers* **2018**, *7*, 13. [CrossRef]
5. Saldivar, A.A.F.; Li, Y.; Chen, W.; Zhan, Z.; Zhang, J.; Chen, L.Y. Industry 4.0 with cyber-physical integration: A design and manufacture perspective. In Proceedings of the 2015 21st International Conference on Automation and Computing (ICAC), Glasgow, UK, 11–12 September 2015; pp. 1–6. [CrossRef]
6. Zhou, K.; Liu, T.; Zhou, L. Industry 4.0: Towards future industrial opportunities and challenges. In Proceedings of the 2015 12th International Conference on Fuzzy Systems and Knowledge Discovery (FSKD), Zhangjiajie, China, 15–17 August 2015; pp. 2147–2152. [CrossRef]
7. Ponis, S.T.; Efthymiou, O.K. Cloud and IoT Applications in Material Handling Automation and Intralogistics. *Logistics* **2020**, *4*, 22. [CrossRef]
8. Cao, Z.; Chen, P.; Ma, Z.; Li, S.; Gao, X.; Wu, R.X.; Pan, L.; Shi, Y. Near-Field Communication Sensors. *Sensors* **2019**, *19*, 3947. [CrossRef] [PubMed]
9. Farsi, M.; Latsou, C.; Erkoyuncu, J.A.; Morris, G. RFID Application in a Multi-Agent Cyber Physical Manufacturing System. *J. Manuf. Mater. Process.* **2020**, *4*, 103. [CrossRef]
10. Lin, C.; Deng, D.; Chen, Z.; Chen, K. Key design of driving industry 4.0: Joint energy-efficient deployment and scheduling in group-based industrial wireless sensor networks. *IEEE Commun. Mag.* **2016**, *54*, 46–52. [CrossRef]
11. Fernández-Caramés, T.M.; Blanco-Novoa, O.; Froiz-Míguez, I.; Fraga-Lamas, P. Towards an Autonomous Industry 4.0 Warehouse: A UAV and Blockchain-Based System for Inventory and Traceability Applications in Big Data-Driven Supply Chain Management. *Sensors* **2019**, *19*, 2394. [CrossRef] [PubMed]
12. Nepa, P.; Motroni, A.; Congi, A.; Ferro, E.M.; Pesi, M.; Giorgi, G.; Buffi, A.; Lazzarotti, M.; Bellucci, J.; Galigani, S.; et al. I-READ 4.0: Internet-of-READers for an efficient asset management in large warehouses with high stock rotation index. In Proceedings of the 2019 IEEE 5th International forum on Research and Technology for Society and Industry (RTSI), Florence, Italy, 9–12 September 2019; pp. 67–72. [CrossRef]
13. Toivonen, A.S. Identifying and Controlling Stray Reads at RFID Gates. Master's Thesis, Aalto University, Espoo, Finland, February 2012.
14. Radio-Frequency Identity Protocols Generation-2 UHF RFID, GS1 Regulation. p. 152. Available online: https://www.gs1.org/sites/default/files/docs/epc/Gen2_Protocol_Standard.pdf (accessed on 26 May 2021).
15. Larionov, A.A.; Ivanov, R.E.; Vishnevsky, V.M. UHF RFID in Automatic Vehicle Identification: Analysis and Simulation. *IEEE J. Radio Freq. Identif.* **2017**, *1*, 3–12. [CrossRef]

16. Stine, R.J.; Markman, H.L.; Markman, J.E. Shielded Portal for Multi-Reading RFID Tags Affixed to Articles. Patent US9760826B1, 12 September 2017.
17. Shao, S.; Burkholder, R.J. Item-Level RFID Tag Location Sensing Utilizing Reader Antenna Spatial Diversity. *IEEE Sens. J.* **2013**, *13*, 3767–3774. [CrossRef]
18. Keller, T.; Thiesse, F.; Kungl, J.; Fleisch, E. Using low-level reader data to detect false-positive RFID tag reads. In Proceedings of the 2010 Internet of Things (IOT), Tokyo, Japan, 29 November–1 December 2010; pp. 1–8. [CrossRef]
19. Keller, T.; Thiesse, F.; Ilic, A.; Fleisch, E. Decreasing false-positive RFID tag reads by improved portal antenna setups. In Proceedings of the 2012 3rd IEEE International Conference on the Internet of Things, Wuxi, China, 24–26 October 2012; pp. 99–106. [CrossRef]
20. Morin, R.B. Method and system for controlling the traffic flow through an RFID directional portal. Patent US8487747B2, 16 July 2013.
21. Goller, M.; Brandner, M.; Brasseur, G. A System Model for Cooperative RFID Readpoints. *IEEE Trans. Instrum. Meas.* **2014**, *63*, 2480–2487. [CrossRef]
22. Oikawa, Y. Simulation evaluation of tag movement direction estimation methods in RFID gate systems. In Proceedings of the 2012 IEEE Radio and Wireless Symposium, Santa Clara, CA, USA, 15–18 January 2012; pp. 331–334. [CrossRef]
23. Jian, Y.; Wei, Y.; Zhang, Y. Estimating the direction of motion based on active RFID. In Proceedings of the 5th International Conference on New Trends in Information Science and Service Science, Macao, China, 24–26 October 2011; Volume 2, pp. 286–290.
24. Wu, J.; Zhu, M.; Xiao, B.; He, W. RFID Based Motion Direction Estimation in Gate Systems. In Proceedings of the 2018 IEEE 22nd International Conference on Computer Supported Cooperative Work in Design (CSCWD), Nanjing, China, 9–11 May 2018; pp. 588–593. [CrossRef]
25. Nikitin, P.V.; Martinez, R.; Ramamurthy, S.; Leland, H.; Spiess, G.; Rao, K.V.S. Phase based spatial identification of UHF RFID tags. In Proceedings of the 2010 IEEE International Conference on RFID (IEEE RFID 2010), Orlando, FL, USA, 14–16 April 2010; pp. 102–109. [CrossRef]
26. Buffi, A.; Tellini, B.; Motroni, A.; Nepa, P. A Phase-based Method for UHF RFID Gate Access Control. In Proceedings of the 2019 IEEE International Conference on RFID Technology and Applications (RFID-TA), Pisa, Italy, 25–27 September 2019; pp. 131–135. [CrossRef]
27. Nikitin, P.V.; Spiess, G.N.; Leland, H.M.; Hingst, L.C.; Sherman, J.H. Utilization of Motion and Spatial Identification in Mobile RFID Interrogator. Patent US9047522B1, 2 June 2015.
28. Buffi, A.; Nepa, P. The SARFID Technique for Discriminating Tagged Items Moving through a UHF-RFID Gate. *IEEE Sens. J.* **2017**, *17*, 2863–2870. [CrossRef]
29. Zhang, Y.; Amin, M.G.; Kaushik, S. Localization and Tracking of Passive RFID Tags Based on Direction Estimation. *Int. J. Antennas Propag.* **2007**, *2007*, e17426. [CrossRef]
30. Tesch, D.A.; Berz, E.L.; Hessel, F.P. RFID indoor localization based on Doppler effect. In Proceedings of the Sixteenth International Symposium on Quality Electronic Design, Santa Clara, CA, USA, 2–4 March 2015; pp. 556–560. [CrossRef]
31. Geigl, F.; Moik, C.; Hinteregger, S.; Goller, M.; GmbH, D. Using Machine Learning and RFID Localization for Advanced Logistic Applications. In Proceedings of the 2017 IEEE International Conference on RFID, Phoenix, AZ, USA, 9–11 May 2017;
32. Hauser, M.; Griebel, M.; Thiesse, F. A hidden Markov model for distinguishing between RFID-tagged objects in adjacent areas. In Proceedings of the 2017 IEEE International Conference on RFID, Phoenix, AZ, USA, 9–11 May 2017; pp. 167–173. [CrossRef]
33. Buffi, A.; D'Andrea, E.; Lazzerini, B.; Nepa, P. UHF-RFID smart gate: Tag action classifier by artificial neural networks. In Proceedings of the 2017 IEEE International Conference on RFID Technology Application (RFID-TA), Warsaw, Poland, 20–22 September 2017; pp. 45–50. [CrossRef]
34. Alfian, G.; Syafrudin, M.; Yoon, B.; Rhee, J. False Positive RFID Detection Using Classification Models. *Appl. Sci.* **2019**, *9*, 1154. [CrossRef]
35. Ma, H.; Wang, Y.; Wang, K. Automatic detection of false positive RFID readings using machine learning algorithms. *Expert Syst. Appl.* **2018**, *91*, 442–451. [CrossRef]
36. Eckstein, E.; Mazoki, G.T.; Richie, W.S., Jr. Article Identification and Tracking Using Electronic Shadows Created by RFID Tags. Patent US7081818B2, 25 July 2006.
37. Alvarez-Narciandi, G.; Motroni, A.; Pino, M.R.; Buffi, A.; Nepa, P. A UHF-RFID Gate Control System Based on a Recurrent Neural Network. *IEEE Antennas Wirel. Propag. Lett.* **2019**, *18*, 2330–2334. [CrossRef]
38. Álvarez Narciandi, G.; Motroni, A.; Pino, M.R.; Buffi, A.; Nepa, P. A UHF-RFID gate control system based on a Convolutional Neural Network. In Proceedings of the 2019 IEEE International Conference on RFID Technology and Applications (RFID-TA), Pisa, Italy, 25–27 September 2019; pp. 353–356. [CrossRef]
39. Hochreiter, S.; Schmidhuber, J. Long Short-Term Memory. *Neural Comput.* **1997**, *9*, 1735–1780. [CrossRef] [PubMed]
40. Costantini, M. A novel phase unwrapping method based on network programming. *IEEE Trans. Geosci. Remote. Sens.* **1998**, *36*, 813–821. [CrossRef]
41. Buffi, A.; Nepa, P.; Lombardini, F. A Phase-Based Technique for Localization of UHF-RFID Tags Moving on a Conveyor Belt: Performance Analysis and Test-Case Measurements. *IEEE Sens. J.* **2015**, *15*, 387–396. [CrossRef]
42. Miesen, R.; Kirsch, F.; Vossiek, M. UHF RFID Localization Based on Synthetic Apertures. *IEEE Trans. Autom. Sci. Eng.* **2013**, *10*, 807–815. [CrossRef]

43. Application Note-Low Level User Data Support. Available online: https://support.impinj.com/hc/en-us/articles/202755318-Application-Note-Low-Level-User-Data-Support (accessed on 12 June 2021)
44. Radio Frequency Identification Equipment Operating in Theband 865 MHz to 868 MHz with Power Levels up to 2 W and in the Band 915 MHz to 921 MHz with Power Levels up to 4 W Harmonised Standard for Access to Radio Spectrum, ETSI ETSI EN 302 208 V3.2.0. Available online: https://www.etsi.org/deliver/etsi_en/302200_302299/302208/03.02.00_20/en_302208v030200a.pdf (accessed on 18 June 2021).

sensors

MDPI

Article

Ranging with Frequency Dependent Ultrasound Air Attenuation

Riccardo Carotenuto [1,*], Fortunato Pezzimenti [1], Francesco G. Della Corte [2,3], Demetrio Iero [1,3] and Massimo Merenda [1,3]

1   Department of Information Engineering, Infrastructure and Sustainable Energy (DIIES), Mediterranea University of Reggio Calabria, 89124 Reggio Calabria, Italy; fortunato.pezzimenti@unirc.it (F.P.); demetrio.iero@unirc.it (D.I.); massimo.merenda@unirc.it (M.M.)
2   Department of Electrical Engineering and Information Technologies (DIETI), University of Naples Federico II, 80125 Naples, Italy; fg.dellacorte@unina.it
3   HWA SRL, Spin-Off Mediterranea University of Reggio Calabria, 89126 Reggio Calabria, Italy
*   Correspondence: r.carotenuto@unirc.it

Citation: Carotenuto, R.; Pezzimenti, F.; Della Corte, F.G.; Iero, D.; Merenda, M. Ranging with Frequency Dependent Ultrasound Air Attenuation. *Sensors* **2021**, *21*, 4963. https://doi.org/10.3390/s21154963

Academic Editor: Edwin C. Kan

Received: 27 May 2021
Accepted: 19 July 2021
Published: 21 July 2021

Abstract: Measuring the distance between two points has multiple uses. Position can be geometrically calculated from multiple measurements of the distance between reference points and moving sensors. Distance measurement can be done by measuring the time of flight of an ultrasonic signal traveling from an emitter to receiving sensors. However, this requires close synchronization between the emitter and the sensors. This synchronization is usually done using a radio or optical channel, which requires additional hardware and power to operate. On the other hand, for many applications of great interest, low-cost, small, and lightweight sensors with very small batteries are required. Here, an innovative technique to measure the distance between emitter and receiver by using ultrasonic signals in air is proposed. In fact, the amount of the signal attenuation in air depends on the frequency content of the signal itself. The attenuation level that the signal undergoes at different frequencies provides information on the distance between emitter and receiver without the need for any synchronization between them. A mathematical relationship here proposed allows for estimating the distance between emitter and receiver starting from the measurement of the frequency dependent attenuation along the traveled path. The level of attenuation in the air is measured online along the operation of the proposed technique. The simulations showed that the range accuracy increases with the decrease of the ultrasonic transducer diameter. In particular, with a diameter of 0.5 mm, an error of less than ± 2.7 cm (average value 1.1 cm) is reached along two plane sections of the typical room of the office considered ($4 \times 4 \times 3$ m^3).

Keywords: ultrasonic ranging; frequency dependent attenuation; ultrasonic signal

1. Introduction

Emerging technologies such as home automation, augmented reality, and gesture interfaces rely on the availability of accurate and fast positioning systems [1,2]. Recently, a large variety of indoor positioning systems (IPS) have proved suitable for many applications, being able to provide cost-effective positioning with sufficiently high speed and accuracy [3,4]. Fast and precise IPS can be used for augmented and virtual reality gestural interfaces [5,6], for navigation in closed places [7,8], for the recognition of human posture and medical rehabilitation [9,10] for the monitoring and care of elderly and disabled people [11], etc. Applications so far recognized for IPS include home automation, robotics, safety, accident prevention through the recognition of dangerous postures and positions of workers, logistics, inventory monitoring, monitoring of body and limb position during sports exercises and training military, game console, monitoring of structures [12], and monitoring of assets and security [13,14]. Certainly, in the near future, positioning systems capable of locating an object with adequate spatial and temporal resolution may enable new possible applications.

Typically, the positioning of a mobile unit or sensor is calculated through a two-step process. First, the distances, or ranges, of the mobile unit from some fixed reference points (RP) are measured. In the second step, these distances are used to geometrically determine the position of the mobile in the reference system defined by the fixed RPs. The ranges necessary for the geometric calculation of the sensor position can be obtained with the desired accuracy and with reasonable cost using the ultrasonic signal time-of-flight technique. With this technique, an ultrasonic traveling signal is emitted from an emitter toward a receiver, and the time of flight (TOF) is measured, which is the time elapsed from the time of emission (TOE) from the emitter to the time of arrival (TOA) at the receiver. In order to estimate this time interval as TOF = TOA − TOE, some technical difficulties must be overcome. First, when the calculation is done by the receiver, then it must know the instant of emission. This implies close synchronization between emitter and receiver, which requires additional hardware, for example, a radio frequency (RF) communication channel. Based on radio frequency channels, several techniques have been proposed in the literature [15,16]. A second difficulty consists in detecting the correct time of arrival (TOA) at the receiver of the traveling ultrasonic signal. Cross-correlation is the most widely adopted technique to have an accurate and robust TOA estimate. Cross-correlation measures the similarity of transmitted and received signals as a function of the time displacement of one relative to the other. The relative displacement that produces the maximum value corresponds to the TOA. Thanks to its integral nature, cross-correlation shows a reduced sensitivity to disturbances [17].

The monotone signal is certainly the easiest to generate and the most suitable for powering commercially available narrow-band ultrasonic transducers. However, the ambient noise makes it difficult to detect the cross-correlation peak corresponding to the TOA since the cross-correlation of a monotone signal shows many adjacent peaks of similar amplitude. Among the different available techniques [18,19], one of the most significant performance improvements is achieved by employing the linear chirp since its cross-correlation shows a very sharp and easily recognizable peak [20–22].

One of the most commonly used methods to derive the sensor position starting from the emitter-sensor distances is trilateration, or multilateration in the case of more than three distance measurements. Multilateration uses the distances between RP and the point to be located as radii of spheres, at the intersection of which is the position sought. In 3D space, the minimum number of spheres, and therefore of RP, is four, which drops to three if only calculating position in a half-space is required. On the other hand, information from additional distance measurements can be used to refine the estimated sensor position, thus making it less susceptible to measurement errors [23].

Some positioning systems do not require any emitter–receiver synchronization; they do not estimate directly the single distance between each RP and the mobile unit, but they measure the time difference between the arrival times of the signals emitted simultaneously by several emitters, also called time difference of arrival (TDOA) [24,25]. From the estimated time differences, the sensor position is calculated as the intersection of three hyperboloids. However, such a mathematical formulation requires at least four RPs for 3D positioning within a half-space, which is unfavorable compared to the intersection of the spheres which only requires three RPs. Furthermore, the hyperboloid intersection-based solution of the TDOA positioning problem is highly nonlinear and much more sensitive to ranging errors than the intersection of the spheres. Moreover, it is worth noting that it is not possible to find the emitter–receiver distance by using only one emitter–receiver pair without having any kind of synchronization. From what has been described, it therefore can be seen that to obtain a reliable distance measurement it is necessary to use a technique that requires shaped signals and a significant computational resource to calculate their cross-correlation [26]. Inevitably, from the realization point of view, this translates into a sensor equipped with a processor capable of performing the cross-correlation at three or four times the positioning rate, since three or four distances are needed to calculate the positioning. In addition, the sensor must also have an RF section (or equivalent) to handle the synchronization signals.

With the aim of reducing the complexity of the measurement process and of the sensor hardware, an entirely new method is proposed here for obtaining the distance measurement between the emitter and the sensor. In fact, the proposed distance measurement does not use the flight time of ultrasonic signals between emitter and receiver, but the new technique exploits the attenuation profile of the signal traveling in the air [27], which is a function, among others, of the distance between the emission point and the point of reception. By measuring the amount of attenuation suffered by signals emitted at different frequencies, the distance between emitter and receiver is obtained with simple calculations.

The paper is structured as follows. Section 2 presents the ranging method in detail, while the simulation set-up and numerical results are described in Section 3. Section 4 draws the conclusions of the work.

2. Ranging Technique Based on the Frequency Dependent Attenuation

The purpose of the proposed technique is to measure the distance between two points in three-dimensional space using an emitter and a receiver of a suitable ultrasonic signal, without any type of synchronization. The acoustic wave that propagates in the air undergoes energy losses due to the molecular frictions that develop in the medium itself, the extent of which depends, in addition to the medium, on the surrounding conditions. However, the attenuation in air depends mainly on relative humidity (RH). In Bass et al. [27], an experimentally obtained absorption curve in air is presented, which relates each RH level and each frequency of the propagating acoustic wave with a value of the absorption or attenuation coefficient.

Consider a sinusoidal signal with pulsation ω, amplitude A, and initial phase β:

$$s = Asin(\omega t + \beta) \tag{1}$$

Furthermore, suppose that there is a line-of-sight (LOS) of length d between the emitter and receiver, which is a direct path without obstacles. The received signal r by the sensor at point P (d, θ, φ) (see Figure 1) first undergoes geometric attenuation, which depends point-by-point on the emission diagram of the emitter:

$$r = D(d, \vartheta, \varphi)s = D(d, \vartheta, \varphi)Asin(\omega t + \beta) \tag{2}$$

where $D(d, \vartheta, \varphi)$ represents the radiation diagram of the emitter including the effect of geometric attenuation. Due to the presence of energy absorption in the propagation medium, an exponential term must be considered in addition [26], included in the following equation:

$$r = D(d, \vartheta, \varphi)Asin(\omega t + \beta)e^{-\alpha d} = Rsin(\omega t + \beta) \tag{3}$$

where $R = D(d, \vartheta, \varphi)e^{-\alpha d}$ is the amplitude of the received sinusoidal signal r and α is the attenuation coefficient, the latter assumed constant throughout the space of interest for all the time necessary for completion of ranging operations. This is an acceptable assumption when considering an air-conditioned home or office without particularly humid or dry areas.

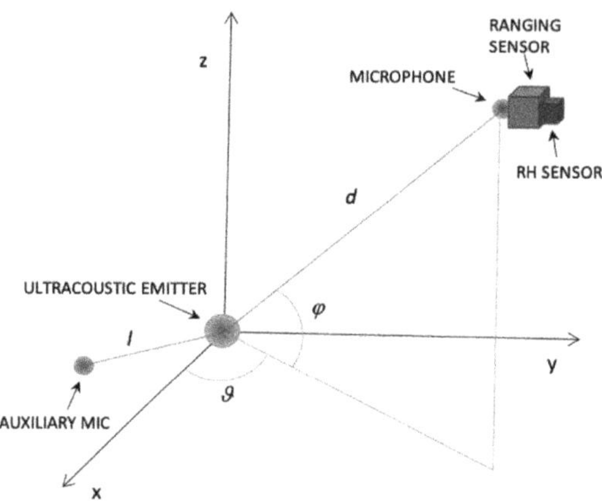

Figure 1. The ranging system: the ranging sensor at point P (d, θ, φ), equipped with a microphone and some processing resources, measures the distance d from the emitter thanks to the frequency dependent attenuation of the air. The attenuation coefficients are measured online along the known distance l using the auxiliary microphone.

Knowing the signal emitted s and the received signal r after propagation along a straight path without obstacles of length d, the latter can be estimated with the relationship:

$$d = \frac{1}{\alpha} \ln \left[\frac{D(d, \vartheta, \varphi)s}{r} \right] \tag{4}$$

Knowing the amplitude of the emitted signal A and that of the received signal R, the estimate of d is still obtained:

$$d = \frac{1}{\alpha} \ln \left[\frac{D(d, \vartheta, \varphi)A}{R} \right] \tag{5}$$

However, for a correct estimate of d it is also necessary to know with a sufficient degree of accuracy $D(d, \vartheta, \varphi)$, which, for any given emitter, depends on the position $P(d, \vartheta, \varphi)$ of the receiver. Since $D(d, \vartheta, \varphi)$ makes the received signal amplitude dependent on the position $P(d, \vartheta, \varphi)$ of the receiver, which is unknown, Equation (5) is not applicable in practice. Furthermore, in general, the actual radiation pattern $D(d, \vartheta, \varphi)$ could be unknown or known with insufficient accuracy. For example, it may depend on the arrangement of reflective surfaces in the space region of interest.

Let us now consider two signals emitted simultaneously by the same emitter, for example, two sinusoids of amplitude A_1 and A_2 with two pulsations ω_1 and ω_2, respectively:

$$s_1 = A_1 sin(\omega_1 t + \beta) \quad s_2 = A_2 sin(\omega_2 t + \beta) \tag{6}$$

The total emitted signal is $s = s_1 + s_2$. It is worth noting that the same reasoning applies more generally to each pair of sinusoids (h, k), with $h \neq k$ and $h, k \in \{1, 2, \ldots n\}$, formed by choosing them two-by-two from a set of n sinusoids with pulsation ω_h and ω_k, respectively. At the receiver, the received signal r is suitably filtered selectively in frequency to yield two signals r_1 and r_2, corresponding to the emitted components s_1 and s_2:

$$\begin{aligned} r_1 &= D_1(d, \vartheta, \varphi) A_1 sin(\omega_1 t + \beta) e^{-\alpha_1 d} = R_1 sin(\omega_1 t + \beta) \\ r_2 &= D_2(d, \vartheta, \varphi) A_2 sin(\omega_2 t + \beta) e^{-\alpha_2 d} = R_2 sin(\omega_2 t + \beta) \end{aligned} \tag{7}$$

Considering the ratio Q of the signals r_1 and r_2 received as r at point P, we obtain:

$$Q(d, \vartheta, \varphi, A_1, A_2, \alpha_1, \alpha_2, \omega_1, \omega_2) = \frac{r_1}{r_2} = \frac{D_1(d, \vartheta, \varphi) A_1 sin(\omega_1 t + \beta) e^{-\alpha_1 d}}{D_2(d, \vartheta, \varphi) A_2 sin(\omega_2 t + \beta) e^{-\alpha_2 d}} \tag{8}$$

where R_1 and R_2 are the amplitudes of the signals extracted from the received signal r, with pulsations ω_1 and ω_2, respectively. The alteration of the signal in the propagation channel consisting of emitter, attenuating propagation medium, and receiver (here for simplicity assumed to have unitary gain) is represented generally for n signals by the product $D_i(d, \vartheta, \varphi) e^{-\alpha_i d}$, with $i = 1, 2, \dots n$. Considering only the amplitudes of the signals involved, we obtain:

$$Q(d, \vartheta, \varphi, A_1, A_2, \alpha_1, \alpha_2) = \frac{R_1}{R_2} = \frac{D_1(d, \vartheta, \varphi) A_1 e^{-\alpha_1 d}}{D_2(d, \vartheta, \varphi) A_2 e^{-\alpha_2 d}} \tag{9}$$

Knowing a priori the ratio between the amplitudes of the two emitted signals $A_1 / A_2 = K$, since it depends on the emission system, we obtain:

$$Q(d, \vartheta, \varphi, K, \alpha_1, \alpha_2) = K \frac{D_1(d, \vartheta, \varphi) e^{-\alpha_1 d}}{D_2(d, \vartheta, \varphi) e^{-\alpha_2 d}} \tag{10}$$

If ω_1 and ω_2 are sufficiently close to each other, then $D_1 \cong D_2$, from which it follows that:

$$Q(d, K, \alpha_1, \alpha_2) \cong K \frac{e^{-\alpha_1 d}}{e^{-\alpha_2 d}} \tag{11}$$

Solving Equation (11) for d, we obtain:

$$d = \frac{1}{\alpha_2 - \alpha_1} \ln\left[\frac{Q(d, K, \alpha_1, \alpha_1)}{K} \right] = \frac{1}{\alpha_2 - \alpha_1} \ln\left[\frac{R_1}{K R_2} \right], \tag{12}$$

where ω_1, ω_2, and K are known constants.

Equation (12) shows that, under the hypotheses made, the distance d between the emitter and the receiver is obtained, known K, ω_1, and ω_2, from the ratio of the amplitudes of the two signals r_1 and r_2, obtained after filtering the received signal r. In practice, r_1 and r_2 can be calculated with an FFT, which has the same computational cost as a cross-correlation, but they can also be estimated with a simple narrowband frequency filter, one for each frequency, which requires significantly less computational effort. Please note that d cannot assume negative values under the assumption that $\omega_1 > \omega_2$ and that the attenuation is monotonically increasing over frequency [27], so that $R_1 > R_2$. Under these hypotheses, the argument of logarithm is strictly positive, and d, too.

Equation (12) can be applied to each pair of sinusoidal signals among n sinusoidal signals emitted simultaneously or in sequence, or by considering n harmonic components of a single signal of arbitrary shape. In practice, n absorption coefficients α_i ($i = 1, 2 \dots n$) can be easily measured for each pulsation of interest α_i ($i = 1, 2 \dots n$) in a continuous manner by having a fixed auxiliary microphone at a known distance l (see Figure 1) placed in the same environment where the system operates, or obtained from data presented, e.g., in Bass et al. [27] having measured the actual RH with a suitable sensor. Note that the calculation can be done at the desired repetition rate, without the limit determined, for example, by the flight time of the signal from the emitter to the receiver. The emission of signals can be continuous over time or for packets of defined duration. In the latter case, by appropriately choosing the length of the signal packet and the repetition frequency, unwanted reflection phenomena typical of closed environments can be mitigated. The approach presented could work, at least in theory, also considering other propagation media, and could be used for underwater ranging, for example. However, this work is focused on indoor positioning in the air.

3. Simulation Setup and Numerical Results

This section provides an overview of the operating principle underlying the simulation software and details on simulation configuration and numerical results.

A   Setup

The realistic acoustic field emitted by a transducer, including diffractive and attenuation effects, was simulated using the academic acoustic simulation tool Field II. It works in the MATLAB^{TM} environment and it is based on the concept of spatial impulse response [28–31]. The ultrasonic field for both the pulsed and continuous wave cases is obtained through linear systems. In a first step, the emitted ultrasound field at a specific point in space is obtained as a function of time using the spatial impulse response by applying to the transducer an excitation in the form of a Dirac delta function. Subsequently, by convolving the spatial impulse response with the excitation signal, the field generated by an arbitrary excitation is computed. Any kind of excitation can be considered, based on the theory of linear systems. This technique owes its name, i.e., "spatial impulse response", to the fact that the impulse response is a function of the spatial position, with respect to the transducer, of the point where the calculated acoustic field is computed [32].

Finally, it is worth noting that, to date, Field II is the only available and reliable acoustic simulator that is not based on a finite element modeling (FEM) approach (e.g., ANSYS, COMSOL, etc.). When dealing with spaces hundreds of times more extended than the typical wavelength considered (less than a couple of cm in the band beyond 18 kHz), as in the case in question, the FEM approach is computationally too expensive. In such cases, the number of nodes is enormous and the calculation becomes very extensive. Instead, the approach used by Field II provides that the calculation of the acoustic field is carried out only in the points considered. This makes the simulation for large spaces very efficient and practically feasible.

However, this approach is partially limited. In fact, the software tool used does not model some important effects in the field of indoor range, such as the phenomenon of reflection. Therefore, it is not possible to easily simulate the reflection of the signal, for example, by acoustically reflective walls, and the phenomena caused by multipath propagation, such as self-interference, typical of even partially reflective environments. Furthermore, the simulator assumes that propagation occurs in free space without considering any obstacles and near-line-of-sight situations. For these reasons, as explained, the simulation results described below are obtained by considering an available line-of-sight between emitter and receiver, and an environment without reflecting walls.

The transducer is represented as follows. The entire surface of the transducer is divided into small rectangles, allowing a transducer surface and field approximations much smaller than the size of the initial element; the smaller the rectangles' size, the lower the field approximation error. In fact, the distance to the field point is large compared to the size of the rectangles. In general, the element size should be much smaller than the wavelength of the signals used. The calculation is made considering that the rectangular elements behave as if they were rectangular pistons, and knowing the exact impulse response of each [32]. The impulse responses produced by each element at each desired field point is the result of the emission of a spherical wave by each of the small elements [33]. The simulation includes diffractive acoustic phenomena, and the tool gives the possibility to modify the shape and dimensions of the transducer, the signal emitted and to test any ranging or positioning technique intended for application.

The effectiveness of the ranging technique proposed here is evaluated in a typical $4 \times 4 \times 3$ m^3 room [34]. The simulation results are computed on a grid of points belonging to a vertical section A and a horizontal section B at an height of 1.5 m from the floor (see plane Sections A and B of the room volume, Figure 2). The grid pitch is 5 cm in all directions. In Figure 2, the boundary lines simply represent the extension of the room; however, walls, ceiling, and floor are not considered, since the simulation tool works as if the emission were in free space. The simulated setup has a disc transducer positioned in the center of the ceiling, in position x = 0, y = 0, and z = 0, with the emitting side facing the floor of the room.

The transducer central frequency is 20 kHz and it is immersed in air at a temperature of 20 °C, a pressure of 1 atm, and a relative humidity of 55%. Air absorption coefficients of 0.416 dB/m @ 18 kHz and of 0.578 dB/m @ 22 kHz are assumed for the simulation [27]. These values are purely exemplary, since in a real room they may vary from moment-to-moment due to the variation, for example, of the RH. Indeed, to cope with this variability, the proposed system measures the value of ω_1 and ω_2, online during its operation via, for example, the auxiliary microphone (see Figure 1). Moreover, it should be noted that in a real environment three or more digits for the attenuation coefficients are not warranted and were used here for demonstration purposes only. Finally, it is assumed that the actual RH, temperature, and pressure of the real room are sufficiently uniform everywhere.

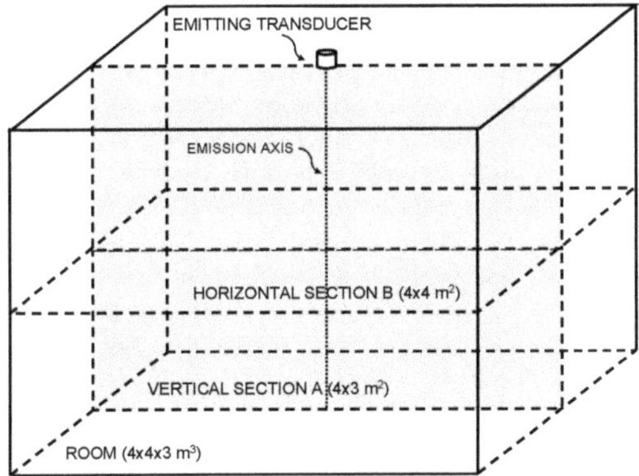

Figure 2. Simulation setup: horizontal and vertical plane sections of the typical $4 \times 4 \times 3$ m^3 room along which the ranging calculations using cross-correlation are computed. The SNR is considered at point P, at distance 1 m from the emitter surface center and on its emission axis.

The shape and size of the emission surface of the transducer determine the emitted and received signals at all points in the space. In this work, circular plane transducers with diameters of 5, 2.5, 1, and 0.5 mm were considered. The circular planar transducers are divided into small square elements with sides 0.025 by 0.025 mm that were used for all the simulations that follow. This element size is a good compromise between the accuracy of the solution and computational resources involved in the simulations.

B.   Numerical Results

A summation signal of two sinusoids at $f_1 = 18$ kHz and $f_2 = 22$ kHz and duration 10 ms was used as emitted signal for the simulations. The simulation was carried out sampling the signal with a sampling frequency $f_S = 10$ MHz, to ensure accurate results. In a first step, the numerical simulation computes the acoustic pressure over time generated by the superimposition of the two excitation signals, for each point of the space considered. Subsequently, an ideal receiver is assumed that linearly transduces the pressure signal into an electrical signal, which is then suitably down sampled to 100 kHz and quantized numerically, to simulate a sampling process that is feasible in a real-world device. Finally, the signal amplitudes at each point and the related ranges are calculated through Equation (12). The coefficients α_1 and α_2 that appear in Equation (12) are calculated starting from the attenuation experienced at point P by the two harmonic components of the signal. The amplitudes A_1 and A_2 of the two components of the emitted signal were set equal using a value of 1. In this first analysis, uniform white noise was added to the signals received with a reference level of SNR 20 dB calculated at 1 m from the transducer on its emission axis (see point P in Figure 2).

The simulation results are shown in the following figures. Figure 3 shows the ranging error committed by using Equation (12) along the vertical section A for four decreasing transducer diameters: 5, 2.5, 1, and 0.5 mm. Figure 4 shows the ranging errors along the horizontal section B for the same transducer diameters. Note the decreasing value ranges reported by the color bars of each subplot when the transducer diameter decreases. Figure 5 shows the cumulative distribution function CDF, i.e., the percent of readings with error less than the value of a given abscissa, for the ranging error along the vertical Section A (blue solid line) and the horizontal Section B (red solid line), respectively. Tables 1 and 2 summarize the ranging results of the four transducer diameters, reporting mean and maximum ranging errors along the vertical and the horizontal sections, respectively. In Table 1 it possible to appreciate the fast decrease of the mean and maximum error from 602.3 and 1919.2 mm down to 12.5 and 27.2 mm, respectively, when the transducer diameter decreases from 5.0 down to 0.5 mm. In Table 2, with the smallest diameter, the mean and maximum errors reach 11.2 and 27.5 mm, respectively.

C.   Discussion

The results obtained clearly show that the proposed numerical method can provide an estimate of the emitter–receiver distance without using the flight time, since the calculation of the ranging through Equation (12) considers only the relative amplitude of the attenuation. The simulation was performed only for two frequencies. By simultaneously using several sinusoids at different frequencies or a broadband signal, it is theoretically possible to obtain a better result as an average of several measurements. The decrease in the ranging error with the decrease in the diameter of the transducer is in agreement with the hypotheses made. In fact, in deriving Equation (12), it was assumed that for frequencies sufficiently close to each other it results $D_1 \cong D_2$, and this is especially true when the emitter is reduced in diameter and approaches the isotropic point-like emitter. In fact, by decreasing the diameter of the transducer, the spatial radiation pattern widens, becoming increasingly smooth and similar for the two pulsations ω_1 and ω_2.

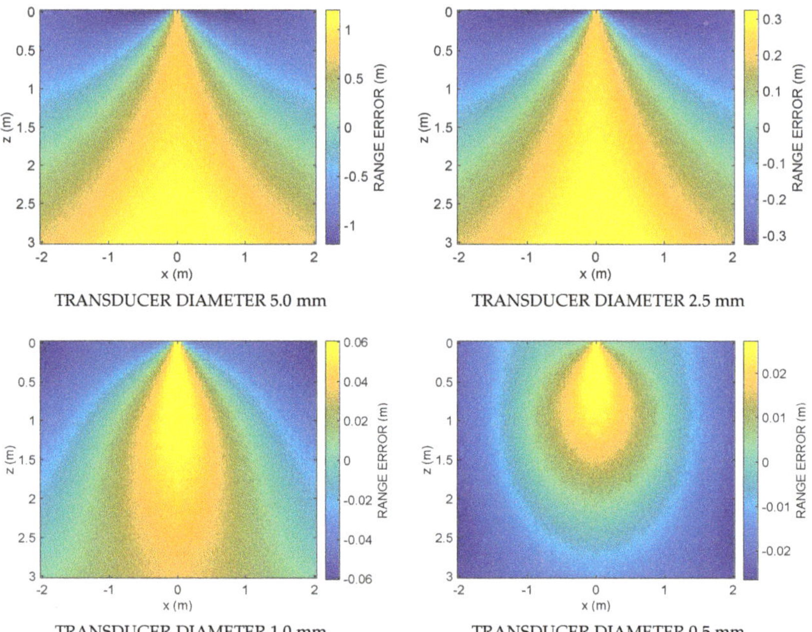

Figure 3. Simulation results: the first column shows the ranging error along the vertical section A for four decreasing values (5.0, 2.5, 1.0, 0.5 mm) of the transducer diameter. Note the decreasing value ranges reported by the color bars when the transducer diameter decreases.

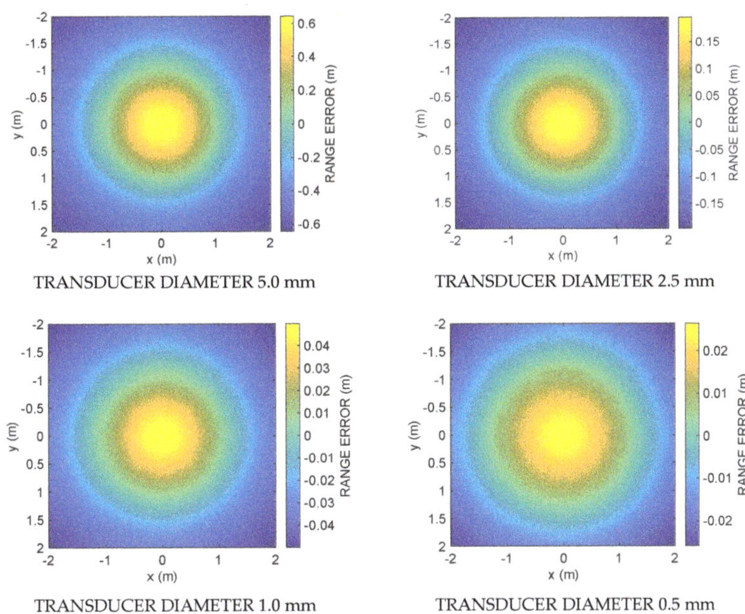

Figure 4. Simulation results: the ranging error along the horizontal section B for four decreasing values (5.0, 2.5, 1.0, 0.5 mm) of the transducer diameter. Note the decreasing value ranges reported by the color bars when the transducer diameter decreases.

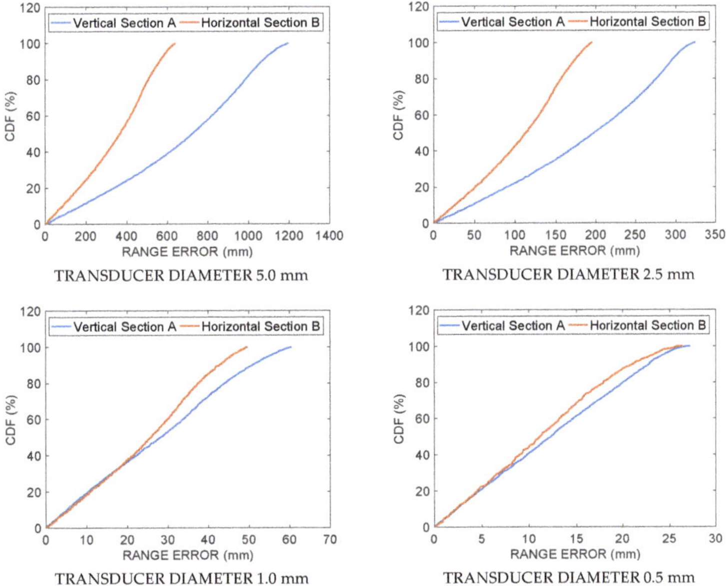

Figure 5. Simulation results: cumulative distribution functions CDF, i.e., the percent of readings with error less than the value of a given abscissa, for the ranging error along the vertical Section A (blue solid line) and horizontal Section B (red solid line) for four decreasing values (5.0, 2.5, 1.0, 0.5 mm) of the transducer diameter.

Table 1. Mean and maximum ranging error as a function of the transducer diameter along the vertical Section A.

Transducer Diameter (mm)	Range Absolute Mean Error (mm)	Range Absolute Maximum Error (mm)
5.0	602.3	1919.2
2.5	171.5	524.4
1.0	30.8	89.9
0.5	12.5	27.2

Table 2. Mean and maximum ranging error as a function of the transducer diameter along the horizontal Section B.

Transducer Diameter (mm)	Range Absolute Mean Error (mm)	Range Absolute Maximum Error (mm)
5.0	603.7	1015.7
2.5	175.7	297.2
1.0	32.4	63.4
0.5	11.2	27.5

In contrast, as shown in Figures 3 and 4, as one goes into the peripheral regions of the acoustic field farthest from the emitter axis, the error increases since D_1 and D_2 differ increasingly. In fact, the region where the error is minimized is the one where D_1 is most similar to D_2, mainly around the axial region. This region, whose three-dimensional shape resembles a cone, widens in space, decreasing the diameter of the transducer.

Certainly, the proposed method does not reach the level of ranging accuracy of many proposed methods that use synchronization, but there are applications that will benefit from the peculiar characteristics of this method, such as personal navigation in malls, airports etc. Even with the accuracy limits discussed, the method still seems to be sufficiently valid for a multiplicity of uses where a not too high accuracy is required, and when the peculiar characteristics of the proposed method take on greater importance: (1) absence of synchronization, which allows the use of a sensor HW of reduced dimensions, since it does not have the RF section, and with less energy consumption compared to sensors that use TOF-based techniques; (2) no limits on the distance measurement rate, since the emitter can emit its signal to the sensor continuously, or with very frequent cycles—from this point of view, the ranging rate is limited only by the onboard computing power; (3) no limitation is imposed by this system architecture on the number of sensors that can coexist in the region of space insonified by the same emitter; (4) the computation of Equation (12) is much less onerous than the computation of a cross-correlation, used by the best ranging techniques based on TOF. On these bases, a wide use of this technique is easily imaginable on mobile devices such as smartphones, tablets, or even notebooks.

4. Conclusions

In this work, a new technique was presented to measure the distance between an emitter and a receiver, which is not based on the time of flight, but is instead based on the different attenuation levels that ultrasonic signals of different frequencies undergo when propagating in the air.

The mathematical derivation of the technique was presented together with the validation of the hypotheses through the use of the Field II acoustic simulator. Simulations were conducted assuming free space propagation, and with room temperature 20 °C, relative humidity 55%, and atmospheric pressure 1 atm. The ranging error was calculated along two sections of a typical $4 \times 4 \times 3$ m^3 room, one vertical and the other horizontal, at an altitude of 1.5 m from the ground. The performance variation of the proposed technique as a function of the diameter of the emitter was shown. Simulation results show that, using a small diameter emitter aperture, 0.5 mm, and with sufficiently isotropic emission, a ranging error less than ± 2.75 cm and a mean error 1.25 cm were achieved along the two room sections considered.

Subsequently, the merits and limitations of the technique were discussed. The technique works in the absence of synchronization, without intrinsic limits on the distance measurement rate, and with an unlimited number of sensors using the same emitter. However, it does not reach, in its first implementation, the level of accuracy of other measurement techniques based on, for example, cross-correlation. In contrast, this allows for the design of sensors with reduced computational power and thus with reduced dimensions, since they do not require RF sections, and with less computational resources and energy consumption than sensors that use correlation-based techniques. Above all, the fact that it does not require synchronization between emitter and receiver makes this technique imaginable on mobile devices such as smartphones, tablets, or even notebooks, and embedded in chips for IoT or RFID.

Author Contributions: Conceptualization, R.C.; methodology, R.C.; software, R.C., F.P., M.M. and D.I.; investigation, R.C., F.P., M.M. and D.I.; resources, R.C.; writing—original draft preparation, R.C.; writing—review and editing, R.C., F.G.D.C., F.P., M.M. and D.I.; visualization, R.C. and F.P., M.M.; supervision, R.C.; project administration, R.C.; funding acquisition, F.G.D.C. All authors have read and agreed to the published version of the manuscript.

Funding: This research received no external funding.

Conflicts of Interest: The authors declare no conflict of interest.

References

1. Mendoza-Silva, G.M.; Torres-Sospedra, J.; Huerta, J. A meta-review of indoor positioning systems. *Sensors* **2019**, *19*, 4507. [CrossRef] [PubMed]
2. Nguyen, C.T.; Saputra, Y.M.; Van Huynh, N.; Nguyen, N.T.; Khoa, T.V.; Tuan, B.M.; Nguyen, D.N.; Hoang, D.T.; Vu, T.X.; Dutkiewicz, E.; et al. A comprehensive survey of enabling and emerging technologies for social distancing—Part I: Fundamentals and enabling technologies. *IEEE Access* **2020**, *8*, 153479–153507. [CrossRef]
3. Zafari, F.; Gkelias, A.; Leung, K.K. A Survey of Indoor Localization Systems and Technologies. *IEEE Commun. Surv. Tutor.* **2019**, *21*, 2568–2599. [CrossRef]
4. Zhang, D.; Xia, F.; Yang, Z.; Yao, L. Localization Technologies for Indoor Human Tracking. In Proceedings of the 5th International Conference on Future Information Technology, Busan, Korea, 21–23 May 2010; pp. 1–6.
5. Przybyla, R.J.; Tang, H.Y.; Shelton, S.E.; Horsley, D.A.; Boser, B.E. 3D ultrasonic gesture recognition. In Proceedings of the 2014 IEEE International Solid-State Circuits Conference Digest of Technical Papers, San Francisco, CA, USA, 9–13 February 2014; Volume 57, pp. 210–211.
6. Carotenuto, R.; Tripodi, P. Touchless 3D gestural interface using coded ultrasounds. In Proceedings of the 2012 IEEE International Ultrasonics Symposium, Dresden, Germany, 7–10 October 2012; pp. 146–149.
7. Paredes, J.A.; Álvarez, F.J.; Aguilera, T.; Villadangos, J.M. 3D Indoor Positioning of UAVs with Spread Spectrum Ultrasound and Time-of-Flight Cameras. *Sensors* **2017**, *18*, 89. [CrossRef]
8. El-Sheimy, N.; Li, Y. Indoor navigation: State of the art and future trends. *Satell. Navig.* **2021**, *2*, 1–23. [CrossRef]
9. Cooper, R.A.; Dicianno, B.; Brewer, B.; LoPresti, E.; Ding, D.; Simpson, R.; Grindle, G.; Wang, H. A perspective on intelligent devices and environments in medical rehabilitation. *Med. Eng. Phys.* **2008**, *30*, 1387–1398. [CrossRef] [PubMed]
10. Voinea, G.-D.; Butnariu, S.; Mogan, G. Measurement and Geometric Modelling of Human Spine Posture for Medical Rehabilitation Purposes Using a Wearable Monitoring System Based on Inertial Sensors. *Sensors* **2016**, *17*, 3. [CrossRef]
11. Marco, A.; Casas, R.; Falco, J.; Gracia, H.; Artigas, J.I.; Roy, A. Location-based services for elderly and disabled people. *Comput. Commun.* **2008**, *31*, 1055–1066. [CrossRef]
12. Fedele, R.; Praticò, F.; Carotenuto, R.; Della Corte, F.G. Instrumented infrastructures for damage detection and management. In Proceedings of the 2017 5th IEEE International Conference on Models and Technologies for Intelligent Transportation Systems (MT-ITSO), Naples, Italy, 26–28 June 2017. [CrossRef]
13. Kang, M.H.; Lim, B.G. An Ultrasonic Positioning System for an Assembly-Work Guide. *J. Signal Process. Syst.* **2021**, 1–12. [CrossRef]
14. Iula, A. Ultrasound Systems for Biometric Recognition. *Sensors* **2019**, *19*, 2317. [CrossRef]
15. Merenda, M.; Iero, D.; Carotenuto, R.; Della Corte, F.G. Simple and Low-Cost Photovoltaic Module Emulator. *Electronics* **2019**, *8*, 1445. [CrossRef]
16. Seco, F.; Jimenez, A.R.; Prieto, C.; Roa, J.; Koutsou, K. A survey of mathematical methods for indoor localization. In Proceedings of the 2009 IEEE International Symposium on Intelligent Signal Processing, Budapest, Hungary, 26–28 August 2009. [CrossRef]

17. Saad, M.M.; Bleakley, C.J.; Dobson, S. Robust High-Accuracy Ultrasonic Range Measurement System. *IEEE Trans. Instrum. Meas.* **2011**, *60*, 3334–3341. [CrossRef]
18. Sabatini, A.; Rocchi, A. Sampled baseband correlators for in-air ultrasonic rangefinders. *IEEE Trans. Ind. Electron.* **1998**, *45*, 341–350. [CrossRef]
19. Aldawi, F.J.; Longstaff, A.P.; Fletcher, S.; Mather, P.; Myers, A. A High Accuracy Ultrasound Distance Measurement System Using Binary Frequency Shift-Keyed Signal and Phase Detection. Available online: http://eprints.hud.ac.uk/id/eprint/3789/ (accessed on 28 April 2020).
20. De Marziani, C.; Urena, J.; Hernandez, Á.; Garcia, J.J.; Alvarez, F.J.; Jimenez, A.; Perez, M.C.; Carrizo, J.M.V.; Aparicio, J.; Alcoleas, R. Simultaneous Round-Trip Time-of-Flight Measurements with Encoded Acoustic Signals. *IEEE Sens. J.* **2012**, *12*, 2931–2940. [CrossRef]
21. Carotenuto, R. A range estimation system using coded ultrasound. *Sens. Actuators A Phys.* **2016**, *238*, 104–111. [CrossRef]
22. Carotenuto, R.; Merenda, M.; Iero, D.; Della Corte, F. Ranging RFID Tags with Ultrasound. *IEEE Sens. J.* **2018**, *18*, 2967–2975. [CrossRef]
23. Filonenko, V.; Cullen, C.; Carswell, J.D. Indoor Positioning for Smartphones Using Asynchronous Ultrasound Trilateration. *ISPRS Int. J. Geo-Inf.* **2013**, *2*, 598–620. [CrossRef]
24. Urena, J.; Hernandez, A.; Garcia, J.J.; Villadangos, J.M.; Perez, M.C.; Gualda, D.; Alvarez, F.J.; Aguilera, T. Acoustic Local Positioning with Encoded Emission Beacons. *Proc. IEEE* **2018**, *106*, 1042–1062. [CrossRef]
25. Carotenuto, R.; Merenda, M.; Iero, D.; Della Corte, F. An Indoor Ultrasonic System for Autonomous 3-D Positioning. *IEEE Trans. Instrum. Meas.* **2019**, *68*, 2507–2518. [CrossRef]
26. Carotenuto, R.; Merenda, M.; Iero, D.; Della Corte, F.G. Simulating Signal Aberration and Ranging Error for Ultrasonic Indoor Positioning. *Sensors* **2020**, *20*, 3548. [CrossRef]
27. Bass, H.E.; Sutherland, L.C.; Zuckerwar, A.J.; Blackstocks, D.T.; Hester, D.M. Atmospheric absorption of sound: Further de-velopments. *J. Acoust. Soc. Am.* **1995**, *97*, 680–683. [CrossRef]
28. Jensen, J.A. Field: A program for simulating ultrasound systems. *Med. Biol. Eng. Comp.* **1996**, *34*, 351–352.
29. Field II Home Page. Available online: https://field-ii.dk/ (accessed on 26 May 2021).
30. Tupholme, G.E. Generation of acoustic pulses by baffled plane pistons. *Mathematika* **1969**, *16*, 209–224. [CrossRef]
31. Stepanishen, P.R. The Time-Dependent Force and Radiation Impedance on a Piston in a Rigid Infinite Planar Baffle. *J. Acoust. Soc. Am.* **1971**, *49*, 841–849. [CrossRef]
32. Stepanishen, P.R. Transient Radiation from Pistons in an Infinite Planar Baffle. *J. Acoust. Soc. Am.* **1971**, *49*, 1629–1638. [CrossRef]
33. Jensen, J.A.; Svendsen, N.B. Calculation of pressure fields from arbitrarily shaped, apodized, and excited ultrasound transducers. *IEEE Trans. Ultrason. Ferroelectr. Freq. Control* **1992**, *39*, 262–267. [CrossRef] [PubMed]
34. Carotenuto, R.; Merenda, M.; Iero, D.; Della Corte, F.G. Mobile Synchronization Recovery for Ultrasonic Indoor Positioning. *Sensors* **2020**, *20*, 702. [CrossRef]

 **sensors**

Article

Comparison of Direct Intersection and Sonogram Methods for Acoustic Indoor Localization of Persons

Dominik Jan Schott [1,*], Addythia Saphala [2], Georg Fischer [3], Wenxin Xiong [4], Andrea Gabbrielli [1], Joan Bordoy [4], Fabian Höflinger [1], Kai Fischer [3], Christian Schindelhauer [4] and Stefan Johann Rupitsch [1]

1   Department of Microsystems Engineering, University of Freiburg, 79110 Freiburg, Germany; and.gabbrielli@gmail.com (A.G.); fabian.hoeflinger@imtek.uni-freiburg.de (F.H.); stefan.rupitsch@imtek.uni-freiburg.de (S.J.R.)
2   Department of Medical Informatics, Biometry and Epidemiology, Friedrich-Alexander-Universität (FAU), 91052 Erlangen, Germany; addythia.saphala@fau.de
3   Fraunhofer Institute for Highspeed Dynamics, Ernst-Mach-Institute (EMI), 79588 Efringen-Kirchen, Germany; georg.fischer@emi.fraunhofer.de (G.F.); kai.fischer@emi.fraunhofer.de (K.F.)
4   Department of Computer Science, University of Freiburg, 79110 Freiburg, Germany; xiongw@informatik.uni-freiburg.de (W.X.); bordoy@informatik.uni-freiburg.de (J.B.); schindel@informatik.uni-freiburg.de (C.S.)
*   Correspondence: dominik.jan.schott@imtek.uni-freiburg.de; Tel.: +49-761-203-7185

Abstract: We discuss two methods to detect the presence and location of a person in an acoustically small-scale room and compare the performances for a simulated person in distances between 1 and 2 m. The first method is Direct Intersection, which determines a coordinate point based on the intersection of spheroids defined by observed distances of high-intensity reverberations. The second method, Sonogram analysis, overlays all channels' room impulse responses to generate an intensity map for the observed environment. We demonstrate that the former method has lower computational complexity that almost halves the execution time in the best observed case, but about 7 times slower in the worst case compared to the Sonogram method while using 2.4 times less memory. Both approaches yield similar mean absolute localization errors between 0.3 and 0.9 m. The Direct Intersection method performs more precise in the best case, while the Sonogram method performs more robustly.

Keywords: presence detection; passive localization; room impulse response; acoustic localization; indoor localization

Citation: Schott, D.J.; Saphala, A.; Fischer, G.; Xiong, W.; Gabbrielli, A.; Bordoy, J.; Höflinger, F.; Fischer, K.; Schindelhauer, C.; Rupitsch, S.J. Comparison of Direct Intersection and Sonogram Methods for Acoustic Indoor Localization of Persons. *Sensors* **2021**, *21*, 4465. https://doi.org/10.3390/s21134465

Academic Editors: Riccardo Carotenuto, Massimo Merenda and Demetrio Iero

Received: 1 June 2021
Accepted: 21 June 2021
Published: 29 June 2021

Publisher's Note: MDPI stays neutral with regard to jurisdictional claims in published maps and institutional affiliations.

1. Introduction

Acoustic localization systems can provide, partly due to the comparably slower wave propagation, a high accuracy indoors similar to radio-based solutions, which are not covered by ubiquitous satellite signals of Global Navigation Satellite Systems (GNSS) [1–3]. For some applications, it may not be desirable to equip persons or objects with additional hardware as trackers due to inconvenience and privacy reasons. Previously, we reported coarsely about indoor localization by Direct Intersection in [4]. In this work, we report in detail on two algorithms for this application and their performances. The proposed system is categorized as a passive localization system [5] and is implemented solely with commercial off-the-shelf (COTS) hardware components.

Echolocation, such as the method used by bats to locate their prey, is a phenomenon where the reflected sound waves are used to determine the location of objects or surfaces that reflect the sound waves due to a change in acoustic impedance. This concept has been extensively used for various investigations in the physics and engineering fields, such as sound navigation and ranging (Sonar) [6,7] and even using only a single transducer for transmission and reception [8].

We draw the approach from bats, which can perceive the incoming reflected wave's direction due to its precise awareness of head angle, body motion, and timing. While the exhaustive echolocation method of bats is not completely understood, one of the more obvious aspects is the back-scattered signals' difference of arrival in time between left and right ears, which can be used to calculate the incoming sound wave's direction [9]. This approach differs from approaches that more generally detect changes in the systems response of a medium, where the responses act like fingerprints. However, in application, insignificant changes in a room may lead to distortions in the response. This makes a better knowledge of the specific room necessary. In contrast, determining times-of-arrival of back-scattered waves is less dependent on the complete impulse response; we therefore chose this approach. We investigate two different algorithms based on the time difference of arrival of the first-order reflection to interpret the returned signals in a small office room of approximately 3 m $\times$ 4 m $\times$ 3 m similar to [10], which are characteristic for the strong multipath fading effects that partially overlap and interfere with the line-of-sight reverberations [11]. The signal frequency employed in our experiment is significantly higher than the Schroeder frequency; therefore, we can assume the sound wave behaves much like rays of light [12]. The physiological structure and the shape of the binaural hearing conformation of bats, together with the natural and instinctive ability to perform head movements to eliminate ambiguities, enhances the echolocation and therefore guarantees excellent objects spatial localization [13]. Our system setup is a fixed structure, and we compensate the adaptive bats head movements by adding two additional microphones to the system. Furthermore, we raise the question of the performance of two approaches and compare the memory consumption and execution time.

The detection of more than one person or object is not investigated in this work.

2. Related Work

Indoor presence detection may be achieved through a variety of different technologies and techniques. For one, radio-frequency (RF)-based approaches have been implemented. In general, these may be classified into two different employed techniques: received signal strength indicator (RSSI)- and radio detection and ranging (Radar)-based approaches. The former offers low-complexity systems with cheap hardware [14,15], whereas with the latter one, higher accuracy may be achieved [16]. The other main concept employed in indoor presence detection is using ultrasonic waves, which are applied in active trackers indoors [17,18] and even underwater [19,20]. An entirely passive approach, as in [21], generally analyzes audible frequencies, which can include speech and potentially violate privacy regulations, similar to vision-based approaches. Acoustic solutions, which operate close to or in the audible range, can be perceived by persons and animals alike, which may cause irritation and in the worst case harm [22]. Therefore, special care has to be invested in designing acoustic location systems. While radio-based solutions are less critical in this concern, due to the fact that most organisms lack sensitivity to radio frequency signals, the frequency allocation is much more restrictive due to licensing and regulations. While LIDAR systems are highly accurate, but comparably costly, other light-based systems have gathered interest again, due to their high accuracy potential, with low systems costs and power consumption [23].

2.1. RF-RSSI

Mrazovac et al. [24] track the RSSI between stationary ZigBee communication nodes, detecting changes to infer a presence from it. In the context of home automation, this work is used to switch on and off home appliances. Seshadri et al. [15], Kosba et al. [14], Gunasagaran et al. [25], and Retscher and Leb [26] analyze different signal strength features for usability of detection and identification using standard Wi-Fi hardware. Kaltiokallio and Bocca [27] reduce the power consumption of the detection system by distributed RSSI processing.

This technique was then improved by Yigitler et al. [28], who built a radio tomographic map of the indoor area. The difference from the previously sampled map of RSSI values is the notification of a presence or occupancy. This general concept is known in the field of indoor localization as fingerprinting. Hillyard et al. [29] utilize these concepts to detect border crossings.

2.2. RF-Radar

Suijker et al. [30] present a 24 GHz FMCW (Frequency-Modulated Continuous-Wave) Radar system to detect indoor presence and to be used for intelligent LED lighting systems. An interferometry approach is implemented by Wang et al. [16] for precise human tracking in an indoor environment. Another promising approach in the RF domain is, instead of using a time-reversal approach (as Radar does), deriving properties of the medium (and contained, noncooperative objects) by means of wave front shaping as proposed by del Hougne et al. [31,32]. This approach would also in principle be conceivable in the acoustic wave domain.

2.3. Ultrasonic Presence Detection and Localization

A direct approach to provide room-level tracking is presented by Hnat et al. [33]. Ultrasonic range finders are mounted above doorways to track people passing beneath. More precise localization can be achieved by using ultrasonic arrays as proposed by Caicedo and Pandharipande [9,34]. The arrays' signals can be used to obtain the range and direction-of-arrival (DoA) estimates. The system is used for energy-efficient lighting systems. Pandharipande and Caicedo [7] enhanced this approach to track users by probing and calculating the position via the time difference of arrival (TDoA). Prior to that, Nishida et al. [35] proposed a system consisting of 18 ultrasonic transmitters and 32 receivers, embedded in the ceiling of a room with the aim to track elderly people and prevent them from experiencing accidents. A time-of-flight (ToF) approach was proposed by Bordoy et al. [36], who used a static co-located speaker-microphone pair to estimate human body and wall reflections. Ultrasonic range sensing my be combined with infrared technology, as has been done by Mokhtari et al. [37], to increase the energy efficiency. In lower frequency regimes, the resonance modes of a room start to dominate the measured signals. This fact may be used to deduce source locations as proposed by Nowakowski et al. [38] (cf. [39,40]).

2.4. Ultrasonic Indoor Mapping

Indoor mapping and indoor presence detection are two views of the same problem. In both instances, one tries to estimate the range and direction for a geometrical interpretation. Ribeiro et al. [41] employ a microphone array co-located to a loudspeaker to record the room impulse response (RIR). The multiple reflections can be estimated from this RIR with the use of l_1-regularization and least-squares (LS) minimization, and a room geometry can be inferred, achieving a range resolution of about 1 m. A random and sparse array of receivers is proposed by Steckel et al. [42] for an indoor Sonar system. In addition to that, the authors use wideband emission techniques to derive accurate three-dimensional (3D) location estimates. This system is then enhanced with an emitter array to improve the signal-to-noise-ratio (SNR) [43]. Another approach, implementing a binaural Sonar sensor, is proposed by Rajai et al. [44]. A sensor was used to detect the wall within a working distance of one meter. In a recent work by Zhou et al. [45], it is shown that a single smartphone with the help of a gyroscope and an accelerometer can be used to derive indoor maps by acoustic probing. Bordoy et al. [46] use an implicit mapping to enhance the performance of acoustic indoor localization by estimating walls and defining virtual receivers as a result of the signals' reflections.

2.5. Algorithms

The first set of methods, which are broadly applied are triangulation algorithms as described by Kundu [47]. In this work we focus on two Maximum-Likelihood approaches, similar to the one proposed by Liu et al. [48]. The first one, Direct Intersection (DI), uses a Look-up-Table (LUT) and spheres inferred from the sensors delay measurements with error margin [49], while the other one, the Sonogram method, populates a 3D intensity map with probabilities to find likely positions of the asset. Since the approaches of the two methods are different, it is likely to expect different outcomes in accuracy, precision, computational complexity, and memory requirements.

3. System Overview

The system consists of a single acoustic transmitter, a multi-channel receiver, a power distribution board, and a central computer to analyze the recorded signals. Four microphones are placed equidistantly around the speaker and connected to the receiver board. The set-up is shown in Figure 1 as it was used for the experiment reported below.

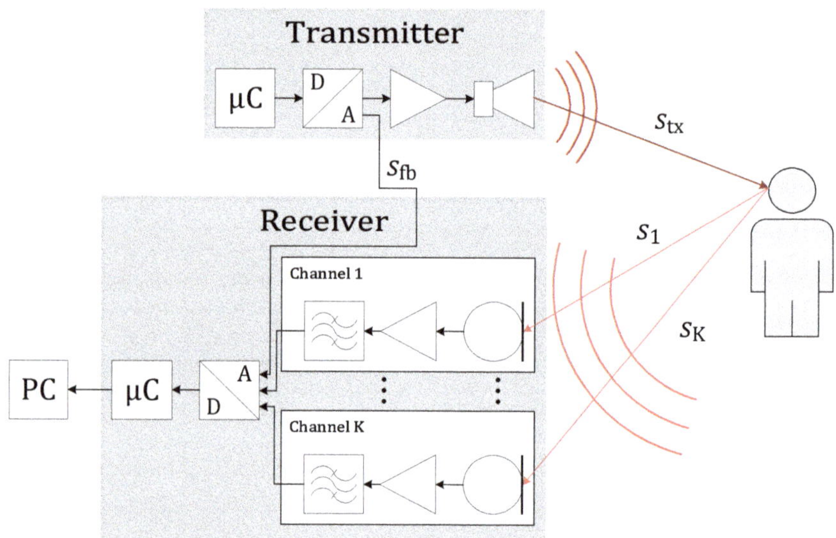

Figure 1. Schematic representation of the system.

3.1. Signal Waveform

Due to their auto-correlation properties and the ability to maximize the Signal-to-Noise-Ratio (SNR) without increasing acoustic amplitude, swept-frequency cosine, i.e., frequency modulated chirp signals, perfectly fit our case-study [50]. Auto-correlated frequency-modulated chirps are able to provide compressed pulses at the correlator output, whose width in time space is defined as follows [51]:

$$P_w = \frac{2}{B}. \tag{1}$$

The frequency-modulated signal employed in our experiments, $x_{Tx}(t)$, is mathematically defined as follows:

$$s_{tx}(t) = \begin{cases} A\cos(2\pi\phi(t)), & \text{for } 0 \le t \le T_s \\ 0, & \text{otherwise} \end{cases} \text{, with} \tag{2}$$

$$\phi(t) = \frac{f_{end} - f_{start}}{2T_s}t^2 + f_{start}t, \tag{3}$$

where A denotes the signal amplitude, f_{start} is the start frequency, f_{end} the end frequency, $B = f_{\text{end}} - f_{\text{start}}$ the frequency bandwidth, T_s is the pulse duration, and $\phi(t)$ the instantaneous phase. The chirp instantaneous frequency is defined as follows:

$$f(t) = f_{\text{start}} + \frac{f_{\text{end}} - f_{\text{start}}}{T_s} t, \quad 0 \leq t \leq T_s. \tag{4}$$

Taking into account the hardware characteristics of our setup, we selected a linear up-chirp pulse with amplitude $A = 1$, $T_s = 5$ ms, $f_{\text{start}} = 16$ kHz, and $f_{\text{end}} = 22$ kHz, which result in a time-bandwidth product of $TB = 30$. The frequency response of a chirp signal directly depends on the Time-Bandwidth (TB) product. For chirps with $TB \geq 100$, the pulse frequency response is almost rectangular [52]. However, due to the hardware limitation of our setup, which do not allow a high (TB) product, the frequency response will be characterized by ripples. In order to mitigate the spectrum disturbances, we consider a window in the time domain the transmitted chirp pulse with a raised cosine window [52]. The frequency band, chirp length, and shaping window were chosen to minimize the system affecting persons and animals in hearing range. We implemented chirps, due to their property of spreading the signals energy over time compared to a single pulse to limit the maximal amplitude and resulting harmonics. While young and highly audio-sensitive people can in principle hear these frequencies, the short signal length of 5 ms compared to the repetition interval of 1000 ms further reduces the occupation of the low ultrasonic channel. Generally speaking, higher amplitudes and lower frequencies potentially increase the operation range of the system, but this comes at a health risk for humans and animals, which we seek to avoid.

3.2. Hardware Overview

To obtain 3D coordinates with static arrangement, a four-element microphone array is sampled, as well as a feedback signal. This array records the incoming echo wave with different time of arrival, depending on the incoming signal direction. Since unsuitable hardware can affect the system's performance [53], both the microphones and speaker were tested for correct signal generation and reception in an anechoic box.

3.3. Data Acquisition

Each microphone's signal was preconditioned before the digitization by the multi-channel analog-to-digital converter, which was chosen to provide each channel with the identical sample-and-hold trigger flank before conversion. Each frame consists of the signal from each microphone and a feedback, which is recorded as an additional input to estimate and mitigate playback jitter. The first layer of digital signal processing is to compress the signal, extracting the reverberated acoustic amplitude over time and removing the empty room impulse response (RIR).

3.3.1. Channel Phase Synchronization

Initially, we calculate the convolution of the feedback channel signal s_{fb} with our known reference signal s_{ref} in its analytic form to obtain the RIR and retrieve the time of transmission from the compressed signal y_{fb}, as shown in Equation (5), where j denotes the imaginary unit.

$$y_{\text{fb}} = |(s_{\text{fb}} \circledast s_{\text{ref}}) + j \cdot \mathcal{H}(s_{\text{fb}} \circledast s_{\text{ref}})| \tag{5}$$

This compressed analytic form y_{fb} of the feedback signal s_{fb} (see Figure 1) ideally holds only a single pulse from the transmitted signal, if the output stage is impedance matched. Searching for the global maximum returns both time of transmission and the output amplitude.

$$a_{\text{out}} = \max_{t \to t_0} y_{\text{fb}}(t) \tag{6}$$

In the following, we refer to the start time of a transmission as t_0, all other channels' time scales are regarded relative to t_0. Therefore, the signals of the microphone channels are truncated to remove information prior to the transmission. The ring-down of small office rooms is in the order of 100 ms, so the repetition interval of consecutive transmissions is chosen accordingly to be larger. This prevents leakage of late echos into the following interval, which would result in peaks being recorded after the following interval's line-of-sight. The remaining signal frames from all microphones are compressed with the same approach as the feedback channel, shown in Equations (5) and (6), to extract each channel's compressed analytic signal y_i and line-of-sight detection time t_i.

3.3.2. Baseline Removal

In the following, we refer to the acoustic channel response after the line-of-sight as the echo profile. An example of such echo profiles is shown in Figure 2. While the line-of-sight signal ideally provides the fastest and strongest response, large hard surfaces, like desks, walls, and floors return high amplitudes, which are orders of magnitude above a person's echo. For a linear and stable channel, we can reduce this interference from the environment by subtracting the empty room echo profile from each measurement, following the approach of [54]. This profile loses its validity if the temperature changes, the air is moving, or objects in the room are moved, e.g., an office chair is slightly displaced. A dynamic approach to create the empty room profile is updating an estimation, when no change is observed for an extended time or alternatively using a very low-weight exponential filter to update the room estimation. In this work, the empty office room was sounded N times directly before each test and averaged into an empty room echo profile $\bar{y}_i^{\circ}$ for each channel i as denoted in Equation (7), to assure unchanged conditions and reduce the complexity of the measurements. The removal itself is then, as mentioned above, the subtraction of the baseline from each measurement, as in Equation (8), under the assumption of coherence.

$$\bar{y}_i^{\circ} = \mathrm{mean}(y_i^{\circ}) \tag{7}$$

$$\tilde{y}_i = y_i - \bar{y}_i^{\circ} \tag{8}$$

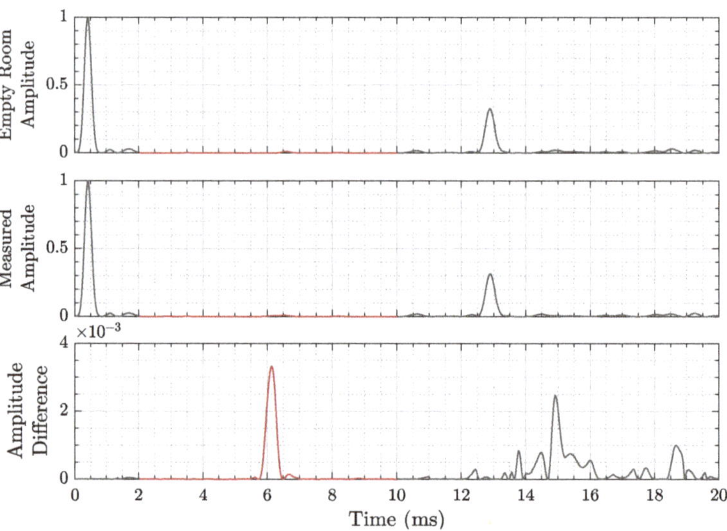

Figure 2. Exemplary magnitude plot of the compressed analytic signal, i.e., RIR, with (**top**) the baseline drawn from an previous recording of the empty room, (**middle**) the room with a person in it, and (**bottom**) the difference of the two above. The red highlighted line in the center marks the area of interest due to geometric constraints. Note the changed scale of the ordinate in the bottom plot.

3.3.3. Time-Gating

For our approach we assume some features of the person, such as being closer to the observing system compared to the distant environment objects, like chairs, tables and monitors, while another area of reverberations is in the close lateral vicinity of the system, consisting, e.g., of lamps and the ceiling. This is exploited by introducing a time gate, which only allows for non-zeros values in the interval of interest as in Equation (9) (also compare Figure 2).

$$\tilde{y}_{tg,i} = \begin{cases} \tilde{y}_i, & \text{for } t_{min} < t < t_{max} \\ 0, & \text{otherwise} \end{cases} \tag{9}$$

Another assumption is that of a small reverberation area on the person. We assume the points of observation from each microphone to be sufficiently close on a person to overlap. The latter assumption introduces an error, which limits the precision of the system in the order of $10\,cm$ [55], which we deem sufficient for presence detection, as a person's dimension is considerably larger in all directions. This estimation is based on the approximate size of a person's skull and its curvature with respect to the distance to the microphones and their spacing. The closer the microphones and the further the distance between head and device, the more the reflection points will approach each other. If we regard a simplified 2D projection, where a person with a spherical head of radius $r_H \approx 10\,cm$ moves in the y-plane only, the position of a reflection point $R = (x_R, z_R)$ on the head can be calculated by

$$x_R = x_C - r_H \sin(\alpha_R), \quad \text{and}$$
$$z_R = z_C - r_H \cos(\alpha_R), \tag{10}$$

where x_C and z_C are the lateral and vertical center coordinates of the head and α_R is the reflection angle. The latter is calculated through

$$\alpha_R = \tan^{-1} \frac{x_C + \frac{d_M}{2}}{z_C}, \tag{11}$$

with the distance d_M between the microphone and sender. The origin is set as the speaker position. By geometric addition, the distance between two such reflection points can be calculated and reach the maximum value if the head moves towards the center. In this case, the reflection points would be on the opposing sides of the head and result in a mismatch of $2\,r_h$. The other extreme is laterally moving to a infinite distance, which increases the magnitude of x_C, while the distance between microphone and speaker stays constant; therefore, the reflection points converge to a single point of reflection. In this work, the distance between head center and speaker remained above $120\,cm$, with a projected error distance of about $1.3\,cm$.

3.3.4. Echo Profile

During the experiment, the reflected signals from the floor, walls, tables, and chairs have a very high amplitude. This interference can lead to masking the echo from the target object. To reduce the effect of the interference, the empty room profile is used to subtract the target impulse response from the input impulse response. If we define the reflection from objects other than the target object as noise, we can increase the signal-to-noise ratio with this method. The empty room impulse response is also called empty room echo profile in this work. In Figure 2, the upper plot is the empty room impulse response, where the experiment room is cleared of most clutter. The middle plot is the room with single static object as target, shown in Figure 3. The lower plot shows the result of subtraction between the the second and first plot, and the scale is adjusted for clarity.

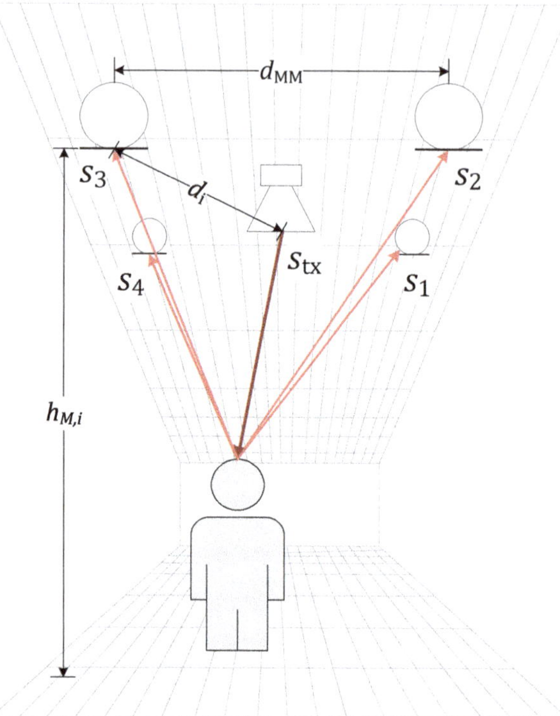

Figure 3. Experimental setup for $K = 4$ receivers spaced by $d_{MM} \approx 0.2\,\text{m}$. The transmitted signal s_{tx} is observed as reflected signals s_i by the system located near the ceiling of the room.

3.3.5. Distance Maps

Look-up tables are calculated before the experiment to estimate the travel distance of a signal from the speaker to each microphone under the assumption of a direct reverberation from a point at position $\vec{x}$ in the room and linear beam-like signal propagation. This grid is formed by setting the center speaker as origin and spanning up a 3-dimensional Cartesian coordinate system of points $\vec{x}$ through the room in discrete steps. We limit the grid to the intervals X_1 to X_3 in steps of 1 cm to decrease the calculational effort and multipath content under the prior knowledge of the rooms geometry as follows:

$$\vec{x} = (x_1, x_2, x_3) \in \mathbb{X}, \text{ where}$$
$$\mathbb{X} = \{X_1 \times X_2 \times X_3\} \subset \mathbb{R}^3. \tag{12}$$

The look-up table approach serves to minimize the processing time during execution. The distance maps provide pointers to convert from binary sampling points to distance points. Each sub-matrix contains the sum of distance between each point in the room to the corresponding ith microphone at the position $\vec{x}_{M,i}$ and to the speaker at position $\vec{x}_S$, which cover the flight path of the echoes, as in Equation (13):

$$M_i(\vec{x}) = \|\vec{x} - \vec{x}_S\| + \|\vec{x}_{M,i} - \vec{x}\|. \tag{13}$$

Therefore, the resultant entries in matrices M depend on the geometric arrangement of speaker and microphones, and the matrix size corresponds to the area of detection, as in Equation (12).

3.4. Data Processing

3.4.1. Direct Intersection

The main assumption for this approach (Algorithm 1) is that the highest signal peak in the observation window of each channel indicates the position of interest, as visualized in Figure 2. Each channels' peak index defines the radius r_i of a sphere around each microphone, which is contained in the point cloud L_i. While ideally those spheres overlap in exactly the point of reverberation, in practical application, where noise, interference, and jitters are present, this is not the case. To compensate this error, we pad the sphere by Δr additional points in the radius until all spheres overlap and the unity of valid estimation points U_L is not empty. The sphere radius widening Δr can be used as an indication of each measurement's quality, as a low error case will require little to no padding, while in high-error cases, the required padding will be large. Another approach is to use a fixed and small padding, which will ensure only measurements of high quality to be successful, but will fail for high error scenarios.

Algorithm 1: Direct Intersection Estimation [56,57].

Input : $\tilde{y}_{tg}$, observed intensity data frames,
$\qquad$ M_i, distance maps,
$\qquad$ K, number of channels,
$\qquad$ Δr_{max}, maximum radius spreading distance.
Output: $\vec{x}_{est}$, estimated 3D-position.

begin
$\quad$ $\Delta r \leftarrow 0$ $\qquad\qquad\qquad\qquad\qquad$ `// initial estimation tolerance`
$\quad$ $N_{OL} \leftarrow 0$ $\qquad\qquad\qquad\qquad\qquad$ `// number of overlapping points`
$\quad$ **for** $i = 1$ **to** K **do**
$\quad\quad$ $r_i \leftarrow \max_{n \to r_i} \tilde{y}_{tg}$ $\qquad\qquad\qquad$ `// get index of peak`
$\quad\quad$ $R_i \leftarrow \{r_i\}$
$\quad$ **while** $(N_{OL} = 0)$ & $(\Delta r < \Delta r_{max})$ **do**
$\quad\quad$ $\Delta r{+}{+}$
$\quad\quad$ **for** $i = 1$ **to** K **do**
$\quad\quad\quad$ $R_i \leftarrow \{r_i - \Delta r, R_i, r_i + \Delta r\}$ $\qquad$ `// recursively add width`
$\quad\quad\quad$ $L_i \leftarrow \text{isMember}(M_i, R_i)$ $\qquad$ `// select points by radius [56]`
$\quad\quad$ $U_L \leftarrow (L_1$ & $\ldots$ & $L_K)$
$\quad\quad$ $\vec{x}_{OL} \leftarrow \text{ind2sub}(\text{size}(U_L), \text{find}(U_L))$ $\qquad$ `// wrap into 3D coordinates [57]`
$\quad\quad$ $N_{OL} \leftarrow \min(\text{length}(\vec{x}_{OL}))$
$\quad$ $\vec{x}_{est} = \text{mean}(\vec{x}_{OL})$

3.4.2. Sonogram

The Sonogram approach (Algorithm 2) leverages available memory and processing power to build a 3D intensity map. This approach utilizes the entire echo profile difference shown in Figure 2 (bottom) and maps them into the 3D distance map explained in Section 3.3.5, with the assumption that the highest peak corresponds to the source of reverberation. The multiplication of impulse amplitude that corresponds to the same coordinates is used as an indication of possible reverberation source. Therefore, the maximum result would have the highest likelihood of being the reverberation source location.

Algorithm 2: Sonogram Estimation [58,59].

Input : $\tilde{y}_{\text{tg}}$, observed intensity data frames,
M_{i}, distance maps,
K, number of channels,
Δr_{max}, maximum radius spreading distance.
Output: $\vec{x}_{\text{est}}$, estimated 3D-position.

begin
$\quad$ $\tilde{y}_{+} \leftarrow \{\tilde{y}_{\text{tg}} > 0\}$ $\qquad\qquad\qquad\qquad$ // remove negative intensities
$\quad$ **forall** $\vec{x}$ **do**
$\qquad$ $I(\vec{x}) \leftarrow \prod_{i=1}^{K} \tilde{y}_{+,i}(M_{\text{i}}(\vec{x}))$

$\quad$ $A_{\text{x1}} \leftarrow \max_{x_1 \in X_1}(I)$ $\qquad\qquad\qquad\qquad$ // 2D matrix
$\quad$ $\vec{a}_{\text{x2}} \leftarrow \max_{x_2 \in X_2}(A_{\text{x1}})$ $\qquad\qquad\qquad$ // 1D vector
$\quad$ $x_{3,\text{est}} \leftarrow \max_{x_3 \in X_3}(\text{smooth}(\vec{a}_{\text{x2}}))$ $\qquad$ // scalar, moving average smoothed [58]

$\quad$ $x_{2,\text{est}} \leftarrow \text{find}(\vec{a}_{\text{x2}}, x_{3,\text{est}})$ $\qquad\qquad$ // select first matching value [59]
$\quad$ $x_{1,\text{est}} \leftarrow \text{find}(A_{\text{x1}}, \{x_{2,\text{est}}, x_{3,\text{est}}\})$

$\quad$ $\vec{x}_{\text{est}} \leftarrow (x_{1,\text{est}}, x_{2,\text{est}}, x_{3,\text{est}})$

---

4. Experiments

4.1. Set-Up

In the experiment, we use a mock-up representing a person's head as the experiment target. The hard and smooth surface of the object is intentional for the sake of usability and to remove unintended movements from our measurements at this early stage. In the set-up shown in Figure 3, the central speaker emits the well-known signal s_{tx}, and the reflected echoes from the target s_1 to s_4 are recorded by the microphone array around the speaker. The depiction in Figure 3 is exaggerated for clarity.

Table 1 shows the spherical coordinates, i.e., radial distance r, azimuth angle θ, and elevation angle ϕ of the target inside the room, with the center of the device as the reference point. The device is positioned on the ceiling, oriented downward. For each position, we measure the distance for the assumed acoustic path with a laser distance meter Leica DISTO^{TM} D3a BT for reference. As mentioned above, the coordinate system's point of origin is set to the center of the device, the x-axis is set perpendicular to the entrance door's wall, and increasing towards the right, the y-axis is parallel to the line of sight from the door and increasing towards the rear end of the room, and the z-axis is zero in the plane of the device (upper ceiling lamp level) and decreasing towards the floor. The two-dimensional depictions are shown in Cartesian coordinates to provide clarity, while the detection results are done in spherical coordinates.

Table 1. Reference Positions.

Position	r (m)	θ (°)	ϕ (°)
①	1.58	77	59
②	1.70	−92	57
③	1.23	−35	54
④	1.26	169	54

4.2. Results

4.2.1. Room Properties and Impulse Response

In preparation for the later experiments, we sounded the room 100 times as described in Section 3.3.2 to record the baseline profiles shown in Figures 4 and 5. This recordings were taken one time and served as a reference for all later experiment runs. During the recordings, the room was left closed and undisturbed.

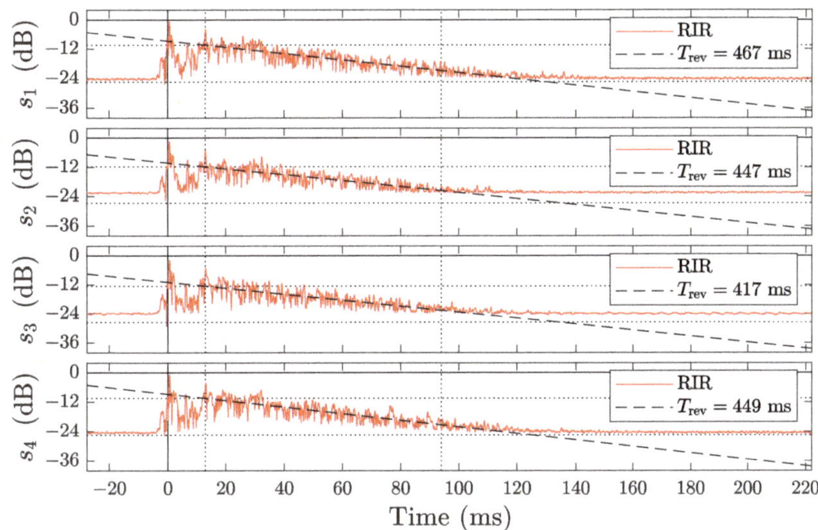

Figure 4. Empty room's impulse response magnitude of a linear chirp ($T_\mathrm{s} = 5\,\mathrm{ms}$, 16 to 22 kHz) in logarithmic scale for all 4 channels s_1 to s_4. The red line indicates the mean response over 100 measurements, with a linear fit indicated by a black dashed line in the interval between 13 to 94 ms (dotted vertical lines) to approximate the reverberation time constant T_rev of the room, given in the legend of each channel's subplot. The upper horizontal dotted line indicate the fit's level at $t = 13\,\mathrm{ms}$, while the lower indicates an additional drop by $-20\,\mathrm{dB}$.

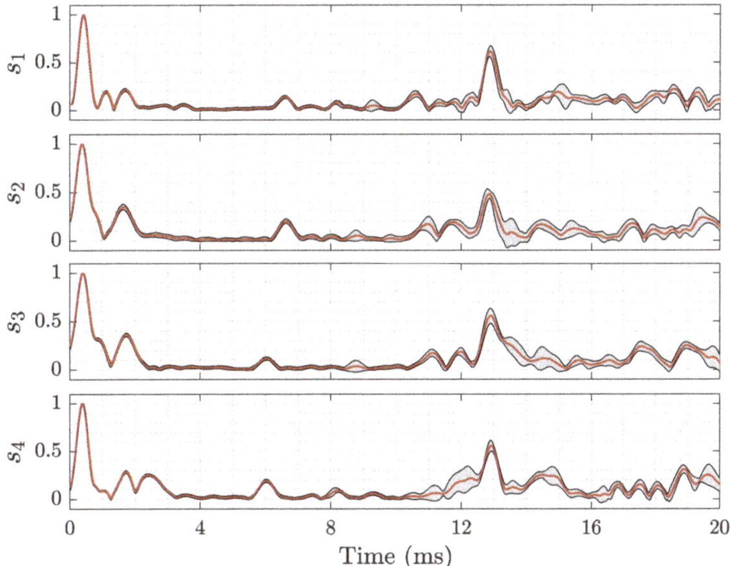

Figure 5. First 20 ms of the empty room's amplitude response for all 4 channels s_1 to s_4. The red line indicates the mean response over 100 measurements, the grey envelope the $\pm 3\sigma$ region. The first peak marks the line-of-sight arrival time and is used for time synchronization.

The room exhibits a different room response for each microphone, as illustrated in Figure 4. We divide the response into four parts: line-of-sight, free space transition, first order echoes, and higher order echoes, i.e., coda [60]. The signal remains in the room for more than 100 ms, before it drops below the noise floor. The definition of the reverberation time from Sabine requires a drop of the sound levels below -60 dB [61,62], for which the low signal-to-noise ratio of less than 24 dB does not suffice. Therefore, we adapted a fractional model and extrapolated the reverberation from a drop of 20 dB. The resulting mean reverberation time of the room is approximately $\bar{T}_{\mathrm{rev}} \approx 445$ ms, which corresponds to a dampening factor $\delta \approx 15.5\,\mathrm{s}^{-1}$ and a Schroeder frequency of approximately $f_{\mathrm{sch}} \approx 230$ Hz, which is far below the transmission band. In this work, we focus on the response in the parts-free space transition and first-order echoes to estimate a person's position. A close-up of the first three parts of the room response is shown in Figure 5.

The recordings still show significant variances in each channel at varying positions, e.g., in the uppermost subplot of Figure 5 from 15 to 16 ms. Below 8 ms, these intervals with increased variances do not occur, indicating a stable channel. The signals' interval close to zero contains strong wall and ceiling echos. Note the very strong reverberation peak at 12.5 to 13.5 ms that is caused by the floor. As our area of interest does not fall within this distance, we omit it for analysis as well. Hence, the time-gate limits as introduced in Section 3.3.3 are $t_{\min} = 3$ ms and $t_{\max} = 8$ ms.

If we transfer the room dimensions into the wavelength space, hence

$$\Lambda = \frac{l}{\lambda_{\mathrm{g}}} = \frac{\ell\, f_{\mathrm{g}}}{c}, \tag{14}$$

with c as the speed of sound and l the room dimensions in the respective Cartesian direction, we can draw an estimator from [63] for the number of modes below the reference frequency f_{g} as

$$N_{\mathrm{mode}} = \frac{4\,\pi}{3}\left(\Lambda_{\mathrm{x}}\,\Lambda_{\mathrm{y}}\,\Lambda_{\mathrm{z}}\right) + \frac{\pi}{2}\left(\Lambda_{\mathrm{x}}\,\Lambda_{\mathrm{y}} + \Lambda_{\mathrm{y}}\,\Lambda_{\mathrm{z}} + \Lambda_{\mathrm{z}}\,\Lambda_{\mathrm{x}}\right) + \frac{1}{2}\left(\Lambda_{\mathrm{x}} + \Lambda_{\mathrm{y}} + \Lambda_{\mathrm{z}}\right). \tag{15}$$

This lets us calculate approximately 15×10^6 modes below 16 kHz and 40×10^6 modes below 22 kHz, which leaves about 25×10^6 modes in the sounding spectrum in-between. If we regard the number of eigenfrequencies below the Schroeder frequency, Equation (15) yields $N_{\mathrm{sch}} \approx 73$ modes that strongly influence the sound characteristics of the room [64].

4.2.2. Direct Intersection

The localization by Direct Intersection from all 100 runs is shown for each of the four reference positions in Figure 6. While the statistical evaluation is performed in spherical coordinates due to the geometric construction during the estimation, this overview plots, as well as those for the Sonogram localization are drawn in Cartesian coordinates that allow for easier verification and intuitive interpretation. The lateral spread of the estimation point cloud in Figure 6 ① is misleading as the points are situated on a sphere around the origin. The projected lateral extent is almost entirely due to the angular errors.

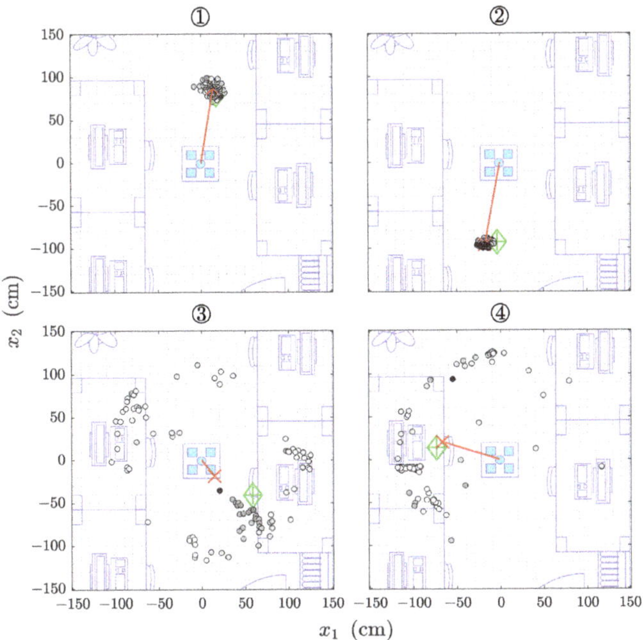

Figure 6. 2D projection of 100 estimations of 3D positions ① to ④ by Direct Intersection. The single estimations are indicated by the black circled markers, the red cross marks the Cartesian averaged position and is highlighted by the red line to the origin, and the green diamond indicates the reference position. The points' infill is proportional to the observed intensity relative to the radius spreading (darker is higher).

Positions ① and ② show a distance estimation deviation of $\sigma_r \approx 10\,\text{cm}$, as well as azimuth and elevation angle errors of $\sigma_\theta \approx \sigma_{CE} < 5°$ for both Direct Intersection and Sonogram localization (compare Tables 2 and 3). For positions ③ and ④, which are situated closer to the desks, the deviation increases to almost 40 cm in distance and almost arbitrary azimuth angles with a $\sigma_\theta \approx 120°$ and more, but a far less affected elevation angle estimation with a $\sigma_\theta < 10°$. The deviations are calculated around the mean estimator for each value. For simplicity of interpretation, the mean error for each dimension is shown in Section 4.2.3.

Table 2. Direct Intersection Estimated Positions.

Position	r (m)	θ (°)	ϕ (°)
①	1.83 ± 0.14	81 ± 4	61 ± 1
②	2.01 ± 0.11	-100 ± 3	61 ± 1
③	1.92 ± 0.37	4 ± 96	59 ± 4
④	2.12 ± 0.25	-58 ± 135	60 ± 3

Table 3. Sonogram Estimated Positions.

Position	r (m)	θ (°)	ϕ (°)
①	1.85 ± 0.10	80 ± 4	58 ± 2
②	2.03 ± 0.11	-100 ± 3	60 ± 2
③	1.77 ± 0.26	-41 ± 69	47 ± 7
④	1.96 ± 0.34	31 ± 119	51 ± 9

The error distributions for each dimension are shown in Figure 7, where each column depicts one of the spherical dimensions (radius, azimuth angle, and elevation angle), while each row represents the results from the reference position indicated to the left of the plot. For the first two positions, the distributions are almost unimodal, but for the latter two, this does not hold true, making the mean value and standard deviation unsuitable estimators.

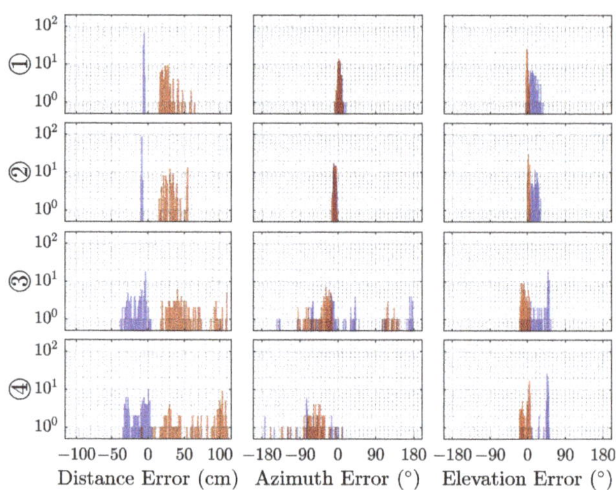

Figure 7. Histograms of the error in estimation compared to the reference over 100 localization repetitions at each position by DI (blue) and Sonogram (red) estimation. Each row depicts the 3 degrees of freedom for each position.

The distribution of the error in the absolute distance between the estimated positions and reference positions (see Figure 8) is likewise a few dozen centimeters for the first two cases, but around 1 m for the latter two. If we recall the reference positions from Table 1, the true distances are between 1 and 2 m, which puts the error in the same order as the expected value.

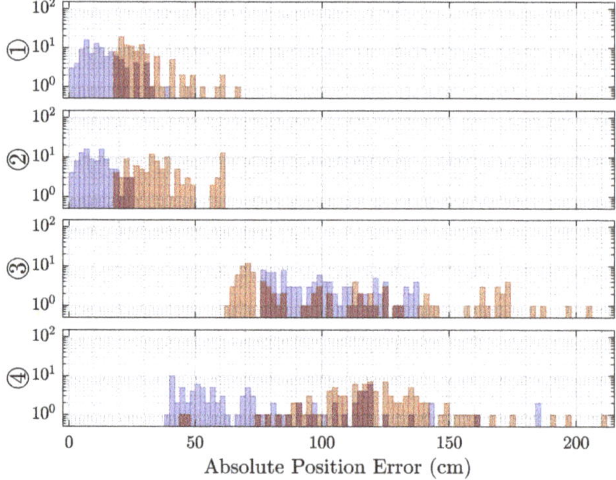

Figure 8. Histograms of the absolute distance error in estimation compared to the reference over 100 localization repetitions at each position by DI (blue) and Sonogram (red). Each row depicts the 3 degrees of freedom for each position.

The Direct Intersection method allows for an investigation into the time variance of the detected maximum peak, which is depicted in Figure 9. In the first two cases, we observe unimodal distributions of around 10 samples in width, while the latter cases show detected peaks all over the interval.

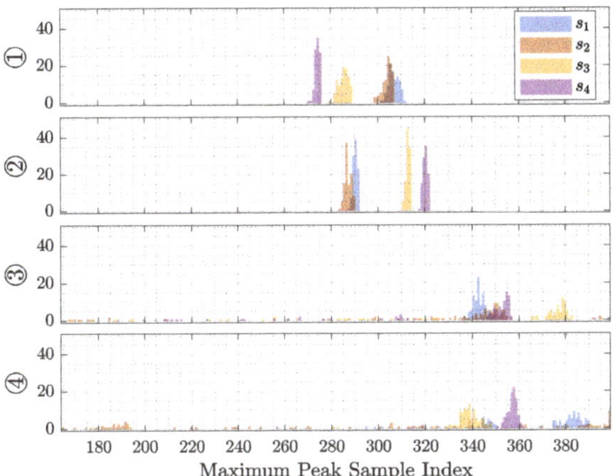

Figure 9. Histograms of the highest peak position of each microphone's channel over 100 localization repetitions at each position by DI. Each row depicts the 3 degrees of freedom for each position.

4.2.3. Sonogram

The Sonogram localization on the same data as before in Section 4.2.2 is shown in Figure 10 for all four cases. The lateral distribution of the estimated locations is not following the spherical shape as closely as is the case for those by Direct Intersection estimations (compare, e.g., Figure 6 ①).

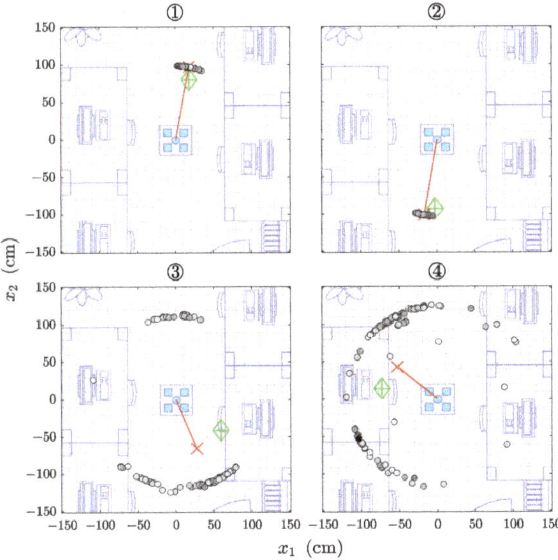

Figure 10. The same position estimation plot as in Figure 6 for positions ① to ④ but by Sonogram. The reference position is given by the green diamond, the averaged estimation by the red cross, and each circle represents a single estimated position. The circles' infill is proportional to the observed intensity.

Similar to before, the method performs well in the cases ① and ②, exhibiting small deviations (see Table 3), but far less precise with the largest deviation increase in the azimuth angle as well. The corresponding mean errors to the reference positions are listed in Table 4.

Table 4. Mean Error for Direct Intersection and Sonogram.

Position	Direct Intersection			Sonogram		
	r (m)	θ (°)	ϕ (°)	r (m)	θ (°)	ϕ (°)
①	0.25	3	2	0.27	2	1
②	0.31	8	4	0.34	8	3
③	0.69	39	5	0.53	6	7
④	0.87	47	6	0.70	138	3

The cases ③ and ④ display two larger clusters of estimated positions, which leads to the bimodal error distributions in Figure 7.

The absolute error is similarly distributed around lower values for the former two cases and widely spread for the two latter cases (see Figure 8). Note that the error distribution plots for the Sonogram are of slightly different horizontal scale, as no errors below 20 cm were observed, while the observed maximal error exceeds 200 cm.

Lastly, the performance of both algorithms with regard to execution time is listed in Table 5 and mean required memory in Table 6. The distribution of those measures is shown in Figures 11 and 12. The Direct Intersection method requires roughly $2.4\times$ less memory than the Sonogram localization. With a best-case mean execution time of 0.66 s, the former algorithm is almost $1.7\times$ faster than the best case mean of the latter method, while the worst-case mean—almost unchanged for the Sonogram approach—is with a factor of 7.1 for the Direct Intersection by far slower than the worst case mean execution time of the Sonogram method.

Table 5. Runtime Performance: Time.

Position	Direct Intersection Time (s)	Sonogram Time (s)
①	0.94 ± 0.17	1.14 ± 0.07
②	0.66 ± 0.13	1.20 ± 0.02
③	6.38 ± 6.60	1.10 ± 0.01
④	8.58 ± 7.12	1.10 ± 0.01

Table 6. Runtime Performance: Memory.

Direct Intersection Memory $(\times 10^8$ bit$)$	Sonogram Memory $(\times 10^8$ bit$)$
1.600 ± 0.004	3.840 ± 0.002

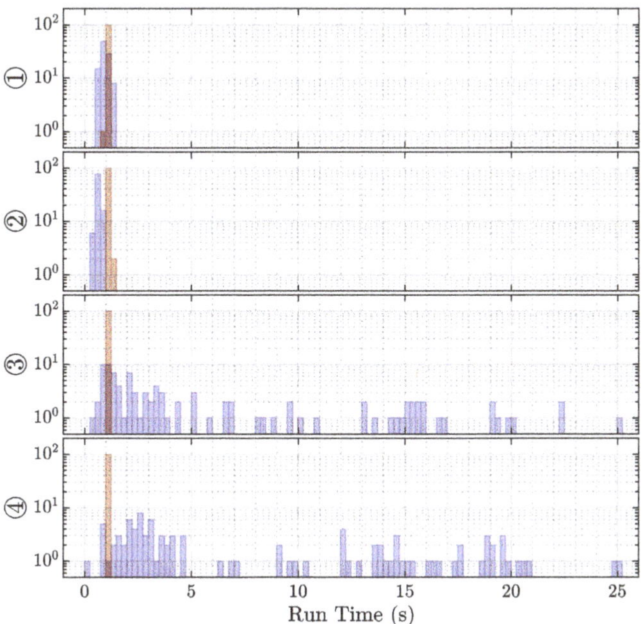

Figure 11. Histograms of the execution time of 100 localization repetitions at each position by DI (blue) and Sonogram (red).

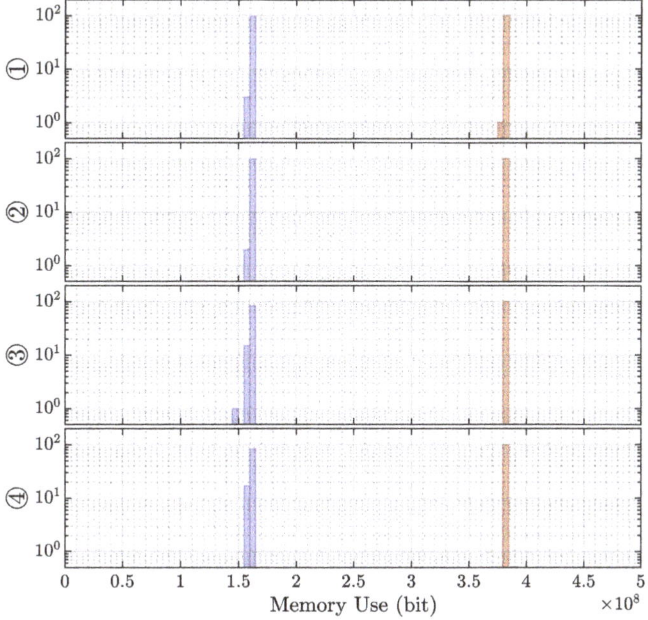

Figure 12. Histograms of the memory allocation during 100 localization repetitions at each position by DI (blue) and Sonogram (red).

The Direct Intersection execution time varies strongly, as we observe it anywhere between 0.25 and 25.0 s; thus, without further limitations, it does not allow for a well-confined prediction of the localization algorithm's execution time.

5. Discussion

5.1. Localization

The localization methods discussed in Section 4 are based on the time of arrival of the line-of-sight reflection from the target. This is possible because the frequency-modulated signal in our experiments is significantly higher than the Schroeder frequency of the room. The Direct Intersection method provides throughout all cases distance estimations that are too short, while the Sonogram-based localization returns distance estimations that are longer than the reference (compare Figure 7). Regarding the absolute error distribution, we observe that the Direct Intersection method performs more accurately, especially in the better cases ① and ②, as well more precise in the first three of the four observed cases, as drawn from Figure 8. The possible cause of the degradation of both methods performance for cases ③ and ④ is in the peak detection algorithm, as Figure 9 shows a wide error range of detected possible peaks. While this was observed specifically for the Direct Intersection method, this also implies the low signal-to-noise ratio of the underlying echo profile, and consequently also affects the Sonogram estimation. Interestingly, the lower estimation errors for cases ① and ② implicate a better performance for the larger distances than the closer ones, which is counter-intuitive from a power perspective, but if we recall the empty room impulse responses shown in Figure 5, where noise is included as the curves' variance, and compare it to the magnitudes of a person's signal in Figure 2, the difference in magnitude is in the same order. For higher distances, the variance increases, as fluctuations in the speed of sound cause phase distortions, but for lower distances, interference effects dominate. The frequency band of the chirp between 16 and 22 kHz sets the wavelength range to approximately 2.2 to 1.6 cm, which is close to the distance between reflection points on a person's head, as shown above in Section 3.3.3. Proximity to objects increases interference as well, which explains the lower performance in the closer positions ③ and ④, where the projected distance onto the sensor system's aperture between the person and the wall, screen, and desk is reduced. If we regard the error distributions of each position in Figure 7 again, the angles and distances roughly fit non-line-of-sight paths, especially for the Sonogram method.

5.2. Performance

The Direct Intersection method requires less than half the memory for its computations compared to the Sonogram method, as the information is very early condensed in the peak selection part of the algorithm. The index look-up is in itself a cheap operation, but due to the sphere-spreading loop to decrease the probability of the algorithm not returning any valid position at all, comes at higher execution duration. The observed worst case for Direct Intersection is with 25 s so high that no real-time tracking is possible anymore. If we look closer at Figure 6 ③, we see that the estimation point gray scale infill is proportional to the inverse spreading factor, so darker colors mean less radial spread before intersecting points could be found. The notion that including strong outliers by allowing the sphere thickness to be spread so far is not confirmed if we consider Figure 6 ④.

6. Conclusions

Both methods show mean distance estimation errors ranging between approximately 0.3 and 0.9 m for objects in distances between 1.2 and 1.7 m, with angular errors between 2° and 138° in azimuth, 1° and 7° in elevation. The Sonogram Estimation allows for analysis of room response in more detail, and the results are more accurate (i.e., average error) in three out of four observed cases, but inversely, the precision (i.e., error variance) of the Direct Intersection is higher in three of the cases. The Direct Intersection method allows for less expensive computation by reducing maximum radius spreading, while the Sonogram

Sensors **2021**, *21*, 4465

method's cost can be reduced effectively by limiting the vertical search interval, e.g., to the clutter free area above the desks. For a full-range sounding of the room, we observed that the locations close to the clutter area are estimated worse regarding both accuracy and precision. For a pragmatic operation on hardware with higher memory limitations the Direct Intersection method will perform faster and with similar precision and accuracy, and can be limited in execution time by restricting the sphere radius spreading at the cost of not being able to estimate the position for several intervals. We esteem further investigation into limiting the degradation of the estimation process by single unreliable channels as most promising for improving passive acoustic indoor localization.

Author Contributions: Conceptualization, D.J.S., A.S., and F.H.; methodology, D.J.S., A.S., and J.B.; software, D.J.S. and A.S.; validation, D.J.S., A.S., G.F., W.X., A.G., and J.B.; formal analysis, D.J.S. and A.S.; investigation, D.J.S. and A.S.; resources, F.H., K.F., C.S., and S.J.R.; data curation, A.S.; writing—original draft preparation, D.J.S., A.S., G.F., W.X., A.G., and J.B.; writing—review and editing, D.J.S., G.F., W.X., and A.G.; visualization, D.J.S.; supervision, J.B. and F.H.; project administration, F.H., K.F., C.S., and S.J.R.; funding acquisition, F.H., K.F., C.S., and S.J.R. All authors have read and agreed to the published version of the manuscript.

Funding: This work was partially supported by the Fraunhofer Gesellschaft and the state of Baden-Württemberg in the Framework of the MERLIN project, and also the German Ministry of Education and Research (BMBF) under the grant FKZ 16ME0023K ("Intelligentes Sensorsystem zur autonomen Überwachung von Produktionsanlagen in der Industrie 4.0 - ISA4.0").

Institutional Review Board Statement: Not applicable.

Informed Consent Statement: Not applicable.

Data Availability Statement: The data presented in this study are available on request from the corresponding author.

Acknowledgments: The authors would like to express their gratitude to the anonymous reviewers for many useful suggestions and support in deepening their understanding of acoustics.

Conflicts of Interest: The funders had no role in the design of the study; in the collection, analyses, or interpretation of data; in the writing of the manuscript; or in the decision to publish the results.

Abbreviations

The following abbreviations are used in this manuscript:

MDPI	Multidisciplinary Digital Publishing Institute
DI	Direct Intersection
DoA	Direction of Arrival
FMCW	Frequency-Modulated Continuous-Wave
LS	Least-Squares
RF	Radio-Frequency
RIR	Room Impulse Response
RSSI	Received Signal Strength Indicator
SONO	Sonogram
SNR	Signal-to-Noise Ratio
TDoA	Time Difference of Arrival
ToF	Time of Flight

References

1. Zafari, F.; Gkelias, A.; Leung, K.K. A Survey of Indoor Localization Systems and Technologies. *IEEE Commun. Surv. Tutor.* **2019**, *21*, 2568–2599. [CrossRef]
2. Billa, A.; Shayea, I.; Alhammadi, A.; Abdullah, Q.; Roslee, M. An Overview of Indoor Localization Technologies: Toward IoT Navigation Services. In Proceedings of the 2020 IEEE 5th International Symposium on Telecommunication Technologies (ISTT), Shah Alam, Malaysia, 9–11 November 2020; pp. 76–81. [CrossRef]
3. Obeidat, H.; Shuaieb, W.; Obeidat, O.; Abd-Alhameed, R. A Review of Indoor Localization Techniques and Wireless Technologies. *Wirel. Personal Commun.* **2021**. [CrossRef]

4. Höflinger, F.; Saphala, A.; Schott, D.J.; Reindl, L.M.; Schindelhauer, C. Passive Indoor-Localization using Echoes of Ultrasound Signals. In Proceedings of the 2019 International Conference on Advanced Information Technologies (ICAIT), Yangon, Myanmar, 6–7 November 2019; pp. 60–65. [CrossRef]
5. Pirzada, N.; Nayan, M.Y.; Subhan, F.; Hassan, M.F.; Khan, M.A. Comparative analysis of active and passive indoor localization systems. *AASRI Procedia* **2013**, *5*, 92–97. [CrossRef]
6. Caicedo, D.; Pandharipande, A. Distributed Ultrasonic Zoned Presence Sensing System. *IEEE Sens. J.* **2014**, *14*, 234–243. [CrossRef]
7. Pandharipande, A.; Caicedo, D. User localization using ultrasonic presence sensing systems. In Proceedings of the 2012 IEEE International Conference on Systems, Man, and Cybernetics (SMC), Seoul, Korea, 14–17 October 2012; pp. 3191–3196. [CrossRef]
8. Kim, K.; Choi, H. A New Approach to Power Efficiency Improvement of Ultrasonic Transmitters via a Dynamic Bias Technique. *Sensors* **2021**, *21*, 2795. [CrossRef]
9. Caicedo, D.; Pandharipande, A. Transmission slot allocation and synchronization protocol for ultrasonic sensor systems. In Proceedings of the 2013 10th IEEE International Conference on Networking, Sensing and Control (ICNSC), Evry, France, 10–12 April 2013; pp. 288–293. [CrossRef]
10. Carotenuto, R.; Merenda, M.; Iero, D.; Della Corte, F.G. An Indoor Ultrasonic System for Autonomous 3-D Positioning. *IEEE Trans. Instrum. Meas.* **2019**, *68*, 2507–2518. [CrossRef]
11. Patwari, N.; Ash, J.; Kyperountas, S.; Hero, A.; Moses, R.; Correal, N. Locating the nodes: Cooperative localization in wireless sensor networks. *IEEE Signal Process. Mag.* **2005**, *22*, 54–69. [CrossRef]
12. Kuttruff, H. Geometrical room acoustics. In *Room Acoustics*, 5th ed.; Spon Press, Taylor & Francis: Abingdon-on-Thames, UK, 2009.
13. Zhang, S.; Ma, X.; Dong, Z.; Zhou, W. Ultrasonic Spatial Target Localization Using Artificial Pinnae of Brown Long-eared Bat. *bioRxiv* **2020**. [CrossRef]
14. Kosba, A.E.; Saeed, A.; Youssef, M. RASID: A robust WLAN device-free passive motion detection system. In Proceedings of the 2012 IEEE International Conference on Pervasive Computing and Communications, Lugano, Switzerland, 19–23 March 2012. [CrossRef]
15. Seshadri, V.; Zaruba, G.; Huber, M. A Bayesian sampling approach to in-door localization of wireless devices using received signal strength indication. In Proceedings of the Third IEEE International Conference on Pervasive Computing and Communications, Kauai, HI, USA, 8–12 March 2005; pp. 75–84. [CrossRef]
16. Wang, G.; Gu, C.; Inoue, T.; Li, C. A Hybrid FMCW-Interferometry Radar for Indoor Precise Positioning and Versatile Life Activity Monitoring. *IEEE Trans. Microw. Theory Tech.* **2014**, *62*, 2812–2822. [CrossRef]
17. Bordoy, J.; Schott, D.J.; Xie, J.; Bannoura, A.; Klein, P.; Striet, L.; Höflinger, F.; Häring, I.; Reindl, L.; Schindelhauer, C. Acoustic Indoor Localization Augmentation by Self-Calibration and Machine Learning. *Sensors* **2020**, *20*, 1177. [CrossRef]
18. Pullano, S.A.; Bianco, M.G.; Critello, D.C.; Menniti, M.; La Gatta, A.; Fiorillo, A.S. A Recursive Algorithm for Indoor Positioning Using Pulse-Echo Ultrasonic Signals. *Sensors* **2020**, *20*, 5042. [CrossRef]
19. Schott, D.J.; Faisal, M.; Höflinger, F.; Reindl, L.M.; Bordoy Andreú, J.; Schindelhauer, C. Underwater localization utilizing a modified acoustic indoor tracking system. In Proceedings of the 2017 IEEE 7th International Conference on Underwater System Technology: Theory and Applications (USYS), Kuala Lumpur, Malaysia, 18–20 December 2017. [CrossRef]
20. Chang, S.; Li, Y.; He, Y.; Wang, H. Target Localization in Underwater Acoustic Sensor Networks Using RSS Measurements. *Appl. Sci.* **2018**, *8*, 225. [CrossRef]
21. Cobos, M.; Antonacci, F.; Alexandridis, A.; Mouchtaris, A.; Lee, B. A Survey of Sound Source Localization Methods in Wireless Acoustic Sensor Networks. *Wirel. Commun. Mob. Comput.* **2017**, *2017*, 3956282. [CrossRef]
22. Hage, S.R.; Metzner, W. Potential effects of anthropogenic noise on echolocation behavior in horseshoe bats. *Commun. Integr. Biol.* **2013**, *6*, e24753. [CrossRef]
23. Rahman, A.B.M.M.; Li, T.; Wang, Y. Recent Advances in Indoor Localization via Visible Lights: A Survey. *Sensors* **2020**, *20*, 1382. [CrossRef] [PubMed]
24. Mrazovac, B.; Bjelica, M.; Kukolj, D.; Todorovic, B.; Samardzija, D. A human detection method for residential smart energy systems based on Zigbee RSSI changes. *IEEE Trans. Consum. Electron.* **2012**, *58*, 819–824. [CrossRef]
25. Gunasagaran, R.; Kamarudin, L.M.; Zakaria, A. Wi-Fi For Indoor Device Free Passive Localization (DfPL): An Overview. *Indones. J. Electr. Eng. Inform. (IJEEI)* **2020**, *8*. [CrossRef]
26. Retscher, G.; Leb, A. Development of a Smartphone-Based University Library Navigation and Information Service Employing Wi-Fi Location Fingerprinting. *Sensors* **2021**, *21*, 432. [CrossRef]
27. Kaltiokallio, O.; Bocca, M. Real-Time Intrusion Detection and Tracking in Indoor Environment through Distributed RSSI Processing. In Proceedings of the 2011 IEEE 17th International Conference on Embedded and Real-Time Computing Systems and Applications, Toyama, Japan, 28–31 August 2011. [CrossRef]
28. Yigitler, H.; Jantti, R.; Kaltiokallio, O.; Patwari, N. Detector Based Radio Tomographic Imaging. *IEEE Trans. Mob. Comput.* **2018**, *17*, 58–71. [CrossRef]
29. Hillyard, P.; Patwari, N.; Daruki, S.; Venkatasubramanian, S. You're crossing the line: Localizing border crossings using wireless RF links. In Proceedings of the 2015 IEEE Signal Processing and Signal Processing Education Workshop (SP/SPE), Salt Lake City, UT, USA, 9–12 August 2015. [CrossRef]

30. Suijker, E.M.; Bolt, R.J.; van Wanum, M.; van Heijningen, M.; Maas, A.P.M.; van Vliet, F.E. Low cost low power 24 GHz FMCW Radar transceiver for indoor presence detection. In Proceedings of the 2014 44th European Microwave Conference, Rome, Italy, 6–9 October 2014; pp. 1758–1761. [CrossRef]
31. Del Hougne, P.; Imani, M.F.; Fink, M.; Smith, D.R.; Lerosey, G. Precise Localization of Multiple Noncooperative Objects in a Disordered Cavity by Wave Front Shaping. *Phys. Rev. Lett.* **2018**, *121*, 063901. [CrossRef]
32. Del Hougne, M.; Gigan, S.; del Hougne, P. Deeply Sub-Wavelength Localization with Reverberation-Coded-Aperture. *arXiv* **2021**, arXiv:2102.05642.
33. Hnat, T.W.; Griffiths, E.; Dawson, R.; Whitehouse, K. Doorjamb: Unobtrusive room-level tracking of people in homes using doorway sensors. In Proceedings of the The 10th ACM Conference on Embedded Network Sensor Systems (SenSys 2012), Toronto, ON, Canada, 6–9 November 2012. [CrossRef]
34. Caicedo, D.; Pandharipande, A. Ultrasonic array sensor for indoor presence detection. In Proceedings of the 2012 20th European Signal Processing Conference (EUSIPCO), Bucharest, Romania, 27–31 August 2012; pp. 175–179.
35. Nishida, Y.; Murakami, S.; Hori, T.; Mizoguchi, H. Minimally privacy-violative human location sensor by ultrasonic Radar embedded on ceiling. In Proceedings of the 2004 IEEE Sensors, Vienna, Austria, 24–27 October 2004. [CrossRef]
36. Bordoy, J.; Wendeberg, J.; Schindelhauer, C.; Reindl, L.M. Single transceiver device-free indoor localization using ultrasound body reflections and walls. In Proceedings of the 2015 International Conference on Indoor Positioning and Indoor Navigation (IPIN), Banff, AB, Canada, 13–16 October 2015. [CrossRef]
37. Mokhtari, G.; Zhang, Q.; Nourbakhsh, G.; Ball, S.; Karunanithi, M. BLUESOUND: A New Resident Identification Sensor—Using Ultrasound Array and BLE Technology for Smart Home Platform. *IEEE Sens. J.* **2017**, *17*, 1503–1512. [CrossRef]
38. Nowakowski, T.; de Rosny, J.; Daudet, L. Robust source localization from wavefield separation including prior information. *J. Acoust. Soc. Am.* **2017**, *141*, 2375–2386. [CrossRef] [PubMed]
39. Conti, S.G.; de Rosny, J.; Roux, P.; Demer, D.A. Characterization of scatterer motion in a reverberant medium. *J. Acoust. Soc. Am.* **2006**, *119*, 769. [CrossRef]
40. Conti, S.G.; Roux, P.; Demer, D.A.; de Rosny, J. Measurements of the total scattering and absorption cross-sections of the human body. *J. Acoust. Soc. Am.* **2003**, *114*, 2357. [CrossRef]
41. Ribeiro, F.; Florencio, D.; Ba, D.; Zhang, C. Geometrically Constrained Room Modeling With Compact Microphone Arrays. *IEEE Trans. Audio Speech Lang. Process.* **2012**, *20*, 1449–1460. [CrossRef]
42. Steckel, J.; Boen, A.; Peremans, H. Broadband 3-D Sonar System Using a Sparse Array for Indoor Navigation. *IEEE Trans. Robot.* **2013**, *29*, 161–171. [CrossRef]
43. Steckel, J. Sonar System Combining an Emitter Array With a Sparse Receiver Array for Air-Coupled Applications. *IEEE Sens. J.* **2015**, *15*, 3446–3452. [CrossRef]
44. Rajai, P.; Straeten, M.; Alirezaee, S.; Ahamed, M.J. Binaural Sonar System for Simultaneous Sensing of Distance and Direction of Extended Barriers. *IEEE Sens. J.* **2019**, *19*, 12040–12049. [CrossRef]
45. Zhou, B.; Elbadry, M.; Gao, R.; Ye, F. Towards Scalable Indoor Map Construction and Refinement using Acoustics on Smartphones. *IEEE Trans. Mob. Comput.* **2020**, *19*, 217–230. [CrossRef]
46. Bordoy, J.; Schindelhauer, C.; Hoeflinger, F.; Reindl, L.M. Exploiting Acoustic Echoes for Smartphone Localization and Microphone Self-Calibration. *IEEE Trans. Instrum. Meas.* **2019**, *69*, 1484–1492. [CrossRef]
47. Kundu, T. Acoustic source localization. *Ultrasonics* **2014**, *54*, 25–38. [CrossRef] [PubMed]
48. Liu, C.; Wu, K.; He, T. Sensor localization with Ring Overlapping based on Comparison of Received Signal Strength Indicator. In Proceedings of the 2004 IEEE International Conference on Mobile Ad-hoc and Sensor Systems, Fort Lauderdale, FL, USA, 25–27 October 2004. [CrossRef]
49. Dmochowski, J.P.; Benesty, J.; Affes, S. A Generalized Steered Response Power Method for Computationally Viable Source Localization. *IEEE Trans. Audio Speech Lang. Process.* **2007**, *15*, 2510–2526. [CrossRef]
50. Cook, C. Pulse Compression-Key to More Efficient Radar Transmission. *Proc. IRE* **1960**, *48*, 310–316. [CrossRef]
51. Springer, A.; Gugler, W.; Huemer, M.; Reindl, L.; Ruppel, C.C.W.; Weigel, R. Spread spectrum communications using chirp signals. In Proceedings of the IEEE/AFCEA EUROCOMM 2000. Information Systems for Enhanced Public Safety and Security, Munich, Germany, 19 May 2000; pp. 166–170. [CrossRef]
52. Milewski, A.; Sedek, E.; Gawor, S. Amplitude Weighting of Linear Frequency Modulated Chirp Signals. In Proceedings of the 2007 IEEE 15th Signal Processing and Communications Applications, Eskisehir, Turkey, 11–13 June 2007; pp. 383–386. [CrossRef]
53. Carotenuto, R.; Merenda, M.; Iero, D.; G. Della Corte, F. Simulating Signal Aberration and Ranging Error for Ultrasonic Indoor Positioning. *Sensors* **2020**, *20*, 3548. [CrossRef] [PubMed]
54. Schnitzler, H.U.; Kalko, E.K.V. Echolocation by Insect-Eating Bats: We define four distinct functional groups of bats and find differences in signal structure that correlate with the typical echolocation tasks faced by each group. *BioScience* **2001**, *51*, 557–569. [CrossRef]
55. Saphala, A. Design and Implementation of Acoustic Phased Array for In-Air Presence Detection. Master's Thesis, Faculty of Engineering, University of Freiburg, Freiburg, Germany, 2019.
56. MathWorks. ismember. Available online: https://de.mathworks.com/help/matlab/ref/double.ismember.html (accessed on 27 May 2021).

57. MathWorks. ind2sub. Available online: https://www.mathworks.com/help/matlab/ref/ind2sub.html (accessed on 27 May 2021).
58. MathWorks. smooth. Available online: https://de.mathworks.com/help/curvefit/smooth.html (accessed on 27 May 2021).
59. MathWorks. find. Available online: https://de.mathworks.com/help/matlab/ref/find.html (accessed on 27 May 2021).
60. Dylan Mikesell, T.; van Wijk, K.; Blum, T.E.; Snieder, R.; Sato, H. Analyzing the coda from correlating scattered surface waves. *J. Acoust. Soc. Am.* **2012**, *131*, EL275–EL281. [CrossRef] [PubMed]
61. Kuttruff, H. Decaying modes, reverberation. In *Room Acoustics*, 5th ed.; Spon Press, Taylor & Francis: Abingdon-on-Thames, UK, 2009.
62. Lerch, R.; Sessler, G.M.; Wolf, D. Statistische Raumakustik. In *Technische Akustik*, 1st ed.; Springer: Berlin/Heidelberg, Germany, 2009. [CrossRef]
63. Lerch, R.; Sessler, G.M.; Wolf, D. Wellentheoretische Raumakustik. In *Technische Akustik*, 1st ed.; Springer: Berlin/Heidelberg, Germany, 2009. [CrossRef]
64. Kuttruff, H. Steady-state sound field. In *Room Acoustics*, 5th ed.; Spon Press Taylor & Francis: Abingdon-on-Thames, UK, 2009.

Article

Clustering-Based Noise Elimination Scheme for Data Pre-Processing for Deep Learning Classifier in Fingerprint Indoor Positioning System

Shuzhi Liu, **Rashmi Sharan Sinha** and **Seung-Hoon Hwang** *

Division of Electronics and Electrical Engineering, Dongguk University-Seoul, Seoul 04620, Korea; shuzhiliu@dongguk.edu (S.L.); rashmisinha@dongguk.edu (R.S.S.)
* Correspondence: shwang@dongguk.edu; Tel.: +82-2-2260-3994

Citation: Liu, S.; Sinha, R.S.; Hwang, S.-H. Clustering-Based Noise Elimination Scheme for Data Pre-Processing for Deep Learning Classifier in Fingerprint Indoor Positioning System. *Sensors* **2021**, *21*, 4349. https://doi.org/10.3390/s21134349

Academic Editors: Riccardo Carotenuto, Massimo Merenda and Demetrio Iero

Received: 9 May 2021
Accepted: 23 June 2021
Published: 25 June 2021

Publisher's Note: MDPI stays neutral with regard to jurisdictional claims in published maps and institutional affiliations.

Abstract: Wi-Fi-based indoor positioning systems have a simple layout and a low cost, and they have gradually become popular in both academia and industry. However, due to the poor stability of Wi-Fi signals, it is difficult to accurately decide the position based on a received signal strength indicator (RSSI) by using a traditional dataset and a deep learning classifier. To overcome this difficulty, we present a clustering-based noise elimination scheme (CNES) for RSSI-based datasets. The scheme facilitates the region-based clustering of RSSIs through density-based spatial clustering of applications with noise. In this scheme, the RSSI-based dataset is preprocessed and noise samples are removed by CNES. This experiment was carried out in a dynamic environment, and we evaluated the lab simulation results of CNES using deep learning classifiers. The results showed that applying CNES to the test database to eliminate noise will increase the success probability of fingerprint location. The lab simulation results show that after using CNES, the average positioning accuracy of margin-zero (zero-meter error), margin-one (two-meter error), and margin-two (four-meter error) in the database increased by 17.78%, 7.24%, and 4.75%, respectively. We evaluated the simulation results with a real time testing experiment, where the result showed that CNES improved the average positioning accuracy to 22.43%, 9.15%, and 5.21% for margin-zero, margin-one, and margin-two error, respectively.

Keywords: fingerprint-based indoor positioning; clustering; RSSI; CNN

1. Introduction

With the increase in demand for location-based services, high-precision indoor positioning for smartphones has acquired importance internationally. While the global positioning system (GPS) can be used for positioning in outdoor environments, the reception of GPS signals is poor indoors. Consequently, indoor positioning is challenging. Scholars at home and abroad have proposed many indoor positioning systems for solving the indoor positioning problem, but problems pertaining to their applicability, stability, and expansion persist. On the basis of technology, indoor positioning methods for smartphones can be classified into wireless-network-based, measurement-based sensor, and vision-based positioning methods [1–3]. In particular, wireless network-based positioning methods mainly use Wi-Fi, Bluetooth, etc. [4–6], among which Wi-Fi positioning is the most widely used positioning method in the literature. There are two main strategies for positioning using Wi-Fi. One is to use a signal propagation model to determine the received signal strength indicator (RSSI), or a channel state information of the Wi-Fi signal to calculate the distance to the access point (AP) for positioning. Another involves constructing a Wi-Fi fingerprint map and using the current Wi-Fi signal to match the fingerprint map to estimate the position [7,8]. This type of fingerprint recognition has notably promoted the development and the usability of the indoor positioning technology.

Many publications [9,10] have reported indoor positioning technology based on the k-nearest neighbor (kNN). Since APs in the environment are significantly displaced from a certain location, APs in certain locations may not be monitored, and the RSSI vector at each location may not include the signals received by all APs. Therefore, the adjacent reference point (RP) may have a similar RSSI vector. Using the kNN algorithm, all RPs on the wireless map consider identifying the nearest neighbor without taking this phenomenon into account. On the other hand, the neighbors found through the kNN algorithm may be scattered in the environment beyond feasible measurement because the signal attenuation of each AP is not only related to distance, but is also affected by many indoor environmental factors. This leads to the minimum signal distance between the RSSI mark position vector and each RP; this distance is not equal to the minimum physical distance between the actual mark position and the recorded RP position. Considering the limitations of the kNN algorithm in indoor positioning technology, indoor positioning technology based on deep learning is an ideal alternative. A previous study [11] experimentally confirmed the possibility of a convolutional neural network (CNN) applied to complex image classification to improve accuracy, particularly in the classification of complex pictures in a dynamic environment. On this basis, another study [12] innovatively utilized the CNN algorithm as an indoor positioning technology framework. Experiments show that CNN can effectively address the limitations of the kNN algorithm and improve the accuracy of indoor positioning.

Hao et al. [13] proposed a fingerprinting technique based on channel state information (CSI). The CSI information of 25 RPs was collected by one transmitter and one receiver with three links of the 121 subcarriers in each link. The size of each RP was 1.1 m $\times$ 0.96 m for all 25 RPs. The density-based spatial clustering of applications with noise (DBSCAN) processing of the CSI data is performed in the offline phase. The noise reduction of the processed dataset was performed with the endpoint-clipping method. This endpoint-clipped dataset was used to train the SVM classifier. The DBSCAN processing of the CSI test data was performed in the online phase. Matching of the test location was done by matching the link dataset of the training dataset. The training dataset was pre-processed for DBSCAN. Unnecessary data, marked as noise, were deleted from the training dataset. The DBSCAN-processed training dataset was augmented and sent to the CNN classifier (deep learning classifier). The test RSSI value was further converted into a 16 $\times$ 16 image and matched with previously trained datasets. The location of the matched image was the location of the test file.

In our previous paper [14], an indoor positioning technology framework based on deep learning classifiers was proposed as shown in Figure 1. In the paper, the positioning system framework was described as two phases: the offline phase and the online phase. The offline phase primarily involves the collection and processing of indoor positioning data. For example, the RSSI data are collected in the test environment, the resulting database is established and trained, and deep learning classifiers are trained. The result at this stage will directly affect the actual positioning accuracy. The online phase primarily involves the actual test, where the real-time positioning test is performed through the deep learning classifier obtained in the previous stage. Although the use of deep learning can improve the accuracy of indoor positioning, deep learning has not been effectively used due to insufficient database capacity. Therefore, we proposed a deep learning indoor positioning framework based on data augmentation in another manuscript [15]. In the offline phase, the program used data augmentation to increase the capacity of the RSSI fingerprint database to improve the training effect of the deep learning model. Experiments demonstrated that this method can further improve the positioning accuracy. In the last step of the online phase, the "majority rule" was used to select the most frequent positioning results returned by the server. This method is termed as the data post-processing algorithm [16]. This method can reduce the error in real-time positioning results to improve the positioning accuracy.

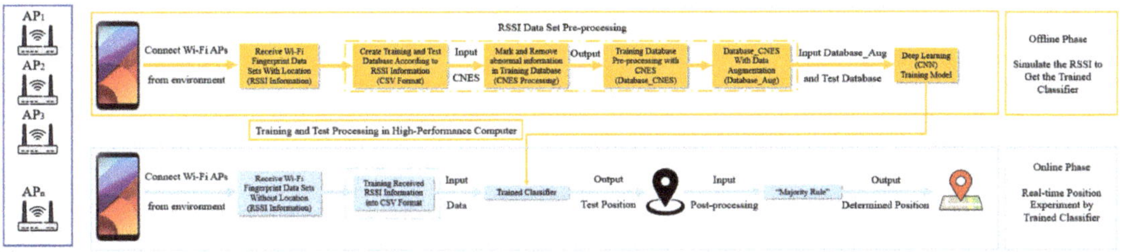

Figure 1. Fingerprint positioning with a deep learning classifier from [15].

A previous study [17,18] showed that in a Wi-Fi fingerprint-based indoor positioning system, changes in a dynamic environment such as the multipath propagation of signals caused by obstacles near the user's location, fading, and the addition or removal of Wi-Fi APs affect the indoor positioning accuracy. Furthermore, another study [19] noted that for an indoor positioning system in a dynamic environment, the error caused by the dynamic environment should be reduced through appropriate methods. Meanwhile, the latest research shows that the interference of moving objects and co-frequency interference in a dynamic environment may cause the Wi-Fi signal pattern to be changed over time, which reduces the positioning accuracy [20]. This paper presents a database pre-processing method based on a clustering-based noise elimination scheme (CNES) to effectively improve the real time positioning accuracy. The proposed CNES scheme is based on the DBSCAN method, and clustering and noise reduction processing were performed for the RSSI fingerprint data at each RP. The pre-processing of RSSI data with a clustering-based noise elimination scheme (CNES) is a novel concept. The proposed method successfully achieved the highest lab simulated positioning accuracy of 92.01% and a real time testing experimental positioning accuracy of 90.42%, which was much higher than the accuracy of the dataset without CNES pre-processing. Furthermore, the proposed method is an infrastructure-free method that does not require any additional infrastructure for implementation. The remainder of this paper is organized as follows. Section 2 presents the background; Section 3 discusses the proposed CNES data pre-processing scheme; Section 4 describes the numerical analysis and presents the laboratory simulation and experiments results; and finally, Section 5 summarizes the conclusions.

2. Background

2.1. Environment Setup

Both data collection and experiment were performed on the seventh floor of the new engineering building at Dongguk University, Seoul, Korea. In Figure 2, the target area with the size of 52 m × 32 m and the roof height of 3 m is divided into 74 grids with 2 m × 2 m squares as the RP. Since each RP was assumed to be the center of the grid, any point in the grid was regarded as the RP. That is, the distance between any two adjacent RPs was considered as 2 m. The RPs such as 1, 25, 36, 40, and 67 were at the corner and their sizes varied between 2 m to 3 m (i.e., +1 m difference). Meanwhile, the RPs such as 10, 11, 18, 50, and 71 were at the ending spot and their sizes varied between 1 m to 2 m (i.e., −1 m difference). The positioning server used in this study was a Dell Alienware Model P31E (Alienware, hardware subsidiary of Dell, Miami, FL, USA), and the smartphone for data collection and testing was a Samsung SHV-E310K (chip fabrication Yongin-si, Gyeonggi-do, Korea). The fingerprint database construction, classification (i.e., position prediction), and online experimental setup were developed with Python.

The data read by an Android device were stored in a buffer. If there was an error in the recorded data, an error message was displayed on a serially connected console. Otherwise, the RSSI data were stored in the buffer, and after a complete scan, they were transferred by the Android console, which was connected by an interface cable to the server, to the server through a Wi-Fi AP. The server determines the Android device's location by comparing the measured RSSI values with the reference data. It was serially connected to the Android console and processed the RSSIs obtained from the surrounding APs with its CPU (Figure 3). The operating frequency of the Wi-Fi device was 2.412–2.480 GHz for the 802.11 bgn wireless standard. Additionally, the input/output sensitivity was 15–93 dBm.

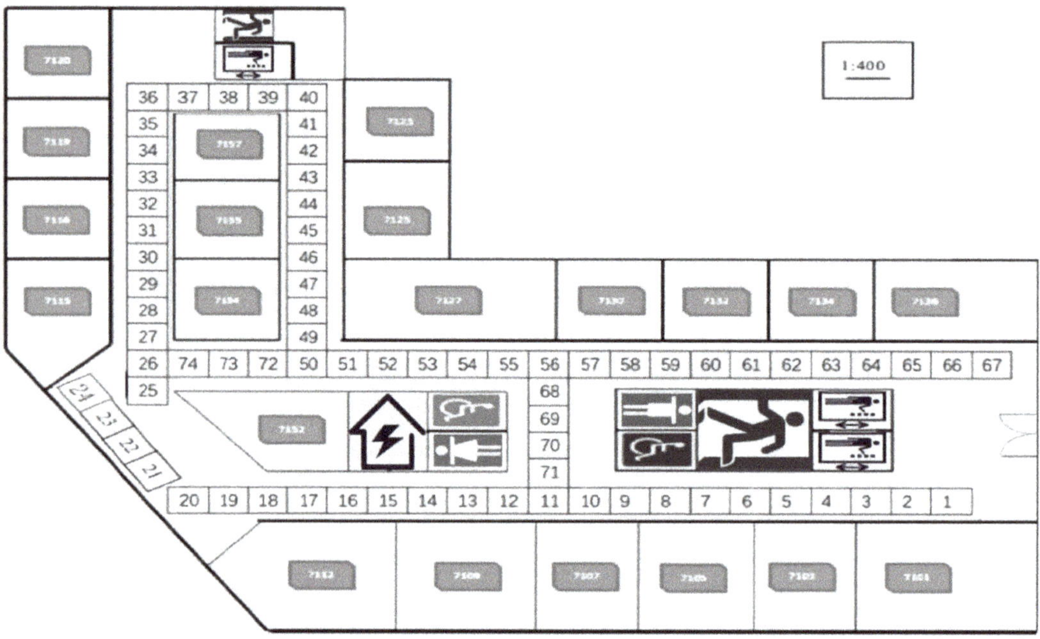

Figure 2. Environment setup for a floor map with 74 reference points.

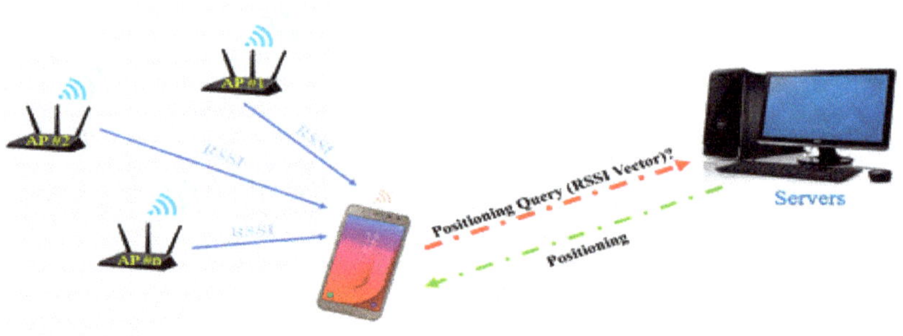

Figure 3. The proposed fingerprint-based Wi-Fi positioning system.

2.2. CNN Model and Data Augmentation

The collected RSSI data were converted into a comma-separated-value (CSV) file and then forwarded to the deep learning model. The structure of the generated CSV file is shown in Figure 4. The CSV file contained all the acquired RSSI information including the media access control (MAC) address from different APs, the RSSI value corresponding to each RP, and the number of RPs. The blue box shows the MAC address information area, which contained a total of 256 MAC information.

MAC Addres	MAC-1	MAC-2	MAC-3	MAC-4	MAC-5	MAC-6	....	MAC-254	MAC-255	MAC-256	256	Total #MAC
RSSI Values	NN-1-1	NN-1-2	NN-1-3	NN-1-4	NN-1-5	NN-1-6		NN-1-254	NN-1-255	NN-1-256	L-1-1	Location Labels
	NN-2-1	NN-2-2	NN-2-3	NN-2-4	NN-2-5	NN-2-6		NN-2-254	NN-2-255	NN-2-256	L-1-2	
	NN-3-1	NN-3-2	NN-3-3	NN-3-4	NN-3-5	NN-3-6		NN-3-254	NN-3-255	NN-3-256	L-1-3	
	NN-4-1	NN-4-2	NN-4-3	NN-4-4	NN-4-5	NN-4-6		NN-4-254	NN-4-255	NN-4-256	L-1-4	
	NN-5-1	NN-5-2	NN-5-3	NN-5-4	NN-5-5	NN-5-6	....	NN-5-254	NN-5-255	NN-5-256	L-1-5	
	NN-6-1	NN-6-2	NN-6-3	NN-6-4	NN-6-5	NN-6-6		NN-6-254	NN-6-255	NN-6-256	L-1-6	
	NN-7-1	NN-7-2	NN-7-3	NN-7-4	NN-7-5	NN-7-6		NN-7-254	NN-7-255	NN-7-256	L-1-7	
				....					....		....	

Figure 4. Input comma-separated-value (CSV) file format [14].

The CNN classifier described in Figure 5 was proposed in our previous study [14], which was composed of five layers. The first layer had input grayscale images of size $16 \times 16 \times 1$, rectified linear unit (ReLU), and dropout. Due to the small size of the input data set, max pooling was not used in the first layer. The second layer consisted of a 16×16 convolution with ReLU and then an 8×8 max pooling layer with a total of 18,496 parameters, which produced the output for the third layer with an 8×8 convolution with ReLU and then a 4×4 max pooling layer. The output was fed directly to a fully connected (FC) layer with 3072 nodes, which led to the next hidden FC layer with 1024 nodes. Finally, the output was calculated using a softmax layer with 74 nodes, which was the total number of RPs in our setup. The inner width was 1024, and the dropout of 0.5 was used for the first four layers. The learning rate was 0.001. The total number of parameters was 2,266,698. This calculated output was the total number of RPs in the current setup. The purpose of data enhancement is to obtain more training data by effectively transforming the existing data, thereby reducing the problem of under-fitting or over-fitting caused by the data quality or the amount of data being too small [21]. The input image was generated from the RSSI values received at 74 RPs during the experiment. At each RP, the RSSI value was recorded for 256 APs, though only a small subset of these APs was visible at each RP. Then, the RSSI values from different APs created a 16×16 image. For example, in Figure 6a, there are a total of nine visible RSSI values between 25 to 70 from 256 APs, with the other values of 0. As shown in Figure 6b, the RSSI values were converted into a grayscale image. The image had different levels of brightness depending on the recorded RSSI values, with higher RSSI values being brighter. The highest RSSI value was 70, which produced the brightest spot in the grayscale image, while the lowest value was 25, which is represented as the darkest nonblack spot. The RSSI values of 0 produced no brightness, so the remaining 247 spots were black. Similarly, the input RSSI files at the other 73 RPs produced different images as an input to the deep learning network.

Before providing the training data to the CNN model, we performed data augmentation for the training database by using the method presented in [21]. The augmentation scheme was operated using only the RSSI values collected at each RP. The RSSI value at each RP was randomly selected and written in a new CSV, which resulted in a large data size compared to the original CSV. The robustness lies in the fact that the pattern of augmented data well mimicked that of the RSSI data before augmentation. For the 24 dataset with 8880 images for each RP after augmentation, the total number of images at each RP was 532,800 with the size of 350 MB. Total number of test files was 1480.

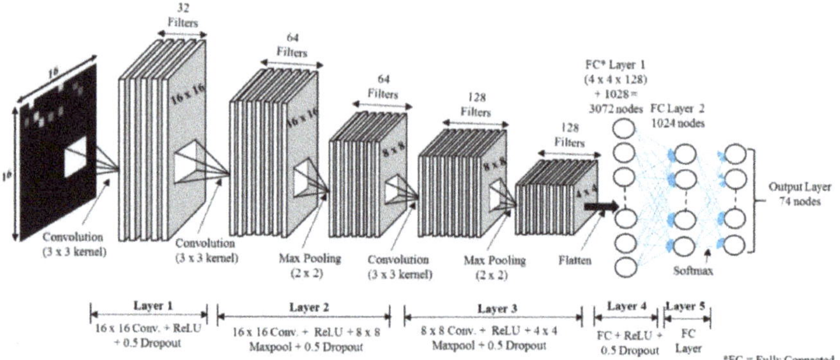

Figure 5. The convolutional neural network (CNN) architecture used in this study [14]. The second layer had a total of 18,496 parameters and the FC layer had 3072 counters, which led to the next hidden FC layer with 1024 converters. Finally, a softmax layer with 74 routines was used.

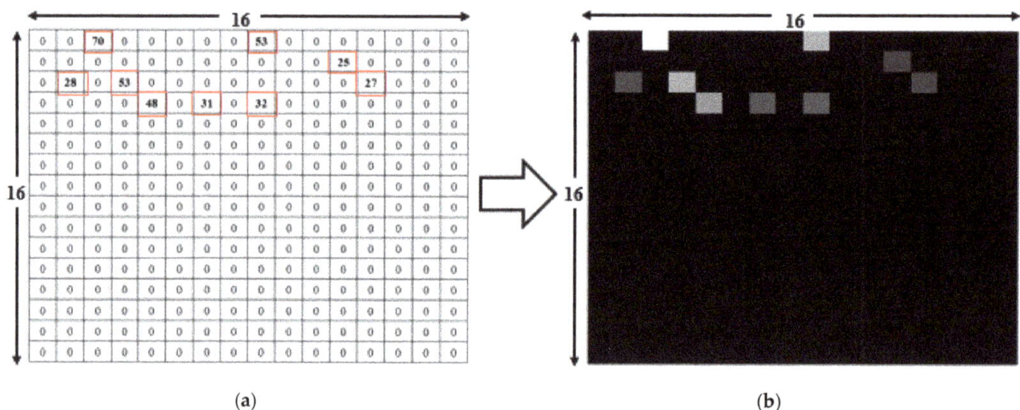

(a)　　　　　　　　　　　　　　　　(b)

Figure 6. Deep learning input file conversion from a CSV file to an image. (**a**) Input CSV readings of the nine visible RSSIs from a total of 256 APs. (**b**) Converted grayscale image with nine bright spots representing APs visible at the RP [14].

2.3. RSSI Dataset Generation

To collect the RSSI data, we used a smartphone in the user's hand and collected data five times on each RP. Each measurement comprised the RP label, the time, the date, the number of available APs, the MAC address, and the corresponding RSSI value at each RP. The RSSI measurement may contain noise, which seriously affects the positioning accuracy due to the time-varying channel characteristics. Moreover, indoor electromagnetic environments are complex and are characterized by multipath fading and other noise. We examined the RSSI fluctuation effects on prediction accuracy in [22]. In this work, the different directions (forward/backward) and times (morning/afternoon) were considered for seven-day data collection. A specific data collection procedure is as follows. The collector holds the smartphone at their waist position (about 1.2 m to 1.3 m height from the ground) and measures the data as stationary at each RP. In the morning, we conducted forward and backward data collections, respectively, which were repeated in the afternoon. Forward refers to the direction to collect the RSSI values from RP1 to RP74 in sequence. Meanwhile, backward means the opposite direction to collect the RSSI value from RP74 to RP1 sequentially. For a seven-day data collection, we collected 28 data files.

Table 1 shows the dataset types. The data collected in the morning and afternoon are denoted by M and A, respectively. The data collected in the forward and backward directions are labelled with F and B, respectively. The number represents the day number when the data were collected, as shown in Table 2. We divided 28 datasets into two parts. One part contained 24 datasets to construct the training database. The other part contained the remaining four datasets to build the test database. The RSSI values were measured five times at each RP in the forward as well as in backward directions. The sampling time for each RSSI measurement was 5 s, which was a total of 25 s for a total of five measurements. For 74 reference points, the total time consumed was 31 min in each direction and 62 min in both directions. For the training data, the measurements were made in the morning and the afternoon for seven days, which resulted in 15 h approximately. For the trial data, the measurements were made for two days in the same manner.

Table 1. Database information and augmentation size.

Database Type	Collection	# of Images	
		Before Augmentation	After Augmentation
Training	24 sets	8880	532,800
Test	4 sets	1480	–

Table 2. Types of dataset.

Dataset	Forward	Backward	Number of Data Files
Morning	MF1, MF2, ..., MF7	MB1, MB2, ..., MB7	14
Afternoon	AF1, AF2, ..., AF7	AB1, AB2, ..., AB7	14
Number of Data Files	14	14	28

3. Proposed Scheme

3.1. Density-Based Spatial Clustering of Applications with Noise (DBSCAN)

Previous approaches to indoor positioning technology focused on the study of small positioning areas. This was because it is necessary for noise samples to appear in the original data as the positioning area expands, especially in a dynamic environment. It has been confirmed [23] that the existence of noise would reduce the waste of computing resources and thus affect the accuracy of indoor positioning. Several studies have [22–25] used a clustering algorithm to cluster and divide the RSSI samples. These studies demonstrated that clustering could reduce the impact of noise in large-area positioning experiments. Density-based spatial clustering of applications with noise (DBSCAN) is a type of clustering that is well-known to the public. It is mainly based on the density of data, and it is highly representative. In this idea, a cluster is a large set, and all objects in it may be densely connected. The algorithm can be unconstrained in the sample database and is able to find clusters of any shape, which are major advantages. The DBSCAN process may be expressed concisely. In simple terms, the core point of a given dataset may be determined arbitrarily. Clustering around this point, all points with reachable density were included in the core point cluster. If many data have not been included, then re-clustering around a new core point is repeated in the cluster. Given a sample dataset, circle the given object with eps radius and count the data objects in this circle. Figure 7 uses a two-dimensional point set to illustrate the concept of core points, border points, and noise points. If there is a point with MinPts or greater in the eps radius around it, the other points will gather around that point, which is called the core point. Points that belong to a cluster but are not core points are called boundary points, and they are primarily points that form the outer edge of the cluster. Points that do not belong to any class become noise points.

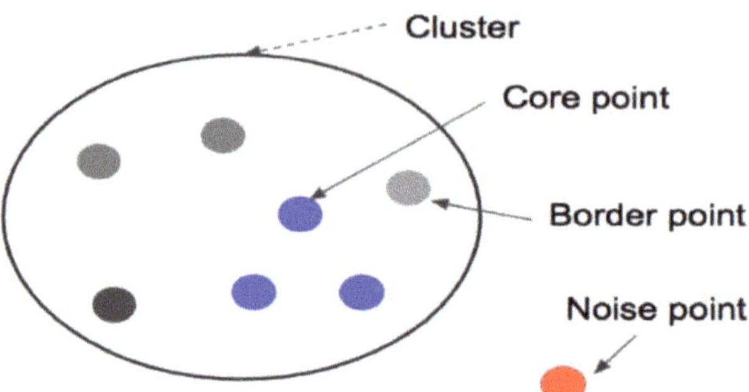

Figure 7. The concept of core points, border points, and noise points.

3.2. Proposed Clustering-Based Noise Elimination Scheme (CNES)

This paper proposed the CNES algorithm based on DBSCAN. The purity of the database was improved by detecting and deleting the noise points of each RP. The algorithm was run in the offline phase of the indoor positioning framework. After using CNES, the training database achieved the results shown in Figure 8. We performed the analysis of effective MinPts value and epsilon 'eps' of DBSAN for RSSI data noise elimination. The effective suitable MinPts value was 4. The value of eps could then be chosen by using a k-distance graph and plotting the distance to the k = minpts − 1 nearest neighbors ordered from the largest to smallest value. Furthermore, the best eps value was analyzed between 60 to 75 points. These points were further used to generate the final eps value.

	MAC Address																				#
	1	**2**	**3**	**4**	**5**	**6**	**7**	**8**	**9**	**10**	**11**	**12**	**13**	**14**	**15**	**16**	**17**	**18**	**19**	**20**	**RP**
	45	40	34	15	52	48	41	41	41	30	26	20	20	18	0	0	0	0	0	0	1
	47	35	43	0	50	0	44	44	44	29	25	0	0	0	44	29	28	19	15	0	1
	45	46	40	0	49	46	43	43	43	30	24	21	0	0	44	0	0	19	15	16	1
	49	38	41	0	51	49	40	40	39	32	20	24	0	23	44	0	0	0	0	18	1
	51	46	42	0	47	48	0	0	43	33	0	0	0	21	0	30	0	10	0	0	1
RSSI Value	43	35	45	21	62	48	32	31	31	0	0	0	0	0	0	0	0	0	0	15	2
	53	30	38	19	59	46	35	34	35	0	0	25	0	0	0	0	0	14	0	0	2
	41	37	0	22	0	44	40	41	41	0	23	0	0	0	45	0	0	0	0	0	2
	55	32	43	21	61	50	34	34	34	0	0	33	0	0	0	0	0	0	0	0	2
	54	34	39	19	57	48	37	38	38	0	0	25	0	0	0	0	0	17	0	0	2
	⋮									⋮								⋮			
	69	0	0	0	0	0	0	0	0	51	0	0	0	0	0	0	0	0	0	0	74
	63	0	0	0	0	0	0	0	0	45	0	0	0	0	0	0	0	0	0	0	74
	61	0	0	0	0	0	0	0	0	45	0	0	0	0	0	0	0	0	0	0	74
	45	0	0	0	0	0	0	0	0	45	0	0	0	0	0	0	0	0	0	0	74
	42	0	0	0	0	0	0	0	0	42	0	0	0	0	0	0	0	0	0	0	74

Figure 8. Clustering-based noise elimination scheme (CNES)-based training dataset with the highlighted and deleted noise points.

Figure 8 shows the RSSI values of MAC addresses ranging from 1 to 20, taking the five sets of RSSI data samples for RP1, RP2, and RP74 as examples. The RSSI sample was the original CSV file generated from RSSI data collected in the experimental environment, which contained noise samples. When CNES is not used, direct use of the database will reduce the accuracy of indoor positioning. The grey, highlighted segments in Figure 8 represent noise samples such as the fifth group of RP1 samples, the third group of RP2 samples, and the fourth group of RP74 samples. The database was first imported into the CNES algorithm for processing; the processing marked and removed noise samples. Post-processing, the new database without noise samples was created. Finally, the new database processed data augmentation and deep learning model training and testing. Figure 9 shows the effect of CNES for eps = 70 and MinPts = 4. The dotted line represents the number of RSSI samples collected at each RP. The solid line represents the number RSSI samples after CNES for eps = 70 and MinPts = 4 at each RP. Figure 10 presents the complete flow graph for the proposed CNES database position estimation. In addition, the pseudocode of the proposed scheme is shown in Algorithm 1.

Algorithm 1: Pseudocode for Clustering-Based Noise Elimination and Position Estimation

---

1.   **Input:** Original CSV fingerprint training dataset
2.   **Define:** CSV dataset:
3.   **for** density calculation
4.   **Define** eps; minpts;
5.   **for** each reference point calculate density 'D'
6.   **if** RP == core point;   \\ Keep the RP data;
7.   **elseif** RP == edge point;   \\ Keep the RP data;
8.   **else** RP ≠ core point | | RP ≠ edge point;   \\ Delete the RP data;
9.   **end if**
10.   **end for**
11.   **end for**
12.   Generate new CSV with density-based noise elimination point;
13.   Augment the output CSV file;
14.   Train the CNN classifier with new CSV file;
15.   Test the file for real time online position estimation;
16.   **end for**

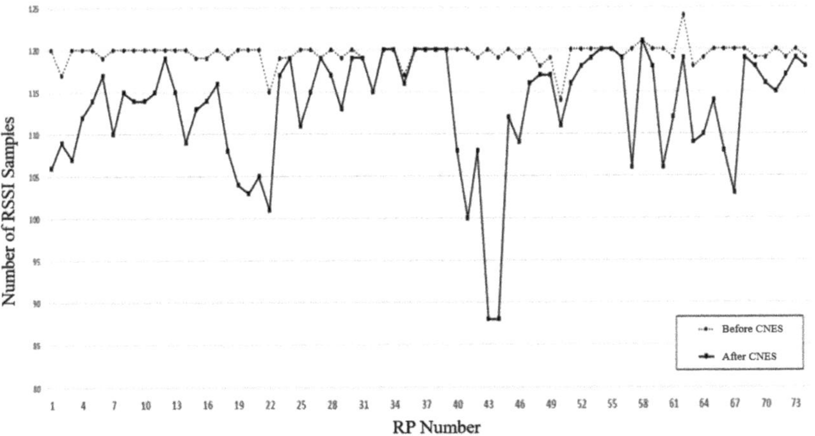

Figure 9. The effect of CNES corresponding to eps = 70 and MinPts = 4 on total number of RSSI samples at each RP.

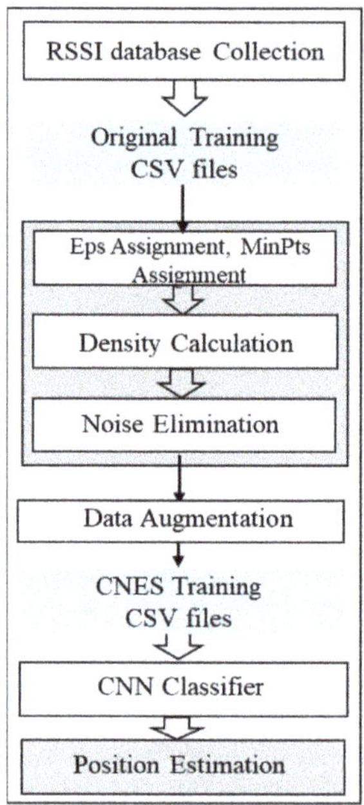

Figure 10. Flow graph for clustering-based noise elimination and position estimation.

4. Numerical Results

A total of 74 RPs was arranged in the positioning environment, as shown in Figure 2. Due to the dynamic environment, uncontrollable factors such as changes in the number of routers, the activation of telecommunication equipment and the movement of pedestrians can generate abnormal information such as noise in the collected RSSI information. DBSCAN recognizes the impact of noise and is robust to outliers. In the experiment, DBSCAN cluster analysis was performed for each RP point to reduce the influence of noise on the data during the data collection process, thereby improving the positioning accuracy. RSSI information was collected five times at each RP point, and a total of 24 datasets were collected in the experiment. Therefore, each RP point had 5×24 kinds of information, as shown in Figure 4. Then, for the total training set (comprising 74 RPs), there were a total of 74 fixed clusters because of the 74 RP labels. However, for each RP, the clustering algorithm marked and eliminated abnormal information, and therefore, there were two clusters for each RP. In the experiment, DBSCAN was used to cluster each RP point. This is because each RP was tagged when collecting RSSI information. The value of eps can be chosen by using a k-distance graph and plotting the distance to the $k = minpts - 1$ nearest neighbors ordered from the largest to the smallest value. Good eps values exist where the plot shows an 'elbow' (i.e., the threshold value above which the number of RSSI samples remains approximately the same), as shown in red circle in Figure 11. For example, eps = 70 in Figure 11. In general, for the suitable eps, a rule of thumb is to select the eps number with only a small fraction of RSSI samples.

4.1. Analysis of Eps

As mentioned, in order to find the best eps value, eps = {60: 75} was used to cluster the training database, and the database was then input into the deep learning model for indoor positioning simulation. In order to accurately verify the simulation accuracy corresponding to different eps values, when performing indoor positioning simulation, we chose the maximum positioning error acceptable in our indoor positioning system for analysis. Assuming the error distance was 4 m, the objective was to choose the eps value most suitable for our indoor positioning system. The simulation results are shown in Table 3, and the indoor positioning accuracy was as high as 94.191% when eps = 70.

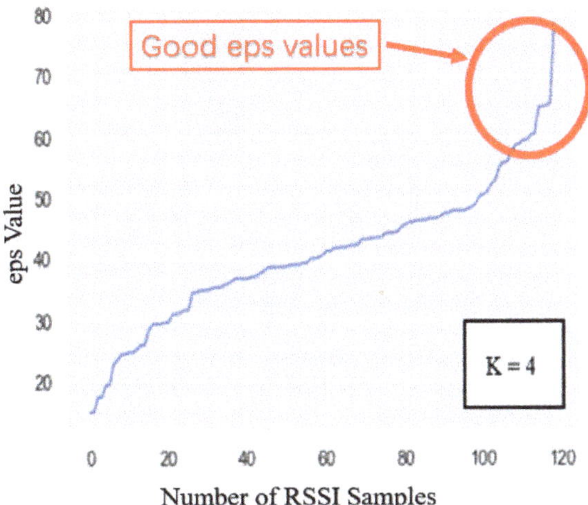

Figure 11. K-nearest neighbor distances to determine eps MinPts for CNES. The red circle represents "elbow", which means there exist good eps values.

Table 3. Simulation accuracies for different eps values (training epochs number = 1000, MinPts = 4).

Eps Value	Lab Simulation Accuracy	Eps Value	Lab Simulation Accuracy
60	93.594%	68	93.491%
61	93.193%	69	92.889%
62	93.293%	70	94.191%
63	92.789%	71	93.189%
64	93.889%	72	92.893%
65	92.593%	73	92.292%
66	92.490%	74	93.093%

In the experiment, DBSCAN was used to cluster the RSSI samples of each RP in the training database, and the RSSI samples outside the core point neighborhood could be eliminated by using the best eps value. The eliminated samples were also the so-called errors. Information samples were not suitable for positioning reference information. Therefore, in the training set clustered by different eps values, the number of RSSI samples retained by each RP was inconsistent, as shown in Figure 12. Among the lines, the top dashed line represents the original training set, that is, with all original RSSI samples retained. As the eps value increases, the curve approaches closer to the original curve. At RP = [10, 11, 19, 20, 21, 22, 41, 42, 43, 44, 45, 56, 64, 65, 66, 67], the range of change becomes larger. In particular, at the point RP = [41, 43, 44], the range of change exceeded 70. This is because these points show the areas where the Wi-Fi signals and people were dense, which may cause RSSI degradation. The red line in Figure 12 denotes eps = 70, which was the best eps value in Table 3. For eps = 70, the average number of removed samples was six, which was lower than 10 for all eps values, which means that more samples were removed in Figure 12. Furthermore, for eps = 70, it was shown that all reference samples were retained at RP = [32, 33, 36, 37, 38, 54, 55, 58] when the environment was better less than the external interference.

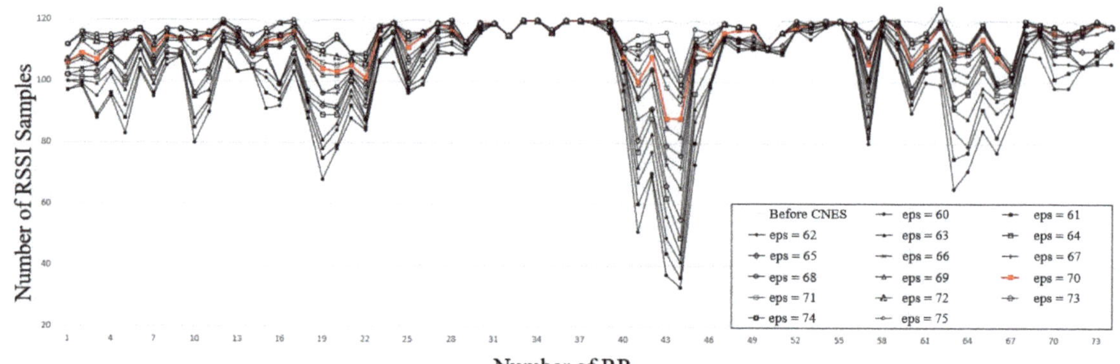

Figure 12. The effect of CNES on the number of RSSI samples at each RP corresponding to different eps values.

4.2. Lab Simulation Results

In the CNN model, the lab simulation results with the highest accuracy were selected for real time testing. In terms of the accuracy of both models, the lab simulation results are shown in Table 4.

Table 4. Summary of lab simulation results.

Lab Simulation Model	Margin (%)		
	0	1	2
CNN	43.50	75.95	87.26
CNES + CNN	61.28	83.19	92.01
Difference	17.78	7.24	4.75

When the RP number is accurately predicted by the CNN trained model, it is called Margin-0 (i.e., 0 m error). When the predicted test RP matches the neighboring RP, it is called Margin-1 (i.e., 2 m error). Similarly, when the test RP matched with difference of two RPs, it is known as Margin 2 (i.e., 4 m error). A comparison of the accuracies of the CNN mode for different margins and for the two techniques is presented in Table 4. As shown in Table 4, the CNES scheme can improve positioning accuracy. Without CNES, the positioning accuracy of Margin-0 was 43.50% only. At the same time, Margin-1 was 75.95%, and Margin-2 was 87.26%. However, the positioning accuracy was significantly improved after using the CNES scheme. The positioning accuracy of Margin-0 exceeded 60%, which was 61.28%. In this way, the positioning effect of Margins-1 and -2 using the CNES solution was also obvious. In particular, the simulation result of Margin-1 reached 83.15%, which was close to the result of Margin-2 without CNES. In addition, the result of Margin-2 exceeded 92%. In addition, we compared the difference between with and without CNES under the same margin. The simulation positioning accuracy after using CNES was improved by 17.78% (Margin-0), 7.24% (Margin-1), and 4.75% (Margin-2). Through a comparison, it can be seen that CNES can significantly improve the accuracy of indoor positioning. In terms of Margin-0, the improvement effect was significant, which means that indoor positioning using CNES can achieve an error of zero-meters in most cases.

The effectiveness of the proposed CNES was also evaluated in terms of positioning accuracy, which is defined as the cumulative distribution function (cdf) of the location error within a specified distance in Figure 13. It was shown that the CNES outperformed that without CNES over the entire range. Note that when the cdf exceeded 94%, the distance error with the CNES was only 4.76 m. Meanwhile, the distance error without CNES was 6.70 m. In addition, it was shown that the cdf with CNES was 73.21% for a one-meter error, but that without CNES was only 58.47%.

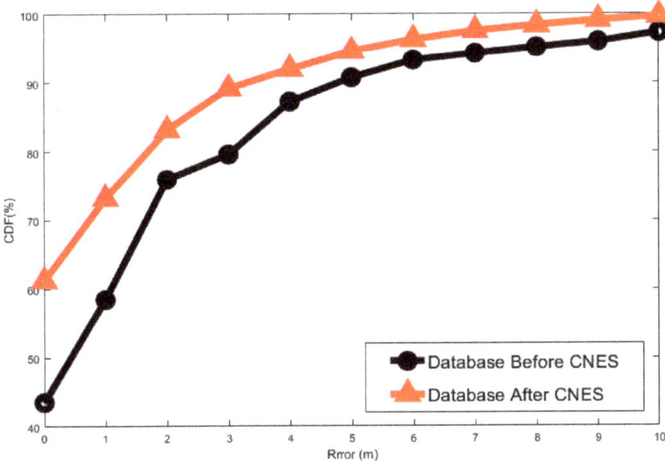

Figure 13. Cumulative distribution function (cdf) vs. distance error. *X*-axis represents positioning errors and *Y*-axis represents the positioning accuracy for different positioning errors.

4.3. PCA

Principal component analysis (PCA) diagrams for the training database and test database for the five cases are shown in Figure 14, and the blue points in the figure represent the RPs in the training data without data augmentation, and the green points represent the unknown location points, that is, the points in the test data. From Figure 14, (a) represents the analysis of data before CNES was applied, and (b) represents the analysis of data after CNES was applied. Furthermore, it is evident that after the CNES was applied, the RP points in the training data were more compact than those in the training data to which the CNES was not applied. This is because the discrete points in Figure 14a are caused by incorrect RSSI fingerprint information, and the use of the CNES reduces wrong information in the training set, thereby improving the accuracy of the deep learning simulation.

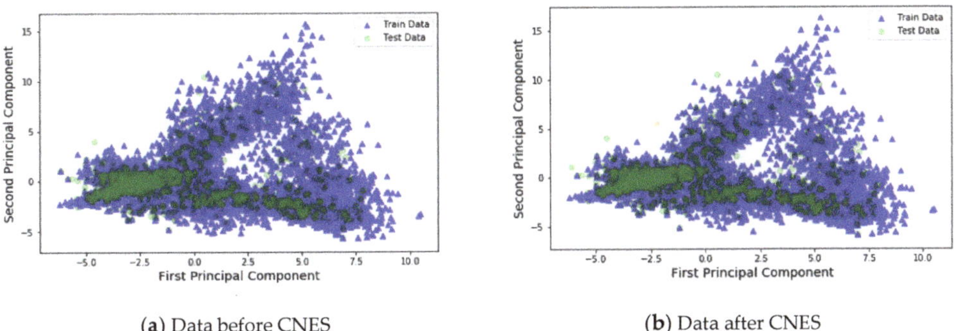

(**a**) Data before CNES   (**b**) Data after CNES

Figure 14. Principal component analysis (PCA) plots for two schemes: (**a**) before and (**b**) after applying the CNES. Green points are test data, and blue points are training data.

4.4. Experimental Results with Real Time Testing

In the experiments, we used the trained classifier for real time testing. The specific process involves passing the measured RSSI values at RPs into the CNN model to obtain the features. Then, the obtained features were compared with those in the trained classifier. The RP with the most similar features was determined as the final position. In addition, we compared two cases for real time testing, namely, CNN and CNES + CNN. "CNN" means

CNN without CNES, and "CNES + CNN" denotes CNN with CNES. For real time testing, we made experiments four times with the plan in Table 5. The experiments were made for two working days (D-1 and D-2), and four times (Test 1, 2, 3, and 4) a day.

Table 5. The experimental plan for real-time testing.

Day	Test 1	Test 2	Test 3	Test 4
D-1	CNN	CNES + CNN	CNN	CNES + CNN
D-2	CNES + CNN	CNN	CNES + CNN	CNN

When processing the experimental results, we stipulated that the test results with measurement errors less than or equal to 2 were successful, which means that only the positioning result and the current position error distance of less than four meters can be used, and the data represent the probability of successful results obtained in the measurement results. An example of the experimental results is shown in Table 6. As shown in Table 6, for each RP positioning test in an experiment, five positioning decisions were performed continuously for the same RP.

Table 6. The example of real time testing experimental results.

RF #	Positioning Decision #					# of Success Decisions		
	#1	#2	#3	#4	#5	Margin-0	Margin-1	Margin-2
1	1	1	2	1	1	4	5	5
2	2	3	2	4	2	3	4	5
3	3	3	2	3	4	3	5	5
●●●●●●								
72	72	73	73	74	72	2	4	5
73	73	73	73	73	26	4	4	5
74	73	74	74	26	26	2	4	5
Experiment Success Rate (%)						61.71	83.35	91.62

In order to facilitate the comparison of the experimental results of CNES, we merged all experimental results into Table 7. Meanwhile, the results were rearranged and are shown in Table 8. It can be seen that the CNES scheme can effectively improve the indoor positioning accuracy, especially in the case of Margin-0 (zero-meter error), the average positioning success rate increased by 22.43%. Without the CNES scheme, the Margin-0 (zero-meter error) positioning success rate was only 39.45%, and after using the CNES scheme, the Margin-0 (zero-meter error) positioning success rate exceeded 60%, which was 61.88%. Meanwhile, the CNES scheme had outstanding positioning accuracy in Margin-1 (two-meter error), and the positioning success rate was 82.77, which was close to the positioning success rate in Margin-2 without the CNES scheme. In addition, in the case of Margin-2 (four-meter error), it exceeded the highest success rate of 90.42% without the CNES scheme. Therefore, the above data show that the CNES pre-processing scheme can indeed greatly improve the accuracy of indoor positioning without changing the hardware.

Table 7. Summary of real time testing of experimental results.

Day	Database (Test Number)	Margin			Database Test Number)	Margin		
		0	1	2		0	1	2
D-1	CNN (Test 1)	38.97	73.78	85.13	CNN (Test 3)	39.12	72.95	84.81
	CNES + CNN (Test 2)	61.71	83.35	91.62	CNES + CNN (Test 4)	62.28	82.47	89.69
D-2	CNES + CNN (Test 1)	62.03	82.73	90.11	CNES + CNN (Test 3)	61.48	82.51	90.27
	CNN (Test 2)	40.23	74.06	85.79	CNN (Test 4)	39.47	73.69	85.14

Table 8. Difference of experimental average results (%).

Database	Average Margin		
	0	1	2
CNN	39.45	73.62	85.22
CNES + CNN	61.88	82.77	90.42
Difference	22.43	9.15	5.21

It is generally known that the multipath effect of the channel will increase the abnormal information (such as noise) in the RSSI dataset. Therefore, the abnormal information in the original training dataset can be eliminated by using the CNES scheme, and then the purity of the training dataset can be improved. According to the simulation and actual experiments, the databases with CNES can adapt to various databases and different environments, thereby improving the positioning accuracy.

5. Conclusions

In this work, a deep learning solution involving a clustering processing scheme was developed. The results showed that the use of pre-processed data along with the CNES could effectively improve the indoor positioning accuracy. The simulation results showed that when the CNES was used as the clustering algorithm, the best effect was obtained for eps = 70. For the indoor positioning simulation, with the CNES RSSI dataset, a positioning accuracy of 92.01% was achieved. The experimental results in the real environment also showed that the CNES pre-processing scheme could increase the positioning accuracy by 22.43%, 9.15%, and 5.21 in Margin-0 (zero-meter error), Margin-1 (two-meter error), and Margin-2 (four-meter error), respectively. Furthermore, the CNES scheme could reduce the effect of interference factors in the dynamic environment on the positioning accuracy and improve the adaptability of indoor positioning accuracy.

Author Contributions: S.L. and S.-H.H. contributed to the main idea of this research work. S.L. and R.S.S. performed the simulations, experiments, and database collection. The research activity was planned and executed under the supervision of S.-H.H. and S.L.; R.S.S. and S.-H.H. contributed to the writing of this article. All authors have read and agreed to the published version of the manuscript.

Funding: The following are the results of a study on the "Leaders in INdustry-university Cooperation+" Project supported by the Ministry of Education and National Research Foundation of Korea.

Institutional Review Board Statement: Not applicable.

Informed Consent Statement: Not applicable.

Data Availability Statement: Not applicable.

Acknowledgments: The authors would like to express their sincere gratitude to Sang Moon Lee, CTO of JMP Systems, Korea, for providing the equipment for database collection and setting up the experimental environment.

Conflicts of Interest: The authors declare no conflict of interest regarding the publication of this article.

References

1. Ashraf, I.; Hur, S.; Park, Y. Indoor positioning on disparate commercial smartphones using Wi-Fi access points coverage area. *Sensors* **2019**, *19*, 4351. [CrossRef]
2. Endo, Y.; Sato, K.; Yamashita, A.; Matsubayashi, K. Indoor positioning and obstacle detection for visually impaired navigation system based on LSD-SLAM. In Proceedings of the 2017 International Conference on Biometrics and Kansei Engineering (ICBAKE), Kyoto, Japan, 15–17 September 2017.
3. Qi, J.; Liu, G.P. A robust high-accuracy ultrasound indoor positioning system based on a wireless sensor network. *Sensors* **2017**, *17*, 2554. [CrossRef] [PubMed]
4. Khalajmehrabadi, A.; Gatsis, N.; Pack, D.J.; Akopian, D. A joint indoor WLAN localization and outlier detection scheme using LASSO and elastic-net optimization techniques. *IEEE Trans. Mob. Comput.* **2016**, *16*, 2079–2092. [CrossRef]

5. Ali, M.U.; Hur, S.; Park, S.; Park, Y. Harvesting indoor positioning accuracy by exploring multiple features from received signal strength vector. *IEEE Access* **2019**, *7*, 52110–52121. [CrossRef]
6. Ashraf, I.; Hur, S.; Park, Y. Smartphone Sensor Based Indoor Positioning: Current Status, Opportunities, and Future Challenges. *Electronics* **2020**, *9*, 891. [CrossRef]
7. Zhang, Y.; Li, D.; Wang, Y. An indoor passive positioning method using CSI fingerprint based on Adaboost. *IEEE Sens. J.* **2019**, *19*, 5792–5800. [CrossRef]
8. Song, Q.W.; Guo, S.T.; Liu, X.; Yang, Y.Y. CSI amplitude fingerprinting based NB-IoT indoor localization. *IEEE Internet Things J.* **2017**, *5*, 1494–1504. [CrossRef]
9. Wu, Z.; Rengin, C.; Shuyan, X.; Xiaosi, W.; Yuli, F. Research and improvement of WiFi positioning based on k nearest neighbor method. *Comput. Eng.* **2017**, *43*, 289–293.
10. Li, C.; Qiu, Z.; Liu, C. An improved weighted k-nearest neighbor algorithm for indoor positioning. *Wirel. Pers. Commun.* **2017**, *96*, 2239–2251. [CrossRef]
11. Chishti, S.O.; Riaz, S.; Zaib, M.B.; Nauman, M. Self-Driving Cars Using CNN and Q-Learning. In Proceedings of the 2018 IEEE 21st International Multi-Topic Conference (INMIC), Karachi, Pakistan, 1–2 December 2018; pp. 1–7.
12. Song, X.; Fan, X.; Xiang, C.; Ye, Q.; Liu, L.; Wang, Z.; He, X.; Yang, N.; Fang, G. A novel convolutional neural network based indoor localization framework with WiFi fingerprinting. *IEEE Access* **2019**, *7*, 110698–110709. [CrossRef]
13. Hao, Z.; Yan, Y.; Dang, X.; Shao, C. Endpoints-clipping CSI amplitude for SVM-based indoor localization. *Sensors* **2019**, *17*, 3689. [CrossRef]
14. Sinha, R.S.; Hwang, S.H. Comparison of CNN applications for RSSI-based fingerprint indoor localization. *Electronics* **2019**, *8*, 989. [CrossRef]
15. Sinha, R.S.; Lee, S.M.; Rim, M.; Hwang, S.H. Data augmentation schemes for deep learning in an indoor positioning application. *Electronics* **2019**, *8*, 554. [CrossRef]
16. Haider, A.; Wei, Y.; Liu, S.; Hwang, S.H. Pre-and post-processing algorithms with deep learning classifier for Wi-Fi fingerprint-based indoor positioning. *Electronics* **2019**, *8*, 195. [CrossRef]
17. Xu, J.; Zhang, H.; Zhang, J. Self-Adapting Multi-Fingerprints Joint Indoor Positioning Algorithm in WLAN Based on Database of AP ID. In Proceedings of the IEEE 33rd Chinese Control Conference, Nanjing, China, 28–30 July 2014; pp. 534–538.
18. Liu, K.; Wang, Y.; Lin, L.; Chen, G. An analysis of impact factors for positioning performance in WLAN fingerprinting systems using Ishikawa diagrams and a simulation platform. *Mob. Inf. Syst.* **2017**, *2017*, 8294248. [CrossRef]
19. Xia, S.; Liu, Y.; Yuan, G.; Wang, Z. Indoor fingerprint positioning based on Wi-Fi: An overview. *ISPRS Int. J. Geo Inf.* **2017**, *6*, 135. [CrossRef]
20. Zhu, Q.; Xiong, Q.; Wang, K.; Lu, W.; Liu, T. Accurate WiFi-based indoor localization by using fuzzy classifier and mlps ensemble in complex environment. *J. Frankl. Inst.* **2020**, *357*, 1420–1436. [CrossRef]
21. Sinha, R.S.; Hwang, S.-H. Improved RSSI-based data augmentation technique for fingerprint indoor localisation. *Electronics* **2020**, *9*, 851. [CrossRef]
22. Park, C.R.; Rhee, S.H. Indoor positioning using Wi-Fi fingerprint with signal clustering. In Proceedings of the 2017 International Conference on Information and Communication Technology Convergence (ICTC), Jeju Island, Korea, 18–20 October 2017; pp. 820–822.
23. Hantoush, R. Evaluating Wi-Fi Indoor Positioning Approaches in a Real-World Environment. Ph.D. Thesis, Universidade NOVA de Lisboa, Lisbon, Portugal, 2016.
24. Li, A.; Fu, J.; Shen, H.; Sun, S. A cluster principal component analysis based indoor positioning algorithm. *IEEE Internet Things J.* **2020**, *8*, 187–196. [CrossRef]
25. Puussaar, A. Indoor Positioning Using WLAN Fingerprinting with Post-Processing Scheme. Ph.D. Thesis, University of Tartu, Tartu, Estonia, 2014.

 *sensors*

Article

Development of a Smartphone-Based University Library Navigation and Information Service Employing Wi-Fi Location Fingerprinting

Guenther Retscher * and Alexander Leb

Department of Geodesy and Geoinformation, TU Wien, 1040 Vienna, Austria; e1326712@student.tuwien.ac.at
* Correspondence: guenther.retscher@tuwien.ac.at; Tel.: +43-1-58801-12847

Abstract: A guidance and information service for a University library based on Wi-Fi signals using fingerprinting as chosen localization method is under development at TU Wien. After a thorough survey of suitable location technologies for the application it was decided to employ mainly Wi-Fi for localization. For that purpose, the availability, performance, and usability of Wi-Fi in selected areas of the library are analyzed in a first step. These tasks include the measurement of Wi-Fi received signal strengths (RSS) of the visible access points (APs) in different areas. The measurements were carried out in different modes, such as static, kinematic and in stop-and-go mode, with six different smartphones. A dependence on the positioning and tracking modes is seen in the tests. Kinematic measurements pose much greater challenges and depend significantly on the duration of a single Wi-Fi scan. For the smartphones, the scan durations differed in the range of 2.4 to 4.1 s resulting in different accuracies for kinematic positioning, as fewer measurements along the trajectories are available for a device with longer scan duration. The investigations indicated also that the achievable localization performance is only on the few meter level due to the small number of APs of the University own Wi-Fi network deployed in the library. A promising solution for performance improvement is the foreseen usage of low-cost Raspberry Pi units serving as Wi-Fi transmitter and receiver.

Keywords: Wi-Fi positioning; navigation; location fingerprinting; RSSI-based positioning; probabilistic approach; information service; book tracking

---

Citation: Retscher, G.; Leb, A. Development of a Smartphone-Based University Library Navigation and Information Service Employing Wi-Fi Location Fingerprinting. *Sensors* **2021**, *21*, 432. https://doi.org/10.3390/s21020432

Received: 19 November 2020
Accepted: 7 January 2021
Published: 9 January 2021

Publisher's Note: MDPI stays neutral with regard to jurisdictional claims in published maps and institutional affiliations.

1. Introduction

In recent years, a number of technologies and methods have been developed and improved for indoor positioning. One of these technologies is based on the use of Wireless Fidelity (Wi-Fi). As such infrastructure is already installed in most public buildings and therefore costs are low, it is one of the most researched technologies for indoor positioning. Thereby positioning can be made either cell-based, as well as using lateration or fingerprinting. In particular, location fingerprinting has proven itself in practice. It is an approach to pattern recognition and based on received signal strength indicator (RSSI) measurements of the surrounding Wi-Fi Access Points (APs) in an off-line training and an on-line positioning phase. During the training phase, the RSSIs of the surrounding APs are measured in the area of interest at reference points to build-up a fingerprinting database, which can be visualized by signal strength radio maps. For the positioning in the on-line phase, the measured fingerprint is then compared at an unknown location with those in the empirically determined radio map. Finally, the position in the radio map that best matches the on-line RSSI measurement is returned. A major disadvantage of the empirical method, however, may be the time required to set up and maintain the database. In addition, the measurements must be carried out again during the installation of new transmitter or other structural changes. Another challenge is the large variation of the observed RSSI values due to signal fluctuations [1]. Despite these disadvantages, fingerprinting is nowadays one of the most popular method for an indoor positioning system (see e.g., [2,3]).

TU Wien (Vienna University of Technology) is the largest scientific-technical research and education institution in Austria. With its four inner-city locations as well as a science center further away from the city center, the University has more than 12,000 rooms in 30 buildings on an available area of approximately 269,000 m^2. With such a large number of buildings and rooms, a positioning and navigation system can be a helpful tool. Especially in the large library building which has six levels with an area of 1160 m^2 the localization and tracking of the books is a challenging task. The motivation of this study is therefore to help students, employees, and visitors of the University to find at least the correct bookshelf. Furthermore, the individual books shall be located and tracked. For that purpose, also the use of Radio Frequency Identification (RFID) is foreseen. It was seen in the tests that not all areas in the library can be covered with Wi-Fi to guarantee the required localization accuracy. Thus, also the integration of Bluetooth is considered for the positioning and navigation system.

The paper is structured as follows: Section 2 provides a comprehensive survey of suitable indoor positioning techniques leading hereto to the chosen technical solution for the library navigation and information system. In Section 3 the specifications and characteristics of the test site and measurement procedures are introduced followed by a description of the analyses carried out for the off-line fingerprinting training phase in Section 4. Section 5 then deals with the impact on the results of different times required for the measurement of a single Wi-Fi scan, referred to as scan duration in the paper, in the kinematic measurement mode. In the following, Section 6 addresses the localization performance and achievable positioning accuracies in the on-line positioning phase. Section 7 pinpoints a useful strategy towards the development of a library navigation and information system. Finally, the paper is concluded and outlook on future work is given in Section 8.

2. Suitable Indoor Positioning Techniques Survey

At the beginning of the study at hand a survey was carried amongst solutions for indoor positioning to be able to select the overall best absolute positioning technology for the library navigation and information service. In this section first the requirements for an indoor positioning system (IPS) in general and in particular for the service at TU Wien are discussed. Suitable technologies are identified and assessed in a compendium where their characteristics and physical properties are analyzed in detail. Moreover, the selection process of the chosen technique is exemplified. An IPS should not only provide a certain required positioning accuracy, but also function reliably and be designed to be user-friendly. Moreover, attention should also be paid to data protection and costs.

2.1. General Aspects

An IPS is a wireless location system used to navigate, locate and position people and objects inside buildings. It usually consists of at least two hardware components, i.e., a transmitter and a receiver. One of the two components is always the mobile device to be located. Depending on the functionality of these two components, a first classification of positioning systems can be made into self- and remote-positioning [3]. In a self-positioning system, the receiver represents the mobile device that measures the signals of transmitters which coordinates are known. The position is then calculated on the mobile device. It is also possible that the measurement results are sent from the receiver to a master station. If the position at the master station is calculated, the positioning mode is referred to as indirect remote-positioning. In a remote-positioning system, the transmitter represents the mobile device and the receivers are fixed at known locations. Then the results of all measurements are collected in a master station and the position of the transmitter is calculated. If the measurement results are sent back from the receiver to the transmitter so that the position is determined on the mobile device, it is called indirect self-positioning.

2.2. Technological Requirements

Due to the complexity of the environment in a building, the limitation of direct line-of-sight (LoS), blocking of signals and multipath effects but also due to reliability, availability and cost of the required equipment, indoor positioning is a big challenge. Both in terms of accuracy, which ranges from the sub-meter level to several meters, and in terms of cost, the systems exhibit a wide range.

The required positioning accuracy of an IPS depends on the application. While positioning accuracy in the meter range is sufficient for many applications, it may be too inaccurate for a warehouse or a library, for example. The achievable accuracy depends, among other things, on the position determination method and technology used. The number of transmitters and receivers plays an essential role. Furthermore, attention should be paid to the dimension in which the position is to be determined. In a multi-story building, in addition to the location (horizontal accuracy), the floor (vertical accuracy) must also be determined. For a library such type of accuracy is needed so that the user is able to find a certain bookshelf. With a bookshelf distance of 2 to 3 m the accuracy should be at least in this range. Since the building has several floors, it should be possible to determine the floor in addition to the location.

In terms of reliability and coverage of the positioning system it should always be available and consist of stable components so that localization can be performed in real-time. In addition, smaller signal failures should be compensated and a seamless transition between outdoor and indoor areas should be possible. This requires a sufficient number and a good distribution of transmitters or receivers. Due to the widely distributed locations of the TU Wien, the positioning system should work on the entire inner-city campus, both outside and inside buildings. The aim is to identify the building in which the user is located or whether he/she is outside the campus.

Any student, staff member or visitor of the University should be able to operate the positioning system easily. With the use of a smartphone being ubiquitous these days, an application for these devices is a useful solution for the navigation and information service. The presentation of information should be adapted to the user and consideration should also be given to physically impaired persons. In addition, care should be taken to ensure that the application consumes as little energy as possible. The power consumption depends, among other things, on whether the position is calculated on the smartphone or externally. If the position is determined on an external server, it can usually be calculated much faster. Positioning should also not take too long, but should be performed in real-time if possible. This means that the latency—the time it takes for the user to see their location—must be also as short as possible.

Privacy is a major challenge for an IPS, as not all people want to share their current location. Therefore, it is also important to consider the privacy and security of IPS users. Therefore, the IPS operator should ask how the user can trust the system [4]. The decisive factor here is where the position is determined, i.e., either on the mobile device or in a master station or on an external server. If the location is determined directly at the mobile device, no forwarding of information to the IPS operator is required, thus ensuring the privacy of the mobile user. However, determining the position on the mobile device again consumes a lot of energy and requires sufficient computing power, so it is significant to reduce the computational complexity of the IPS [5]. In [6], different IPS are compared and it was found that most of the commonly employed technologies present data protection problems. Only remote positioning technologies, such as the use of inertial sensors, provide a high level of data protection. Their disadvantages, however, are their low precision due to accumulation of errors leading to high sensor drift rates. Furthermore, they require an absolute positioning method in addition for determining the start location and an update of the only relatively determined positions.

Last but not least, the cost of an IPS is one of the basic decision criteria and depends on several factors, such as available money, time, infrastructure and energy. The time factor is related to installation and maintenance. Also the costs for software, server and maintenance of databases have to be considered. The number of transmitters or receivers are thereby considered as infrastructure costs. However, it is not always obvious what the actual cost of an IPS will be. For example, it can be assumed that a Wi-Fi-based IPS does not incur hardware costs because the required Wi-Fi APs are already available. This is valid, however, for RSSI-based solutions at TU Wien only. New approaches based on measurement of the ranges to the Wi-Fi APs require the installation of new hardware.

Power consumption is also a decisive cost factor of a positioning system. Some devices are completely energy passive, such as passive RFID tags. These devices only respond to external fields and can therefore have an unlimited lifetime. Other mobile devices, on the other hand, only have a few hours of battery life without recharging [3]. For TU Wien the positioning system should not cause any costs for the user, so he/she should not have to buy new hardware. For indoor applications, technologies should be used that are not too expensive to purchase and install. For outdoor use, GNSS positioning can be used if enough satellite signals are available.

2.3. Compendium of Common Technologies

Different signals and technologies can be used for positioning of a mobile device. In order not to go beyond the scope of this work these technologies are presented briefly in the following, with a closer look at their characteristics and physical properties as well as their advantages and disadvantages. A first division can be made into optical, sound-based and radiofrequency technologies, magnetic fields and inertial sensors.

Optical technologies work with signals that are in the visible or infrared range. This corresponds to electromagnetic radiation with wavelengths from 380 nm to 10 μm. Unlike radio signals, infrared signals and visible light cannot penetrate walls and other obstacles, limiting positioning to enclosed spaces. Transmitter identification can be modulated on the infrared signal, which allows different transmitters to be distinguished. Therefore, the position determination can be carried out with all common measuring principles. One of the first IPS to use infrared signals is Active Badge [7]. Technologies that use visible light for data transmission are also called Visible Light Communication (VLC) [8,9]. For instance, light-emitting diodes (LEDs) can be used. The transmission of data using LED is possible because the light source can be switched on and off again in very short intervals. This flickering can be so fast that it cannot be perceived by human eyes. A variety of modulation methods can be used. The principle for VLC is that each fixed LED lamp has a different flicker coding, so that the mobile sensor (e.g., the smartphone camera) receives the light and compares the modulation with the known coding scheme [10].

The position of a user can also be determined by using acoustic signals. The time of the broadcast can be determined by simultaneously sending a radio and an acoustic signal. Since the radio signal arrives earlier than the acoustic signal at the receiver, the difference between these two times can be used to calculate the range. For this process, however, an exact synchronization of the transmitter and receiver clocks is necessary [11]. Due to the fact that acoustic signals travel with slower speed than infrared or radio frequency signals the acoustic signals travel time measurement allows for higher accuracy. The propagation velocity thereby depends on the energy of the signal as well as on the density and temperature of the medium it penetrates. The sound-based technologies can be divided into audible [12] and non-audible sound (ultrasonic signals) [13]. The position can be determined by all range-based measurement principles (see Section 2.4). Nakashima et al. [14] determine the position using a digital watermark that they inserted into the audio track of each speaker. However, this technology is difficult to implement in reality, as it can be assumed that there is a lot of noise in a building, making it difficult to determine the position. In addition, audible noise—especially in Universities libraries—can be very annoying in everyday life. Thus, ultra-sonic signals are usually employed for sound-based positioning. With this technology, the travel time of emitted ultrasonic pulses is usually measured and the position subsequently determined by means of lateration. For example, in [13] the position is determined using Time of Arrival (ToA), while in [15] localization is performed using Time Difference of Arrival (TDoA). The signals can be either transmitted by the mobile device and received by permanently installed receivers or vice versa. The mobile device can also be a transmitter and receiver at the same time, so that the distance can be determined by Round Trip Time (RTT) measurements, which means that no additional infrastructure is required. In addition to the multipath effect, another problem with this technology is that ultrasonic systems are very sensitive, since even a low noise already generates ultrasonic waves and thus interferes with the system. Among the commonly known systems are the Active Bat [16], Dolphin [17] and Cricket [18]. Lopes et al. [15] proposed a reliable acoustic indoor positioning system fully compatible with a conventional smartphone. Thereby acoustic ranging in the audio band based using non-invasive signals is carried out using the smartphone audio I/O. In order to support the positioning system a Wireless Sensor Network (WSN) of synchronized acoustic beacons is used for TDoA ranging. They achieved an absolute positioning error of less than 10 cm on the 95% significant level in their tests.

Unlike infrared signals and visible light, radio signals can penetrate walls and other obstacles, making positioning not limited to enclosed spaces. Each station that transmits radio frequency (RF) signals has a unique address by which it can be identified. The most popular RF technologies are Wi-Fi, Bluetooth, Radio Frequency Identification (RFID) and Ultra-wide Band (UWB).

A Wireless Local Area Network (WLAN) transmits electromagnetic signals over the free 2.4 and 5 GHz Industrial, Scientific and Medical (ISM) band with wavelengths of 12.5 and 6.0 cm, respectively. WLAN or Wi-Fi is based on the IEEE 802.11 standard. Indoors the signals have a range of 20 to 100 m [19] and can also pass through walls. Positioning can be performed using time-based methods (ToA and RTT), Angle of Arrival (AoA) or RSSI. The specifics of Wi-Fi are further discussed in Section 2.6.

Bluetooth is also an electromagnetic signal, which has a wavelength of approximately 12.5 cm in the frequency range between 2.402 and 2.480 GHz. The latest Bluetooth version (5.1) now has a range of approximately 200 m. In addition, from this version onwards, the direction angle of the received or transmitted signal can also be measured using AoA [20]. Position determination via Bluetooth can also be time-based or RSSI-based. However, since Bluetooth transmits in the same spectrum as Wi-Fi, it is susceptible to interference. Advantages of Bluetooth are availability as it is supported by most smartphones, low cost, and low power consumption, which allows transmitters to run on battery for several months or even years [21]. It has been considered by the IT Department of TU Wien to use Bluetooth Low Energy (BLE) Beacons in addition on the campus for areas with limited or no Wi-Fi coverage. However, one must also take into account the signal attenuation, the multipath effect and the fluctuations in signal strength while using Bluetooth. Zhao et al. [22] demonstrate that BLE is more accurate than Wi-Fi for localization when lateration approaches are employed where RSSI values have to be converted into ranges.

Typical RFID frequency ranges are 125–134 kHz for low frequency, 13.56 MHz for high frequency and 860–890 MHz for ultra-high frequency (UHF) [23]. In RFID, a reader communicates with one or more transponders (so-called tags) using radio waves. If a tag comes near a reader, the two start communication and can exchange information with each other, such as the location of the permanently mounted component. The tags or readers are then mounted at strategically important locations inside the building (e.g., at entrances). Communication takes place either by inductive coupling or electromagnetic waves [11]. RFID tags can be classified according to whether they are passive, semi-passive or active [24]. Passive tags do not have their own power supply and respond with the energy the reader releases with the help of a small antenna. They are much lighter, smaller and cheaper than active tags, so they only have a range of 1 to 2 m. A passive tag can be attached to a mobile device or a book, for example, which is read by permanently installed readers. Thus, they can be used for cheap book labelling, e.g., also to replace barcodes. In the TU Wien library it is thought to use this technology. RFID can then also serve as a security function by installing readers at the building exits, which give an alarm signal if a book leaves the library unauthorized.

UWB is based on the transmission of electromagnetic waves by a sequence of very short pules (less than 1 ns) with a wide bandwidth of larger than 500 MHz. This allows reflected signals to be better filtered, minimizing the multipath effect and improving positioning accuracy, which is one of the major advantages of this method [25]. Unlike other RF-based technologies, UWB devices can also transmit signals in multiple frequency bands simultaneously. Another advantage is the lower power consumption. However, UWB devices are still more expensive to purchase and install. Position determination can be performed using lateration (ToA and TDoA) and/or angulation (AoA) [11,19].

Other RF-based localization technologies include Zigbee and Long-range Wide-area Network (LoRaWAN). Zigbee is a common low power technology, often used in Internet of Things (IoT) applications with same ranges and coverage as the aforementioned technologies. LoRaWAN, on the other hand, can reach ranges of up to 15,000 m transmitting at 915 MHz, which may allow a significant reduction of transmitters in order to cover an environment. In the tests conducted by Sadowski and Spachos [26] it was seen, however, that this technology showed the worst performance for indoor localization. Zigbee has a similar low energy requirement to LoRaWAN, while its performance is much higher. The authors further mention that BLE is a low power and cost efficient solution for IoT localization in small crowded areas. Wi-Fi, however, consumes the most power out of all the examined technologies, but is advantageous because of its high ubiquity. It also achieved the highest positioning accuracies in their tests. These results confirm that a localization service based on Wi-Fi is the way to go ahead for our application.

An indoor position determination can also be performed by measuring the magnetic field strengths. Both the geomagnetic field [27] and an artificially generated magnetic field [28] can be used. Although there are some approaches using the later, most modern systems use the Earth's magnetic field strength [11]. Using the embedded magnetometer in a smartphone magnetic field fluctuations can be measured as the magnetic field shows local anomalies, which are caused by objects, such as electrical devices and cables and building structures, such as concrete reinforcement. Assuming that these anomalies within a building are nearly static and have sufficient variability, they provide a unique magnetic fingerprint such that localization can be carried out with fingerprinting (see Section 2.5) [29]. A drawback, however, is that magnetic fields are already disturbed by small changes in the environment, e.g., caused by people, which complicates localization.

Global Navigation Satellite Systems (GNSS) chipsets are an integral part of every smartphone. At least the US Navstar Global Positioning system (short GPS) is supported in smartphones. In addition, more and more smartphones include the measurements from the Russian GLONASS, the European Galileo and the Chinese Beidou. The GNSS signals are in the L-band, i.e., in the frequency range from 1.164 to 1.300 MHz with a wavelength from 25.7 to 23.1 cm and 1.559 to 1.610 MHz with 19.2 to 18.6 cm wavelength. In this work GNSS, is used only for outdoor positioning, but it is out of the scope of this paper to further discuss its application in the TU Wien localization service.

The users' location can also be determined relatively from a given start position with the help of inertial sensors, which are embedded in every smartphone. They include Micro-Electro-Mechanical Systems (MEMS)-based accelerometers and gyroscopes. The measurements of the accelerometer can be used to obtain the distance travelled, such as from step counting, and the gyroscope is used to estimate the direction of movement of the user. The employed location technology is dead reckoning and in case of pedestrian navigation it is referred to as Pedestrian Dead Reckoning (PDR). Due to the large error drift of MEMS-based sensors, which accumulates over time, a combination with absolute positioning techniques is required to update the measurements (see e.g., [30,31]). Using filtering, such as with a Kalman or particle filter, is another popular approach to reduce sensor drift rates and estimate the current users' location (see e.g., [32]). In this work, the use of inertial sensors is not foreseen at this stage as the sole use of an absolute positioning technology is evaluated.

Additionally, it is worth mentioning that more and more smartphones also have a pressure sensor built in, which can be used to measure the air pressure and thus determine the altitude. This can be particularly useful in multi-storey buildings as the sensor can be used to determine the current floor (see e.g., [33,34]).

Smartphone cameras provide visual information in addition to the position, but computational-intensive image recognition software is required for localization. One approach for smartphone positioning is scene analysis. Similar to the fingerprinting method, a database is first filled with images of the environment and linked to the respective location. After that, a photo can be taken from the smartphone user's point of view and this is then compared with the images from the database to determine the position [35]. The big advantage thereby is that no signals are used for position determination and therefore the effects of signal propagation do not play a role. Moreover, no additional infrastructure needs to be installed. A disadvantage, however, is the large amount of time required to set up and maintain the database, which has to be updated when structural changes occur. If the position is not determined on the mobile device, the image must first be transferred to the master station, which means a large data transfer. On the other hand, if the position is to be computed on the mobile device, it requires a large amount of RAM since the image recognition software packages are computationally intensive. However, due to the recent technical developments with improved image recognition algorithms and computational capabilities as well as greater data transmission rates, these drawbacks have been minimized [36]. Another method is called visual odometry [37,38] where the self-motion (translation and orientation) of a person or object is determined using single or multiple cameras attached to the mobile object. Thereby the images must contain sufficient meaningful information, such as color, texture, shape, etc., to estimate the movement of the camera. Visual odometry offers a good trade-off between cost, reliability, and implementation complexity [39] and is widely used in mobile robotics [40].

Table 1 provides a comparison of the most commonly employed technologies in terms of their advantages and disadvantages as well as respective costs and Table 2 summarizes the characteristics of different suitable positioning methods. As with these technologies, however, it is somewhat difficult to give an exact positioning accuracy figure, as it depends heavily on the measurement principle, method and infrastructure used, and is therefore not mentioned. Table 3 provides rough ranges of achievable positioning accuracies for the different useable positioning methods. The cost of each technology here depends mainly on the infrastructure already in place. From the user's point of view, it is assumed that a smartphone is used for positioning and therefore no costs are incurred. The costs in Table 1 therefore relate only on the installation and maintenance of the infrastructure in the building. Since an IPS is used in various applications, there is no one method or technology that is superior to the others, but only one that best meets the requirements set, both technically and economically. As each of the techniques presented has different advantages and disadvantages, a combined solution is the best way to overcome the drawbacks of each individual method and reduce measurement errors. If several positioning technologies are combined, the system is also referred to as hybrid IPS.

Table 1. Positioning technologies comparison.

Technology		Advantages	Disadvantages	Costs
optical	infrared	low power consumption	do not penetrate walls; susceptible to interference; low range	low
	visible light	low power consumption; high accuracy	do not penetrate walls; susceptible to interference	low
acoustic-based	audible	no costs	disruptive in everyday life	none
	ultra-sonic	high accuracy	susceptible to interference	medium
radio frequency	Wi-Fi	use of available infrastructure; accuracy on the m-level	susceptible to signal fluctuation; high power consumption	medium
	Bluetooth	low power consumption	susceptible to signal fluctuation and interference	low
	RFID	cheap passive tags can be mounted everywhere	low range and accuracy	medium
	UWB	multipath resistant; low energy consumption; high accuracy	expensive hardware	high
magnetic	natural	no costs	susceptible to interference	low
	artificial	low fluctuations	susceptible to interference	high
smartphonesensors	GNSS	freely available	not useable in buildings	none
	inertial	work independently	high sensor drift	none
	camera	works independently; visual information	computationally high costs	none

Table 2. Characteristics of positioning methods.

	Measurement Principle	Advantages	Disadvantages	Positioning Accuracy
CoO	cell-based	simple algorithm	relative positioning to location of transmitter	cell size dependent
Lateration	ToA, TDoA, RTT, RSSI-based	no off-line training phase	susceptible to multipath; LoS requirement	dm–m
Angulation	AoA	no off-line training phase	susceptible to multipath; LoS requirement; antenna array needed	dm–m
Fingerprinting	RSSI	no multipath influence; no LoS requirement	off-line training phase	m
Scene Analysis	-	no multipath influence	off-line training phase; high data transfer rates required; computationally intensive	dm–m
Dead Reckoning	-	only smartphone sensors used	relative positioning; large drift rates	m

Table 3. Characteristics and properties of range-based localization techniques for lateration.

	ToA	RTT	TDoA	RSSI-Based
signal propagation	does not matter	does not matter	does not matter	matters
LoS	required	required	required	not required
multipath	matters	matters	matters	partially matters
time synchronisation	transmitter and receiver	transmitter and responder	transmitter and receiver	not required
positioning accuracy	dm–m	dm	dm–m	m

From Table 2 the following conclusions can be drawn: Cell-of-Origin (CoO) based positioning is only suitable for determining a first approximate solution. Unlike location fingerprinting and scene analysis, lateration and angulation do not require an off-line training phase. With these two methods, however, care must be taken that the receiver measures the LoS signal and not a reflected signal. Angulation also requires an antenna array or a directional antenna to measure the incident angles. The major advantage of fingerprinting is that it is more resistant to multipath than lateration and angulation. In addition, there is no need LoS between the transmitter and receiver. The major disadvantage, however, is the time required to set up and maintain the training fingerprint database. The advantages and disadvantages of scene analysis are similar to those of fingerprinting. An additional disadvantage is the large amount of data that has to be transmitted if the position is to be calculated on an external server. If the position is determined on the smartphone, however, sufficient computing power is required, since the image recognition software packages are computationally intensive. Dead Reckoning (DR) is the only method based on relative positioning included in the Table. The big advantage over the other methods is that it only requires a smartphone whit its embedded inertial sensors. As already aforementioned, smartphone sensors show high drift rates leading to low positioning accuracies which are steadily increasing in time.

2.4. Range-Based Localization Operational Principle

The most common measuring principles for position determination using ranges between a transmitter and a receiver are briefly reviewed in the following. Mostly it is assumed, however, that only the horizontal position coordinates must be determined. As aforementioned it is also important to locate the user on the correct floor in a multi-storey building. The techniques have in common that the mobile device can be either the transmitter or the receiver. These measurement principles form the basis for the methods of localization.

To derive the ranges between a transmitter and receiver several methods are applicable, such as Time of Arrival (ToA) (also referred to as Time of Flight (ToF)), Time Difference of Arrival (TDoA) and Round Trip Time (RTT) measurements [36]. In addition, ranges may be derived from RSSI measurements using so-called path loss models (see e.g., [41] for examples of common path loss models). These models are describing the relationship between RSSI values and distance by assuming that the RSSI decreases with increasing range from the transmitter.

For lateration, the individual ranges between the transmitters and the mobile device are first estimated. The methods ToA, RTT, TDoA and RSSI can be used for this purpose. At least three ranges must be measured for unambiguous localization in 2D. This is referred to as trilateration. If more ranges are used it is called multi-lateration. For the position determination the intersection is calculated from the distance radii. The location of the transmitter is in the center of a circle in 2D and a sphere in 3D (see e.g., [1]).

Table 3 summarizes the main properties of range-based localization techniques. The main disadvantage of the time-based methods (ToA, RTT and TDoA) are that an accurate time synchronization is required and such an error would have a large impact on the position determination. In addition, care must be taken not to measure the reflected signals, but rather the LoS signals, in order to obtain accurate results. The largest disadvantage of the RSSI-based lateration method is that the measured signal strengths can vary by a large extent. However, there is no need for a LoS between transmitter and receiver and that the multipath does not play a major role. Furthermore, no time synchronization between the two components is necessary. The levels of achievable positioning accuracies provided in the Table are also representative for the use of Wi-Fi in RSSI- and RTT-based lateration solutions.

2.5. Location Fingerprinting

Fingerprinting is a pattern recognition approach based on the RSSI measurement principle. The method consists of the training phase (or off-line phase) and the positioning phase (on-line phase). During the training phase the RSSIs of the surrounding transmitters are measured at several reference points in space and saved in a multi-dimensional database which can visualized in radio maps. The radio maps can be stacked into so-called datacubes as proposed by Retscher [1]. The radio map datacubes are the 3D arrays of the radio maps of the sensed APs at a certain location. The datacube has two spatial axes and a vertical AP axis. It is created by stacking radio maps vertically onto each other allowing the examination of the interrelations of the three quantities easily. For positioning in the on-line phase, the measured fingerprint at an unknown location is then compared with those in the empirically determined radio map datacubes. Finally, the position in the radio map that best matches the on-line measurement is returned. The radio map can also be created using simulated models taking into account the signal propagation in the area of interest; but this can be very complex.

In contrast to lateration, fingerprinting uses the signal attenuation and the multipath effect to determine the position. In addition, there is no need for a direct LoS between transmitter and receiver. One disadvantage of this method is the time required to set up and maintain the fingerprint database. In addition, training measurements must be carried out again when a new transmitter is installed or when structural changes are made [42]. Another challenge is the large variation in the observed RSSI values due to signal fluctuations. Despite these drawbacks, fingerprinting is now one of the most popular methods for an IPS. The specifics regarding the use of Wi-Fi fingerprinting are reviewed in the following section.

2.6. Specifics of Wi-Fi Positioning

Smartphone-based positioning using Wi-Fi plays a dominant role in the indoor positioning field and thus, it has become increasingly popular. This section presents briefly the Wi-Fi specifications and the properties of the two most commonly employed techniques in Wi-Fi positioning, i.e., the location fingerprinting and lateration-based approaches.

A Wi-Fi AP broadcasts small packets, i.e., the beacons, containing the Service Set Identifier (SSID) and the Media Access Control (MAC) address approximately every 100 ms. This ensures continuous data transmission. An AP can also transmit several signals simultaneously, with each signal belonging to its own Wi-Fi network. Furthermore, it is important that different channels are assigned to the APs, otherwise interference will occur [43]. The mobile device receives the signal and can identify the AP by the MAC address. The RSSI can additionally be sensed with various smartphone applications. Since the RSSI, the SSID and the corresponding MAC addresses can be accessed without any authenticated connection, this information is freely available. This allows wireless positioning to be performed autonomously, avoiding also privacy concerns that typically arise with other positioning technologies [31]. The size of the covered radio cell depends on the transmission power and the spatial conditions of the environment. Here, fluctuating influences such as the humidity in the air and in the building structure play a major role.

Wi-Fi is based on the IEEE 802.11 standard, which was developed by the Institute of Electrical and Electronics Engineers (IEEE). Since its introduction, several extensions have been developed, each with its own characteristics, such as the frequency band used or the range. Two frequency bands, i.e., the 2.4 and 5 GHz band, are available for Wi-Fi. The frequency range in the 2.4 GHz band (2400–2483.5 MHz) is divided into 14 channels, with only the first 13 channels used in Austria. Although the channel spacing is 5 MHz, a radio connection requires a bandwidth of 20 MHz (or at 802.11b 22 MHz). In order to avoid interference, therefore, in the case of spatially overlapping cells, overlapping-free frequency ranges with a distance of four channel numbers must be selected. The legally regulated maximum transmission power for the 2.4 GHz band in Austria is 100 mW. A total of 19 channels are freely available in the 5 GHz band. The frequency range 5150–5350 MHz may only be used with a maximum transmission power of 200 mW. The lower frequency range of 5150–5250 MHz may also be used with automatic power control, i.e., Transmit Power Control (TPC). TPC reduces the transmission power depending on the need. For example, if there is good connection between the devices, the transmission power is reduced. The 5470–5725 MHz frequency range may only be used outside buildings using TPC and Dynamic Frequency Selection (DFS) and with a maximum transmission power of 1000 mW. With the help of DFS, the AP automatically detects other radio systems and can switch to another frequency. This ensures that radar installations, satellite positioning services are not interfered [44]. The combination of TPC and DFS thus allows APs to determine the channels with the best availability and to use the lowest possible transmission power. The user therefore only receives the transmission power required for the current distance to the AP.

Each frequency band has its own advantages and disadvantages. In principle, the higher the frequency, the shorter the range (due to the higher signal attenuation). The 2.4 GHz band thus theoretically has a larger range, as it overcomes shielding materials with a lower loss. However, it has the disadvantage that the frequency band is compatible with other electronic devices or needs to be shared with other radio techniques, such as Bluetooth, microwave ovens, radio remote controls, etc., making it more prone to interference. The advantage of the 5 GHz band is the significantly higher data transmission rate, which does not matter with an IPS, since no data is transmitted as only the RSSI are measured. The big disadvantage of the 5 GHz band is that the signal is more shielded by walls. In the conducted tests it is seen that the 5 GHz Wi-Fi signals have a shorter range than the 2.4 GHz signals due to these properties (see Section 4.3). The Wi-Fi antennae in the APs bundle the electromagnetic waves and can thus influence the signal. Depending on the design of the antenna, the range and direction of the signals can be controlled and thus also the size of the radio cell. Commercially available APs have usually a range of 20 to 100 m within a building.

Due to signal damping and attenuation as well as signal fluctuations and noise, Wi-Fi positioning is normally not robust against dynamic changes in the environment. Thus, location fingerprinting is most commonly employed localization technology. For fingerprinting deterministic and probabilistic approaches can be employed. On the other hand, for lateration methods based on the measurement of the RSSI and RTT can be used. In the case of RSSI-based lateration, however, lower positioning accuracies are achievable in comparison to the RTT measurements with the new Wi-Fi IEEE 802.11mc standard. In Retscher [1] a comprehensive review of these techniques and the common mathematical models may be found. In this study, the available Wi-Fi AP hardware in the library of TU Wien supports only RSSI-based solutions. Thus, the chosen localization technique in this work is fingerprinting. A probabilistic fingerprinting approach is employed as they usually provide higher positioning accuracies than deterministic methods (see Section 6.1).

3. Test Site and Measurement Procedures

3.1. Test Site

The University library of TU Wien is a multi-storey building and is connected with another large office building referred to as 'campus Freihaus'. The chosen trajectory in the library with its waypoints on the ground and second floor is shown in Figure 1. It starts from outdoors in front of the main entrance (waypoints 1 to 5, not shown in the Figure) and has a length of around 379 m. Partly on the ground floor (points 10 to 19) and on the second floor (points 27 to 42) the trajectory runs along bookshelves. The reading room on the second floor is an open space of a size of approximately 830 m^2. The layout of the bookshelves is illustrated with grey lines in Figure 1. The trajectory waypoints, also referred to as checkpoints, were placed at important passages and at the bookshelves on every second row. The distances between the checkpoints are therefore approximately 3 to 8 m. The number of visible APs was quite low as on the ground floor only two APs and on the second floor only four APs of the University Wi-Fi network are located (see yellow stars in Figure 1 for their location). Throughout the whole library, only four APs which are almost at the same location in the reading rooms as on the second floor can be found on each of the five floors. That is why, the second floor was chosen as major study area as this floor is representative for the whole library building apart from the ground floor. The low number of APs results in challenging conditions for matching of the fingerprints in the on-line positioning phase due to the small number of AP in close proximity.

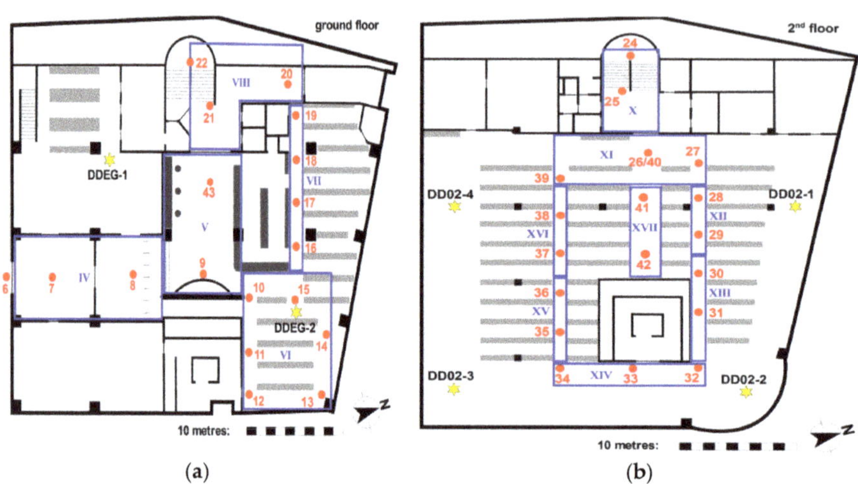

Figure 1. Ground floor (**a**) and reading room (**b**) of the library on the second floor showing the location of the APs (yellow stars), the checkpoints (red points) and the cells (blue Roman numbers). The bookshelves are represented as grey lines.

Measurements were carried out during normal opening hours of the library with many people around. The users walked along the trajectories with an average walking speed of $1~\text{ms}^{-1}$ in both ways back and forth taking around four minutes each for the whole trajectory. Apart from measurements in kinematic mode, also stop-and-go and static observations were carried out along the trajectory and on the checkpoints. For the analyses, also cells (denoted with blue Roman numbers in Figure 1) were defined consisting of different numbers of checkpoints in dependence of the local spatial conditions. If several checkpoints are part of these cells a higher localization performance can be achieved as demonstrated in Section 6.2.

3.2. Wi-Fi Signal Availabilities

Apart from the Wi-Fi University network in total six APs on the ground floor and 41 APs on the entire second floor providing signals could be sensed. Figure 2 shows the average number of visible Wi-Fi signals per scan on all checkpoints of the trajectory leading from outside through the ground floor to the reading room on the second floor. Note, that these numbers represent the different MAC addresses per scan and not the physical APs. In the Figure, the ratio of the University owned APs of the TUnet network (orange bars) and all visible signals (blue bars) is shown. On average, 19 stationary AP signals from the University network per scan could be measured in the library and outside. At checkpoints 1 to 6 the difference between the signals from the TUnet and other signals is the largest. These points are located outdoors and many other external signals are received. When considering the frequency distribution, checkpoints 5 and 6 stand out, where an above-average number of signals per scan is observed. These two checkpoints are located directly in front of the library entrance, which is why many signals of the TUnet from the adjacent office building can also be received here. The low number of signals at checkpoint 19 results from the fact that this point is located at a corner of the room on the ground floor. Checkpoints 20 to 25—with the exception of point 23—also show a lower number of signals per scan. These checkpoints are located in the staircase, where there are no APs. Especially at checkpoint 24, very few signals with an average of 4.7 signals per scan were received.

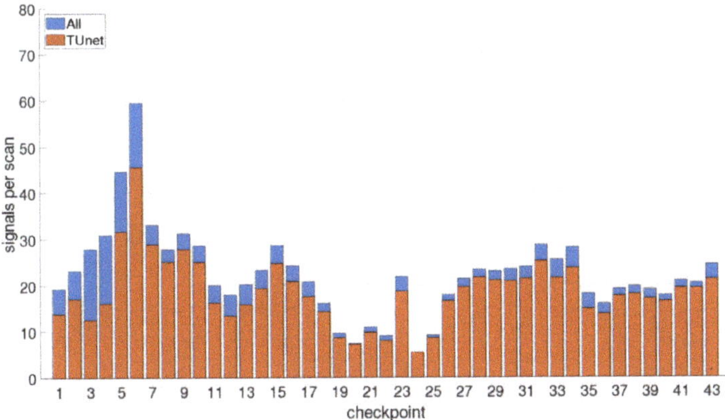

Figure 2. Average number of visible signals per scan on all checkpoints of the trajectory leading from outside through the ground floor to the reading room on the second floor.

3.3. Test Measurement Procedures

Test measurements were carried out in three different modes, i.e., static, stop-and-go and kinematic. In the case of static measurements, individual signal strengths measurements were carried out in several user orientations at the checkpoints. The necessary orientation measurements were performed in the possible movement directions. For example, only two orientations were measured in the corridors and four orientations at nodes (where two corridors intersect). At least 50 single Wi-Fi scans with several different smartphone models were measured at each checkpoint.

In the kinematic measurement mode, the Wi-Fi RSSIs were continuously recorded along the defined trajectory while the user walked along with a usual walking speed of $1\ \mathrm{ms}^{-1}$ back and forth. The obvious advantage of this mode is that the time required for the off-line training phase is much shorter than for the static measurements, in which a measurement cycle took approximately 40 min compared to 4 min only in kinematic mode. However, this measurement procedure does not exactly carry out a Wi-Fi scan on every checkpoint due to time taken for a single Wi-Fi scan. Thus, the result significantly depends on the scan duration [1,4]. For the creation of the radio maps (see Section 4.2), however, the signal strengths on the checkpoints must be known, which is why the RSSI values of a measurement run have to be interpolated in time. Therefore, a timestamp was set on each checkpoint while passing by. The linearly interpolated RSSI values on the checkpoints can then be saved in the fingerprint database. Figure 3 illustrates a kinematic measurement process and the linear interpolation of signal strengths. If the signal of an AP is not measured during the scan, then a RSSI value of -102 dBm was assigned for the missing value. This value was chosen as, since the lowest value measured in the test site was -101 dBm. As already mentioned, each smartphone takes a certain amount of time to perform a Wi-Fi scan (compare Table 4). In Figure 3, therefore, the two smartphones with the shortest and the longest scanning time are presented, whereby the two smartphones performed the measurement simultaneously. Although higher number of scans, i.e., 201 scans, can be performed during the same time interval with the OnePlus 5T smartphone due to the shorter scanning time than with the Sony Z3 (115 scans), there is a great similarity between the two signal series. The signal strengths are derived from the 5 GHz Wi-Fi signal of the AP DDEG-2 in the case shown.

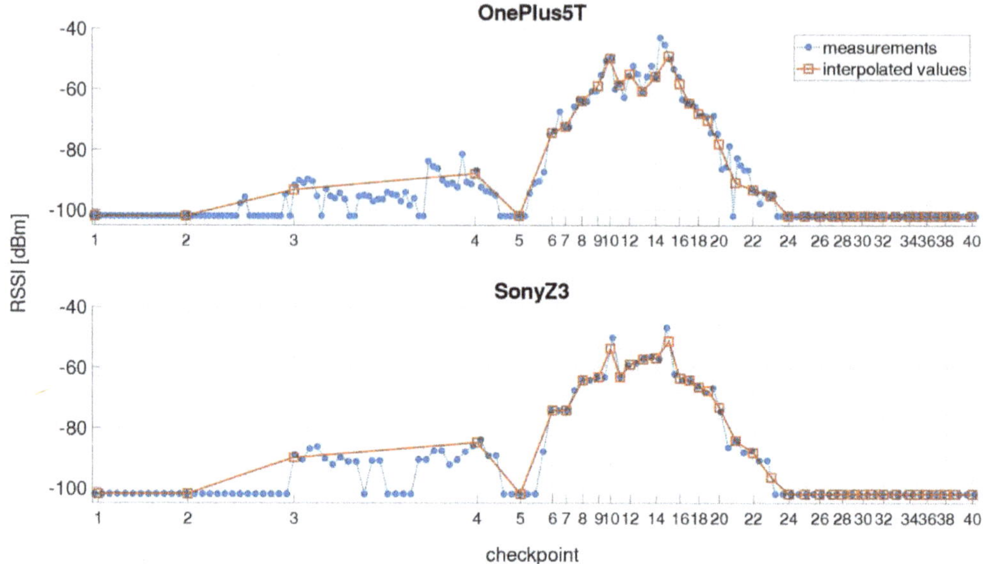

Figure 3. Linear interpolation of the kinematic measurements for the two smartphones (**top**) OnePlus 5T and (**bottom**) Sony Z3 which measured simultaneously.

Table 4. Average scan durations and sensed AP signals per scans for the six available smartphones.

	Scan Duration [s]	Average AP Signals per Scan
Nexus 5X	3.8	40.8
OnePlus 5T	2.4	42.6
Samsung S3A	3.5	33.7
Samsung S3B	3.5	27.5
Samsung S7	2.5	39.5
Sony Z3	4.1	39.3

In the stop-and-go mode, measurements were carried out at each checkpoint for a certain period of time of approximately 20 s so that at least five Wi-Fi scans are available. Thus, in contrast to the kinematic mode no interpolation must be performed. As an example, Figure 4 shows a measurement run for the same two smartphones and AP as in Figure 3. Again a great similarity between the signal sequences of the two smartphones can be seen. For a more detailed analysis, the scans on selected checkpoints are magnified in Figure 5. Checkpoints 3 and 24 are the ones where the signal could be at first or at last received along the trajectory. These two checkpoints are also those with the lowest RSSIs. However, only with the OnePlus 5T smartphone the signal could be measured with RSSI values −94.9 dBm or −92.9 dBm, respectively. At checkpoint 10 the highest RSSI values can be measured with average signal strength of −44.7 dBm with the OnePlus 5T and −43.2 dBm with the Sony Z3. At checkpoint 19 it is noticeable that the signal could not be measured during a scan, although it was measured shortly before and after with approximately −62 dBm. Furthermore, it can be observed at all checkpoints that the signal strengths are not always stable, but differ slightly from scan to scan. In order to investigate how the RSSI values behave within this short period of time, the standard deviations were calculated. The largest signal fluctuations occurred at checkpoints 4 with ±4.3 dBm and 6 with ±3.3 dBm. Viewed over all measurements and checkpoints, the average standard deviation of the signal strengths during the stopping phase is only ±1.5 dBm.

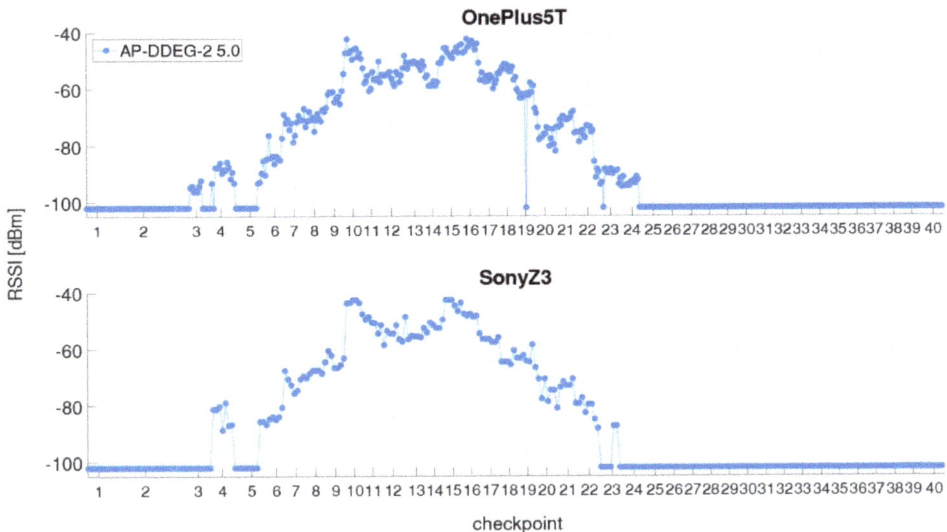

Figure 4. Examples of a stop-and-go measurement run for the two smartphones (**top**) OnePlus 5T and (**bottom**) Sony Z3 which measured simultaneously.

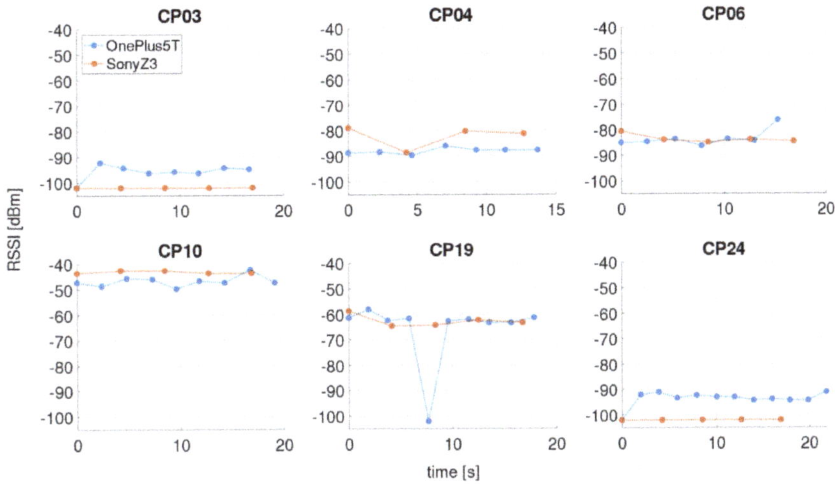

Figure 5. Zoomed views of the stopping phases for selected checkpoints for the two smartphones.

4. Analyses of the Off-Line System Training Phase

Training measurements were carried out in front of the main building of the University, in the library and the Freihaus building along predefined trajectories with reference waypoints at decision points, such as trajectory crossings, and at irregular intervals depending on the local conditions. During the kinematic measurements, a time stamp was set at the waypoints when the user passed, in order to be able to interpolate the RSSI values at these points.

4.1. Measurement Mode Comparison

The fingerprint databases were created from the RSSI measurements of all smartphones either separately or combined. However, not every measurement is used separately, but the average RSSI values are collected for each checkpoint in a vector. In order to obtain a combined device-independent fingerprint database, a calibration with a multivariate linear regression as in [1] was carried out in the form of:

$$y_{RSSI} = a_S \cdot x_S + b_S \tag{1}$$

where x_S is the measured RSSI value from the smartphone S which should be calibrated, y_{RSSI} the averaged reference vector calculated from all average RSSI values that are estimated with the linear regression model and a_S and b_S are the calibration coefficients. To obtain a gradient which is equal for each smartphone, $a_S =$ const. is assumed in the linear regression model. The adjusted RSSIs using these calibration coefficients can then be used for a combined fingerprint database. Applied to these datasets, the variation range could be reduced from 27 to 16 dBm using the calibration. Overall, the average standard deviation of all measurements could be reduced from 4.2 to 3.0 dBm (for further details see [1]). In order to compare these measurement modes and the databases, the differences between the mean RSSI values were calculated for each checkpoint and AP.

Figure 6 shows the mean signal strengths and their standard deviations of one AP where the largest differences between the database values were found in the library for the three databases from the static, stop-and-go and kinematic off-line training measurements. Thereby the 2.4 GHz signal from the AP DD02-2 showed the largest difference at checkpoint 34. The average difference for this AP resulted in 0.8 dBm between the databases derived from static and kinematic training measurements, 1.0 dBm between the static and stop-and-go database and 1.5 dBm between the kinematic and stop-and-go database. The largest difference between the stop-and-go and the kinematic database reached 11.3 dBm. For further comparison, the correlation coefficients between pairs of the same APs were calculated. For this purpose, on the one hand, the database from the averaged RSSI was used, and on the other hand, their variances. The results for the mean correlation coefficients and differences between the databases are presented in Table 5. With regard to the RSSI values, the databases show nearly no differences and are highly correlated. The average difference between pairs of the same AP is also very low. In the case of the variances, the correlation with the kinematic measurements is somewhat weaker. This is probably due to the lower number of off-line measurements, i.e., 60 scans per checkpoint, in this measurement mode. All in all, however, the databases are very similar, which is why the databases are now combined for the subsequent creation of the radio maps and the analyses of the positioning results.

Table 5. Mean correlation coefficients and variances for the comparison of the different fingerprint databases.

	RSSI		Variances	
	$\bar{r}$	$\bar{d}$ [dBm]	$\bar{r}$	$\bar{d}$ [dBm]
static—kinematic	0.96	0.3	0.93	3.9
static—stop-and-go	0.99	0.3	0.97	2.8
kinematic—stop-and-go	0.96	0.4	0.94	3.4

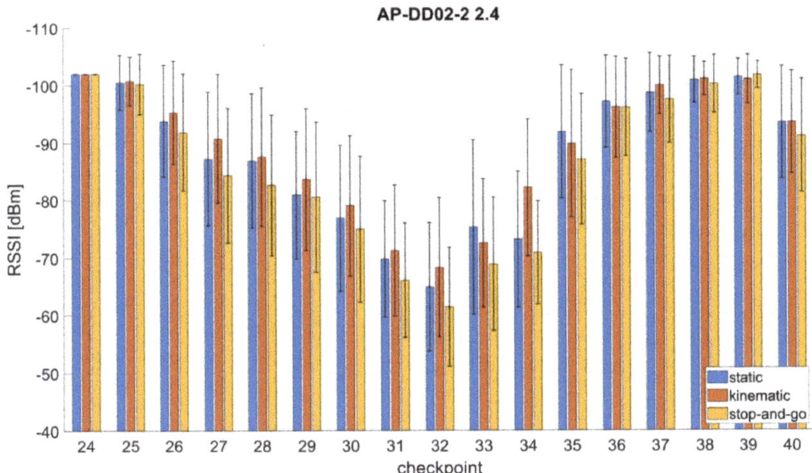

Figure 6. Mean RSSI and their standard deviations for comparison of the databases of the three measurement modes static, kinematic and stop-and-go.

4.2. Radio Map Generation

In order to know the RSSIs and variances of the APs not only at the waypoints, but also in the whole test site, an area-wide interpolation is carried out for each AP for both the RSSI values and the variances. Different interpolation methods can be used for this purpose (see e.g., [45]). An interpolation by natural neighbors, also referred to as Voronoï interpolation, is used in this work [46–48]. The grid width of the interpolated radio maps is set to 1 m, which results in that positioning can be carried out within meter accuracy. In a multi-storey building, when creating the radio maps, it must be kept in mind that a separate radio map for each AP is created for each floor, always using only the RSSI measurements on those checkpoints that are located on the respective floor. The different radio maps of a floor can be combined into a three-dimensional array in the form of datacube (see Section 2.5), with the first two dimensions resulting from the extent of the floor and the third dimension from the number of APs.

The creation of an empirically determined radio map starts with the classification of reference points on the basis of a building map. Care should be taken to ensure that the reference points are well distributed throughout the building. In the off-line training phase, the signal strengths—derived from different APs—are then measured at each reference point. A fingerprint respective scan $s_{RP_i,t}$, which was carried out at the reference point RP_i at time t, is thus composed of the measured RSSI values $RSSI_{AP_j}$ of the N APs [1]:

$$s_{RP_i,t} = \begin{bmatrix} RSSI_{AP_1} \\ RSSI_{AP_2} \\ \vdots \\ RSSI_{AP_N} \end{bmatrix} \tag{2}$$

The measured signal strengths are then assigned to the corresponding APs in the fingerprint database. To do this, however, it first has to be determined which APs are to be used for localization. If several scans are performed at a reference point, then the database consists of all the scans at each reference point. It can happen that the number of signals strengths received for each scan is different, because for example an AP temporarily does not broadcast a signal, or the signal is too weak to be sensed. This leads to problems in determining the position when RSSI values of different APs occur in the observed fingerprint and in the fingerprint in the radio map. Therefore, a constant RSSI value is used for the missing fingerprint in this work, which means that the signal of the AP was not measurable. As aforementioned here a constant minimum value of -102 dBm was used, since the lowest value ever sensed in the area was -101 dBm.

Not every scan is used separately for localization, but the RSSI averages of the measurements are collected in the vector given in Equation (2). A suitable reference value for the sensed RSSI values must be found for this purpose. If the measured values are assumed to be normally distributed and contain only random errors, then the mean value is an optimal reference value. The database's fingerprint f_{RP_i} consists of all the mean RSSI values sensed at that reference point in the form:

$$f_{RP_i} = \left(f_{AP_j} \right) = \frac{1}{N} \sum_t^N RSSI_{t,AP_j} = \begin{bmatrix} \overline{RSSI}_{AP_1} \\ \overline{RSSI}_{AP_2} \\ \vdots \\ \overline{RSSI}_{AP_N} \end{bmatrix} \tag{3}$$

Each reference point generally has a unique characteristic and therefore acts like a RSSI fingerprint (therefore also the name fingerprinting for the localization approach). Thereby each RSSI is measured with a certain precision at each reference point. The value for the precision of a measurement series is the variance or standard deviation. This information can also be used for fingerprinting by providing each fingerprint with its covariance matrix. The measurements are assumed to be uncorrelated, which means that the empirical covariance matrix C_{ff} contains only the variances $s^2_{RSSI_{AP_j}}$ of the APs AP_j in the diagonals as given in:

$$C_{ff_{RP_i}} = \begin{bmatrix} s^2_{RSSI_{AP_1}} & 0 & \cdots & 0 \\ 0 & s^2_{RSSI_{AP_2}} & \cdots & 0 \\ \vdots & \vdots & \ddots & \vdots \\ 0 & 0 & \cdots & s^2_{RSSI_{AP_N}} \end{bmatrix} \tag{4}$$

The fingerprinting database thus contains a fingerprint f_{RP_i} for each reference point and is thus a two-dimensional array with the reference points as columns and the APs as rows. To determine the position, either the database can be used directly, or a radio map for each AP can be created by means of surface interpolation of the RSSI values. For that purpose, however, the coordinates of the reference points must be known. The interpolated radio map of an AP then contains the mean RSSI values on the reference points as well as the interpolated RSSI values between them. Since an individual radio map is created for each AP, the radio maps can be combined into a three-dimensional array, with the first two dimensions resulting from the length and width of the building and the third dimension from the number of APs. For complex buildings, however, it is not necessary to create a single radio map array for the entire building, but it is also possible to create single arrays for certain areas (e.g., floors). The size of the radio maps depends on the grid size as well as the spatial conditions and influences the quality and duration of localization. The larger a radio map is, the longer it takes to determine the position of the user. The accuracy of the position determination depends mainly on the density of the APs and the quality of the radio maps. The quality of the radio maps deteriorates over time due to the fluctuations in the AP signals and changes in the environment, such as changes in the position of furniture or other objects. Therefore, it is important that new fingerprints are collected regularly in order to keep the radio maps up-to-date. In a previous study, the authors [45] have developed a continuous kinematic training of the fingerprint database where measurements along walked trajectories are used to update the database. An important finding in the investigation of the radio maps is that the database created either from static, stop-and-go and kinematic measurement modes show a great similarity in both RSSIs and variances. For future work, this also means that continuous kinematic system training can be carried out, which means that the training phase is much shorter.

As examples, Figures 7 and 8 show the radio maps of the two APs on the ground floor and the four APs on the second floor, respectively. Thereby only the 5 GHz frequency band of the Wi-Fi signals is presented. In addition, own radio maps were created for the outdoor area. Since there is only one checkpoint on the first floor, no surface interpolation can be performed, which is why the radio map on this floor consist only of the fingerprint of checkpoint 23. The signals of 77 APs were used in total, which is why the radio map datacubes have the sizes presented in Table 6. From Figure 7 can be seen that the Wi-Fi signals can be received at all checkpoints on the ground floor. AP DDEG-1 can be sensed at checkpoints 7, 8 and 43 with the highest signal strength of approximately −60 dBm. Surprisingly, however, the RSSI values are quite low under this AP. This is most likely not the case, but is due to the fact that no measurements were carried out in this area around the AP. The signals of the AP DDEG-2 can be well received in the entire area. Since checkpoint 15 lies directly under this AP, the signal is received with a high average value of −48.6 dBm. The radio maps for the 5 GHz Wi-Fi signals of all four APs on the second floor are shown in Figure 8. All of them can be received at all checkpoints. However, the signals in the diagonally opposite corners are very weak, i.e., only −91 to −98 dBm. While cross-comparing Figures 7 and 8 it is seen that the RSSI values are much lower on the second than on the ground floor. From Figure 8 can also be seen that the signal strengths were only interpolated between the checkpoints, which is why localization can only be performed in the inner area of the reading room between these bookshelves. If one would extrapolate the RSSI values outside this area, this would result in none realistic RSSI values as the signals could be determined either very strong or weak with high or low RSSI values, respectively. In order to have a radio map for the whole reading room, measurements not only on the selected checkpoints along the trajectory but also at additional points at room boundaries need to be carried out. The investigation in this paper, however, dealt with measurements in kinematic mode along the predefined trajectories to minimize the workload for system training. From the results presented in Section 4.3 one can see that the similar low signal strength values from all four APs led to lower positioning accuracies.

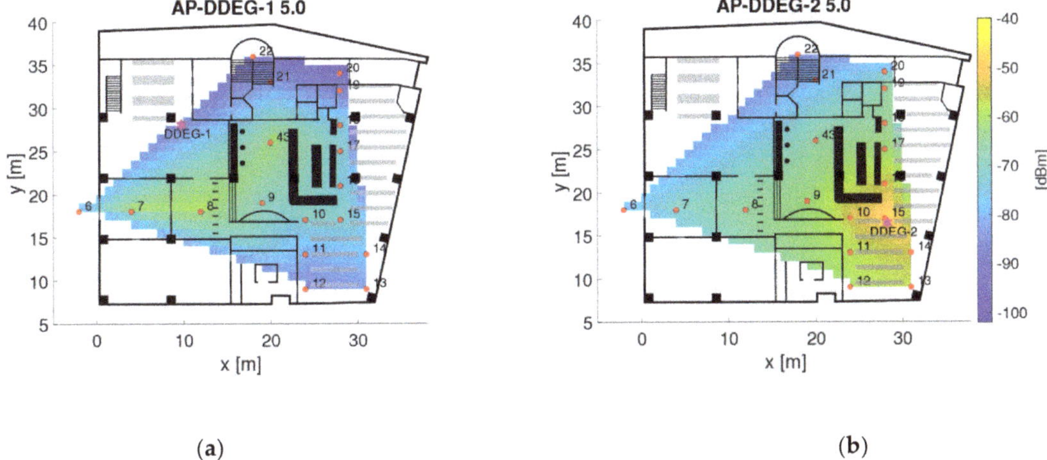

Figure 7. Radio maps for the two APs (**a**) AP-DDEG-1 and (**b**) AP-DDEG-2 on the ground floor of the library.

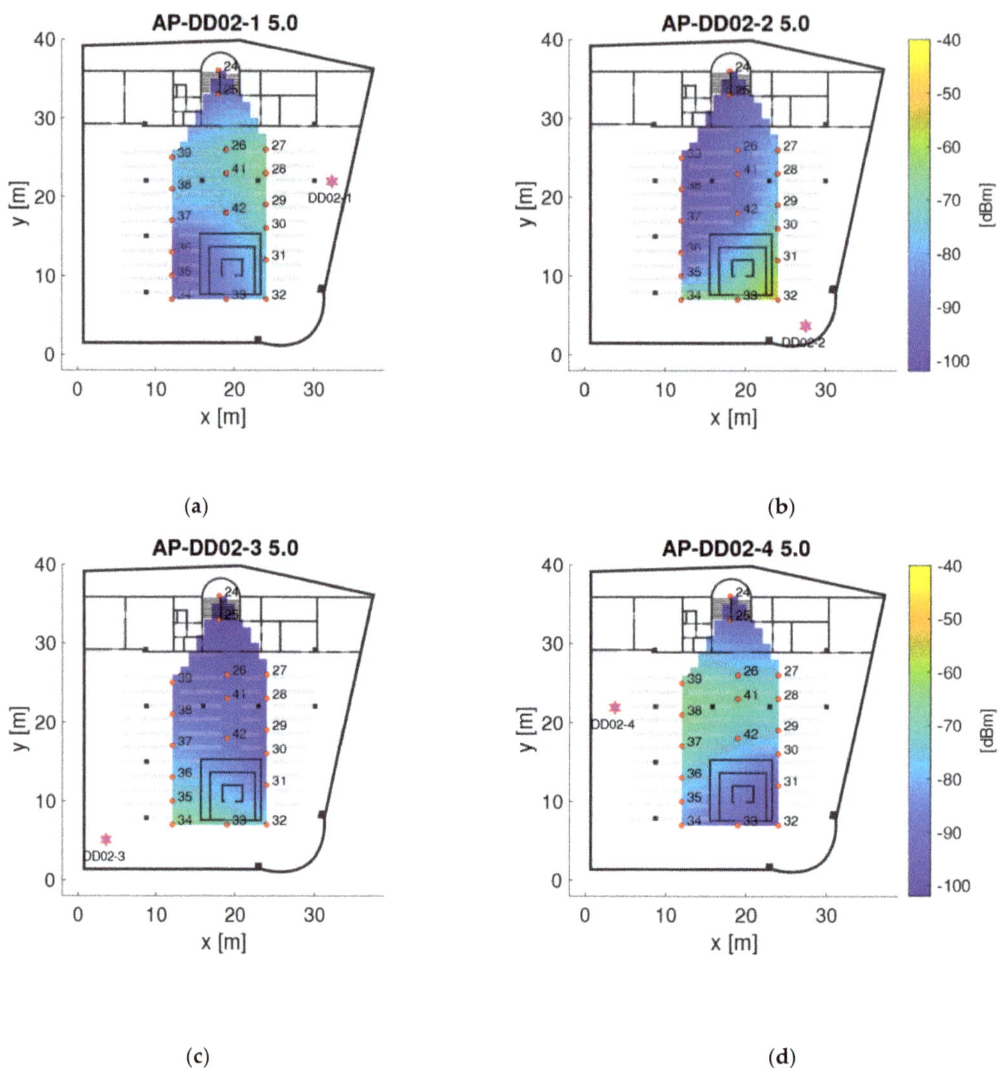

Figure 8. Radio maps for the four APs (**a**) AP_DD02-1, (**b**) AP-DD02-2, (**c**) AP-DD02-3 and (**d**) AP-DD02-4 on the second floor of the library.

Table 6. Sizes of the radio map datacube arrays.

Outdoor	$32 \times 172 \times 77$
Ground floor	$34 \times 28 \times 77$
1st floor	$1 \times 1 \times 77$
2nd floor	$13 \times 30 \times 77$

4.3. Visibility and Range of the Wi-Fi Signals

As aforementioned, not all APs are detected at every scan. Figure 9a,b show, for example, the signal strengths and visibility of the APs at those checkpoints where to lowest and highest number of APs could be sensed. At checkpoint CP20 (Figure 9a), the signals of nine different APs were measured, of which only two were visible more than in 75% of the observations. CP20 is located quite remotely in a room corner on the ground floor of the staircase (compare Figure 1a). The two frequency bands of the AP DDEG-2 were most visible and were also received with higher signal strength. The 5 GHz signal was visible in 96% of the scans and had and average RSSI value of −77.4 dBm. Although the 2.4 GHz signal has a slightly lower visibility, i.e., 94%, it could be received with average RSSI values of −75.2 dBm. At checkpoint CP06 (Figure 9b), 47 different APs could be sensed at least once, of which many APs were visible in more than 75% of cases. CP06 is directly located at the library entrance. Here, the 5 GHz signals of the APs DCEG-3 and DC01-3 were the most visible with 99%. The 2.4 GHz signals of DCEG-3 have a visibility of 98% with the highest signal strength on average of −65.0 dBm. The 5 GHz signal, on the other hand, showed a −5.2 dBm lower mean signal strength of −70.2 dBm. Thus, a correlation between the visibility and the RSSI values is obvious. Therefore, the correlation coefficient of these two measured values was determined for each checkpoint; it resulted in 0.96. This means that the higher the signal strength of an AP, the more often this AP is also visible.

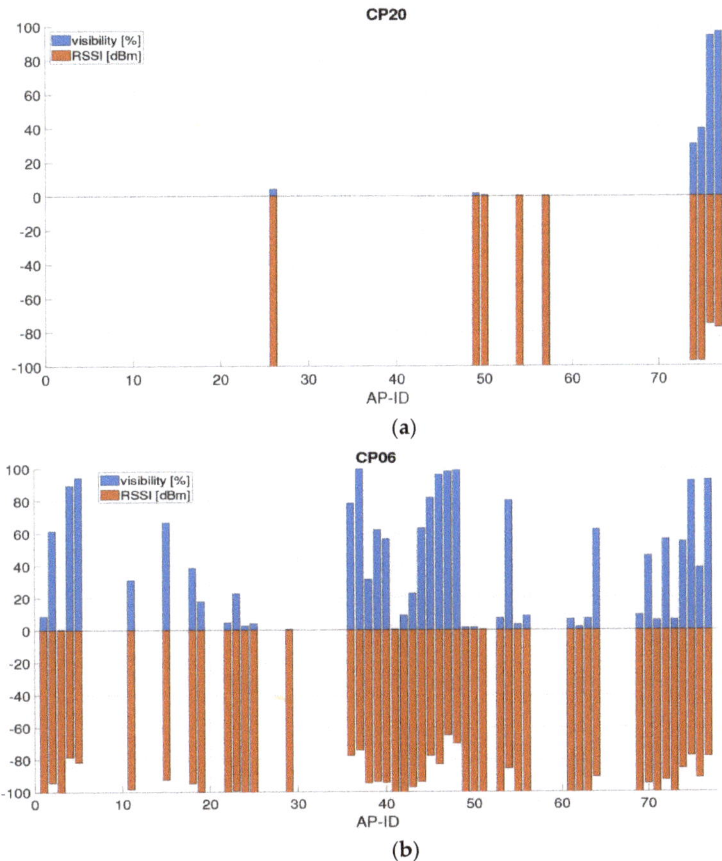

Figure 9. Checkpoints with (**a**) the lowest (CP20) and (**b**) highest (CP06) visibility.

A further correlation exists between the two frequency bands 2.4 and 5 GHz that all APs provided. Across all checkpoints, the 2.4 GHz signals were on average 3.6 dBm higher than the 5 GHz signals. The average standard deviations resulted in ±4.5 dBm for the 2.4 GHz and ±3.5 dBm for the 5 GHz frequency band. In terms of signal range, the 2.4 GHz has a longer range from an AP compared to the 5 GHz band as it penetrates shielding materials with less loss and also has less free space path loss (FSPL), although the 5 GHz band has a

3.0 dBm higher transmitting power. The FSPL describes the reduction of the power density of an electromagnetic wave in free space, i.e., without interference from damping media, such as air or interference caused by reflections. As shown in [1], the attenuation thereby depends on the signal frequency and the signal weakens with increasing distance from the transmitter, also in terms of the signal-to-noise ratio. The FSPL in the unit dB is usually described on a logarithmic scale by means of the Friis transference equation [49]:

$$FSPL\ [dB] = 10 \cdot \log_{10}\left(\frac{4 \cdot \pi \cdot d \cdot f}{c}\right)^2 \tag{5}$$

where d is the distance between the transmitter and receiver in [m], f is the frequency of the signals in [Hz], and c is the propagation speed in [ms^{-1}]. Thus, the power decreases with the square of the distance to the transmitter. This applies for direct LoS signals. In practice, an empiric logarithmic distance model can be derived from Equation (5), because also with LoS signals, reflections and damping due to physical objects occur. Thereby, a Wi-Fi signal is already considerably attenuated within a few meters from an AP and the attenuation increases with the increasing frequency [1]. This mathematical relationship proves that the 5 GHz Wi-Fi signals have a shorter range that the 2.4 GHz signals.

Furthermore, the use of an AP from a different floor in the building was analyzed. For this purpose, one AP from the first floor of the library (between the two test areas) visible on most checkpoints was selected. The RSSI values of the AP DD01-2 for both the 2.4 and 5 GHz frequency bands are presented in Figure 10. They were received on 36 checkpoints with varying RSSI values. With an average of 79.7 dBm for the 2.4 GHz band and 88.0 dBm for the 5 GHz band the highest Wi-Fi signals were measured at checkpoint CP32 which is located directly above the AP on the second floor. This proves again that that there is a significant difference between the two frequency bands in terms of signal strength and range.

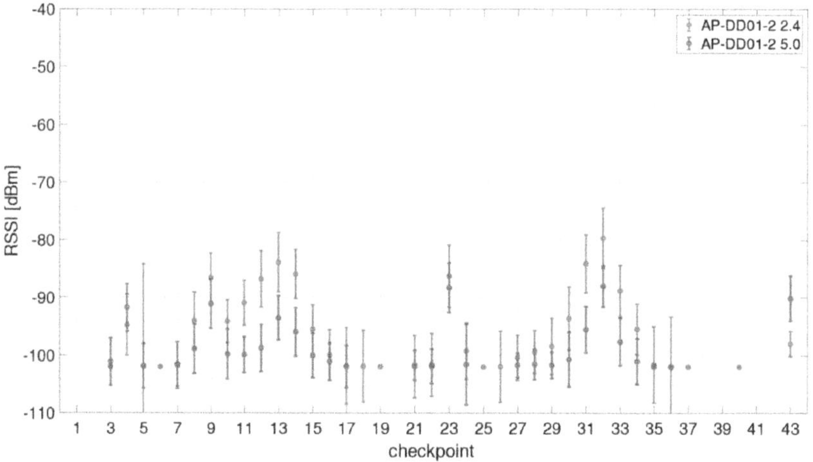

Figure 10. Sensed RSSIs of the AP DD01-2 from the first floor on all checkpoints for both the 2.4 and 5 GHz frequency bands.

4.4. Kinematic System Training

Retscher and Hofer [50] introduced the checkpoint concept for Wi-Fi positioning. System training for fingerprinting is usually carried out in static mode on reference points distributed in a regular grid in the area of interest. The main disadvantage of this training procedure is therefore the required high workload. With trajectory checkpoints, the time needed for system training can be reduced by three quarters as shown in [50]. In the following steps of development, static training was replaced by kinematic measurements while walking along the trajectories [45]. Without stopping at reference points the user walks along predefined trajectories throughout the building. These kinematic measurements, however, pose much greater challenges than the usual static training measurements. As discussed in Section 3.3, the RSSI values on the checkpoints need to be interpolated from the whole RSSI time series. The reason for the required interpolation is that the RSSIs are continuously recorded and a single Wi-Fi scan takes some time. The scan duration depends heavily on the number of sensed APs and mostly on the hardware of the smartphone. In the following Section 5 this impact is discussed in further detail. Table 4 shows average scan durations for the employed smartphones in the test. One might think that the long times occurred only because of the fact that the smartphones used in the tests are quite old. The obtained different range of scan durations, however, is representative for a great variety of smartphones which are available on the market. They also cover a wide range of different hardware. As can be seen from Table 4, the scan durations varied between around 2.5 to over 4 s. The average number of sensed APs per scan was in the range of around 30 to 40 AP signals. The analysis of the system training measurements showed that there are sufficiently stable signals available everywhere on the campus to carry out a position determination using Wi-Fi fingerprinting. Retscher and Leb [45] could demonstrate that the achieved positioning accuracies for the kinematic system training are not much worse than with static measurements. The big advantage, however, is that the training phase is much shorter and continuous system training can also be carried out if needed.

5. Impact of Different Scan Durations on the Positioning Results

Every smartphone needs a certain amount of time to perform a single Wi-Fi scan. These can be very different in length, as has been the case with the six different smartphones used (see Table 4). In Figure 3 the series of the two smartphones with the shortest and longest scan duration were presented in Section 3.3. A great similarity between the two time series can be observed although more scans along the trajectory can be performed with the OnePlus 5T smartphone than with the Sony Z3. As shown in this section, however, the scan duration has a significant influence for kinematic positioning in the on-line phase. If one looks at the whole collected dataset irregular scan durations were found for individual smartphones. They can deviate quite significantly from the mean scan durations presented in Table 4. Figure 11 shows such a case where two smartphones, i.e., the Nexus 5X and the Sony Z3, are compared. It can be seen that the Sony Z3 can have very long scanning times of even up to 15 s. The Nexus 5X, on the other hand, performs many scans with a measuring time of only a few milliseconds. These irregular scanning times are examined in more detail in the following.

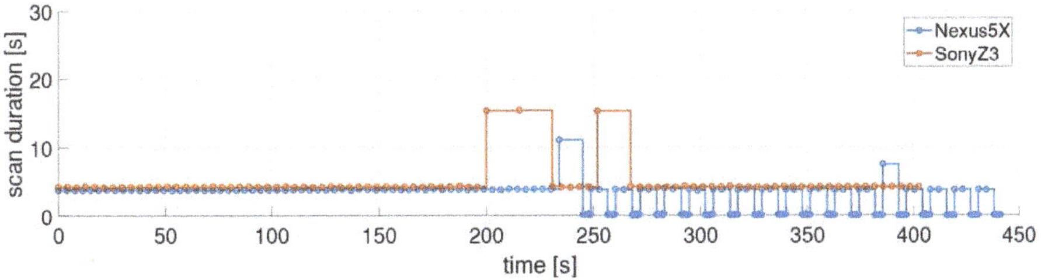

Figure 11. Irregular scan durations of the two smartphones Nexus 5X and Sony Z3.

The short scan durations of the Nexus 5X are shown in Figure 12 together with the measured signal strengths. Again the results of the 5 GHz signal of the same AP as in Figure 3 are presented. As shown in Figure 12a, the irregular scanning periods start between the checkpoints 6 and 7. The pattern is always similar as first slightly longer scan duration occurs followed by a series of scans with a short scan duration, whereby the total duration of these scans corresponds to the average scan duration. After that, two scans occur with average scan duration and then a series of short scans starts again. A closer look indicated that during these short scan durations the RSSI values do not change which causes then problems in localization of the user. To reduce their effect, these scans were eliminated from the dataset. However, as a result a gap of one scan is present in the dataset as indicated in the Figure 12b. The reason for this effect, however, could not be clarified. It was only found with the Nexus 5X.

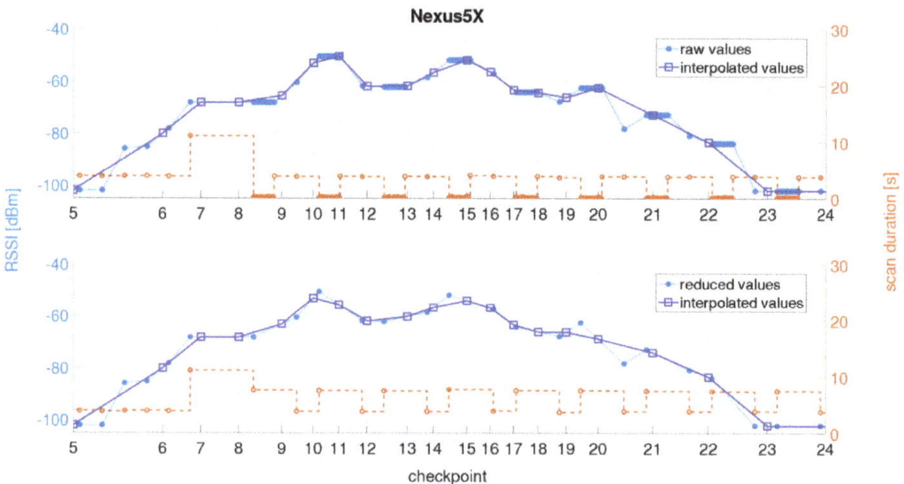

Figure 12. Kinematic measurements with the Nexus 5X smartphone with (**top**) raw and (**bottom**) interpolated RSSI values in dependence on the scan duration.

Figure 13a shows the long scan durations of the Sony Z3 together with one of the Samsung S3, i.e., the S3A (Figure 13b). Both smartphones carried out the measurements at the same time. The Sony Z3 showed the longest scan durations near checkpoints 7, 9 and 16, which results that no Wi-Fi scan was performed along a distance 15 m while walking with an average speed was 1 ms^{-1}. This leads to the fact that no scans are performed near checkpoints 8, 10, 11, 17 and 18. The interpolation can still provide similar values for the kinematic measurements as with the Samsung S3A phone. However, this does not apply in general. If, for example, no Wi-Fi scans were carried out between checkpoint 11 and 14, the interpolation would estimate too high RSSI values for the checkpoints in between. It was found that the smartphone tried to connect automatically to known Wi-Fi networks although it was first disconnected from the network. The connection function was disabled in order to have no influence from the signal strength changes while trying to connect on the positioning result. In the following, the maximum allowable scanning time for a meaningful interpolation was investigated. If there is long scan duration between two checkpoints then it has no influence on the interpolation. The maximum allowable scanning duration therefore depends on the spatial conditions, i.e., essentially on the distance between the checkpoints. If two checkpoints are located close to each other, it can be assumed that the signals show similar high values. If they are several meters apart, the RSSI values can vary significantly depending on the environment and the interpolation may no longer provide meaningful values. Since the fingerprint database in this work consists of many scans and these irregular scan durations only occurred in a few measurement runs, these scanning delays have no significant effect on the presented positioning results. If a long scan occurs in the on-line positioning phase, it is clear that no positioning can be carried out during this time, as no Wi-Fi RSSI values are available.

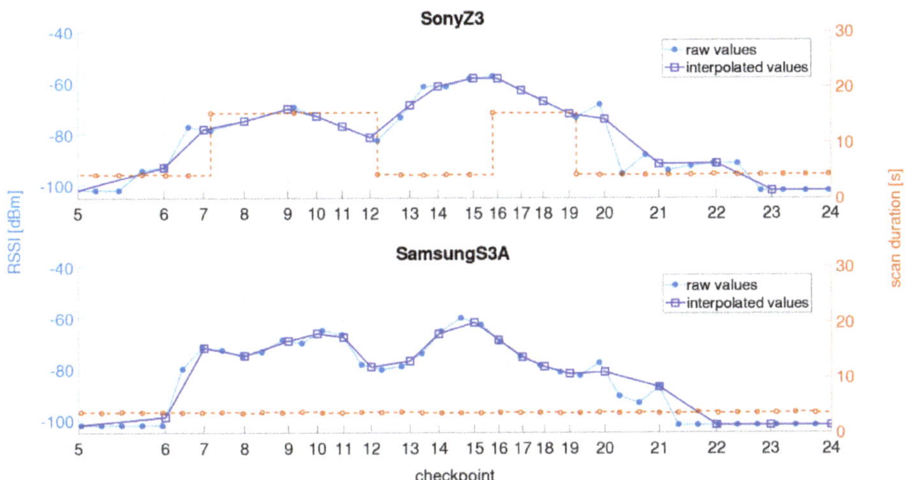

Figure 13. Raw and interpolated RSSI values of the two smartphones (**top**) Sony Z3 and (**bottom**) Samsung S3A in dependence on the scan duration.

6. Localization in the On-Line Positioning Phase

For localization in the on-line positioning phase, RSSI measurements are carried out and matched to the fingerprint database. Most commonly either deterministic or probabilistic fingerprinting techniques based on pattern recognition are employed [2]. In this study, a probabilistic approach is applied as it provides, in general, higher positioning accuracies than deterministic methods in indoor positioning [51–54]. The main reason for this is that probabilistic fingerprinting accounts better for signal fluctuations. For the analyses of the achievable positioning accuracies, on-line measurements were carried out with all three measuring modes, i.e., in static, stop-and-go and kinematic mode. In the following, the operational principle of a simple and straightforward probabilistic fingerprinting approach is briefly reviewed and then the results for static and kinematic positioning modes are presented.

6.1. Probabilistic Fingerprinting Approach

A probabilistic fingerprinting approach was selected where the basic idea is to compute a conditional probability density function (PDF) of the unknown position (see e.g., [1,55]). Starting from Bayesian filtering, a dynamic system with measurement noise can be dealt with. The posterior PDF of the unknown positions can be derived using Bayes' theorem (see e.g., [56,57]) and the measurements because of the fact that the fingerprints contain information about the signal characteristics. In this work, a probabilistic approach based on the derivation of the Mahalanobis distance is applied [58]. The Mahalanobis distance d^M has the form [1]:

$$d^M\left(f_{map}^i,\ f_{obs}\right) = \left(f_{obs} - f_{map}^i\right)^T C_{ff_{map,i}}^{-1} \left(f_{obs} - f_{map}^i\right) \tag{6}$$

where f_{obs} is the current on-line RSSI measurement at the position f_{map}^i in the fingerprint database (or radio map) and $C_{ff_{map,i}}$ its empirical covariance matrix.

Equation (6) means that the estimated reference point with the highest probability density is the point at which the Mahalanobis distance d^M between the observed fingerprint and the fingerprint of the corresponding point in the fingerprinting database is the smallest. The advantage of using the Mahalanobis distance is the additional use of the covariance matrix $C_{ff_{map,i}}$, since the distance metric is adjusted using the covariance matrix. This is also a distance criterion for the fingerprint matching. As the inverse of the covariance matrix is the weight matrix, the weighted square sum of the RSSI differences (between off-line training and on-line positioning phase) is calculated for the Mahalanobis distance. Then the weights are inversely proportional to the variances of the corresponding fingerprints. In fact, the Euclidean vector distance most commonly used in the deterministic fingerprinting approach (see e.g., [58,59]) is a special case of the Mahalanobis distance, which occurs when the covariance matrix becomes the unit matrix. In a previous study of the authors of this contribution it was seen that the simple and straightforward calculation of the Mahalanobis distance while using kinematic system training yielded to comparable results as algorithms where a static or stop-and-go mode for localization is applied.

Its principle of operation is reviewed by giving a simple example. Figure 14 illustrates the position estimation for five on-line measurements. For each measurement at a checkpoint (CP), the Mahalanobis distance is estimated for each individual CP in the fingerprint database. The CP with the shortest distance is then the desired position. As shown in the Figure, the positions at CP01, CP02, CP04 and CP05 have been correctly determined. The on-line measurement at CP03, however, has its minimum at CP05, which means that the position in this on-line measurement has been indirectly determined. If one defines a so-called matching success rate (MSR) it would be zero for this checkpoint. This example shows the advantages of using the Mahalanobis distance for probabilistic fingerprinting. It is based on the knowledge of its covariance matrix. Standard deviations of each fingerprint must be known. In general, however, not all APs can be received anywhere in the measuring area, which can lead to problems with distance calculation if RSSI values are of different APs in the on-line and off-line fingerprint. As already mentioned, in the case a value of -102 dBm is used for the non-receivable AP. In the event that the signals of an AP could not be received at a single off-line measurement at a certain checkpoint—i.e., only values of -102 dBm are set at the corresponding location in the database—the variance is zero. However, the variance must not be zero, otherwise the determinant of the covariance matrix is also zero and thus the covariance matrix is singular and not invertible. However, the matrix must be inverted when calculating the Mahalanobis distance, see Equation (6). To avoid this problem, a variance of 0.0001 dBM is used in this case. If a signal from an AP can now be received in the on-line measurement at a point that could not be measured in the off-line phase, then the weighting becomes very large as the weighting is inversely proportional to the variance, which also increases the distance between the two fingerprints. As a result, then the likelihood that this point is the location one is looking for decreases. If the position is determined using interpolated radio maps, then the deviations of the calculated positions from the true position can be specified. The Mahalanobis distance between each point in the radio map and the on-line fingerprint is calculated first. Then, from the position in the radio map where the shortest distance was calculated (the nearest neighbor), the deviation from the true position is calculated using the Euclidean distance. Ideally, the Mahalanobis distances near the respective checkpoint are very short and grow with increasing distance. However, a set of K-smallest distances can also be selected to determine the position. This is referred to as K-nearest neighbor (KNN) approach. The searched position is then derived from the center of gravity of the K-nearest neighbors. Therefore, the static measurements were used to determine at which K the smallest deviations from the true position occur. In the library, the arithmetic mean of all deviations for K = 1 is 2.9 m and the median 2.0 m. As shown in Figure 15, the arithmetic mean and the median of all deviations in the measuring area increases the more neighbors for position determination are included. Therefore, only the nearest neighbor approach with K = 1 was applied in the further evaluation.

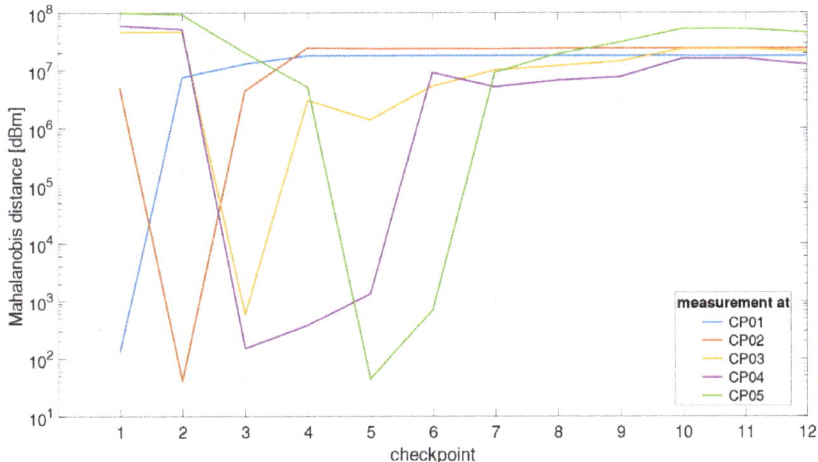

Figure 14. Positioning using the Mahalanobis distances at five different checkpoints (CP01 to CP05).

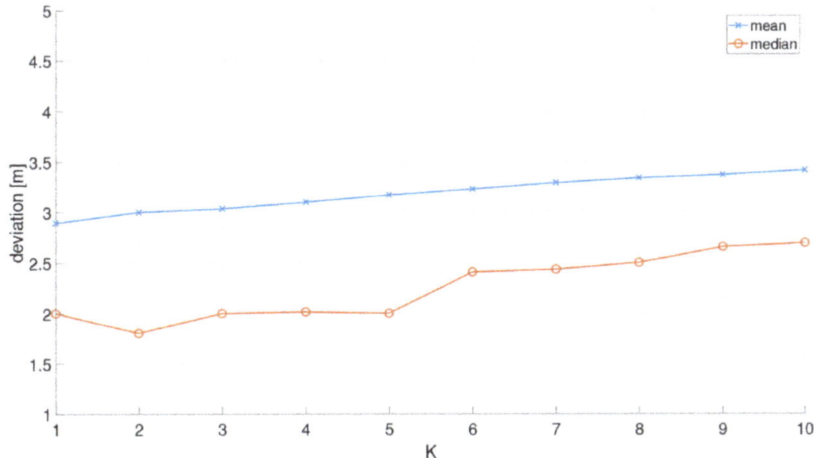

Figure 15. Mean and median deviations for static positioning in dependence of the K value for the nearest neighbor approach.

6.2. Static Positioning

For static localization, five RSSI scans were carried out in two different orientations—usually in the possible direction of movement along the trajectory—at each checkpoint. For the analysis, a matching success rate (MSR) was defined, i.e., how often the correct checkpoint in the on-line positioning phase was assigned. The achieved MSR was quite low of only 61% on average for all 43 checkpoints. Especially checkpoints at room borders and edges were determined with low MSRs. In addition, most of the incorrectly matched points were assigned to neighboring checkpoints. The test site was then divided into cells for cell-based localization (Figure 1 shows the cells with blue Roman numbers). If cell-based positioning is carried out then the MSRs can be significantly increased as indicated in Table 7. The two worst results were achieved in cells X and XI, which are either in the staircase or at the entrance to the second floor near the staircase. Here the adjacent cells are determined frequently. The two cells VI and VII located on the ground floor also showed low MSRs. In fact, this is caused by the usage of only two APs from the network of the University.

Table 7. Matching success rates (MSR) in the cells in the library (for the location of the cells see Figure 1).

Cell	Checkpoints	Location	MSR
I	1, 2	outdoor 1	100.0%
II	3, 4	outdoor 2	100.0%
III	5, 6	entrance area (outdoor)	100.0%
IV	7, 8	entrance area (indoor)	66.7%
V	9, 43	ground floor lobby	87.5%
VI	10–15	ground floor area 1	56.9%
VII	16–19	ground floor area 2	54.2%
VIII	20–22	ground floor staircase	77.8%
IX	23	first floor staircase	100.0%
X	24, 25	second floor staircase	50.0%
XI	26, 27, 39, 40	second floor area 1	39.6%
XII	28, 29	second floor area 2	91.7%
XIII	30, 31	second floor area 3	66.7%
XIV	32–34	second floor area 4	100.0%
XV	35, 36	second floor area 5	70.8%
XVI	37, 38	second floor area 6	87.5%
XVII	41, 42	second floor area 7	66.7%

For the following analyses, the positions were estimated on the basis of the interpolated radio maps, allowing the deviations of the calculated location to the ground truth to be determined. On the ground floor of the library, the average deviations from the ground truth resulted in 3.4 m and the median in 2.2 m. In particular, checkpoints CP11 to CP19 show above-average deviations of up to 5.0 m, as was also seen when looking at the MSR. This is caused again by the fact that there are only two APs on the ground floor and that there are no building structures that influence the Wi-Fi signals in such a way that the RSSI varies significantly on each checkpoint. On the second floor of the library, the positioning accuracies are better with mean deviations of 2.2 m and a median of 2.0 m. The largest deviations of 8.3 m resulted on CP42 in one measurement run with the Samsung S3B. Figure 16a shows the worst and Figure 16b the overall best result in localizing of this checkpoint. The estimated location resulted in a deviation of 1.4 m from the ground truth for the best solution. Here the difference of the Mahalanobis distance between the true location and the estimated position is approximately only 0.3 dBm. In the worst case, the Mahalanobis distances differ with values as large as 20 dBm. Further significant average deviations on this floor were seen at the two checkpoints CP27 and CP40. These two checkpoints have already achieved poor results when one looks at the MSR. CP27 and CP40 are located in the entrance area of the second floor near the staircase.

If one looks at the achieved results of the different smartphones, differences can be seen. Table 8 shows the statistical values for each of the six used smartphone in two orientations in the possible movement directions. As can be seen, the orientation of the user does not always play a major role for the resulting positioning accuracies. This is because of the fact that several orientations were measured for the off-line training measurements and thus the influence of the human body could be minimized. Table 8 also shows no major differences between the smartphones as they were calibrated with a linear regression model using the coefficients a_S and b_S obtained from Equation (1) (see Section 4.1).

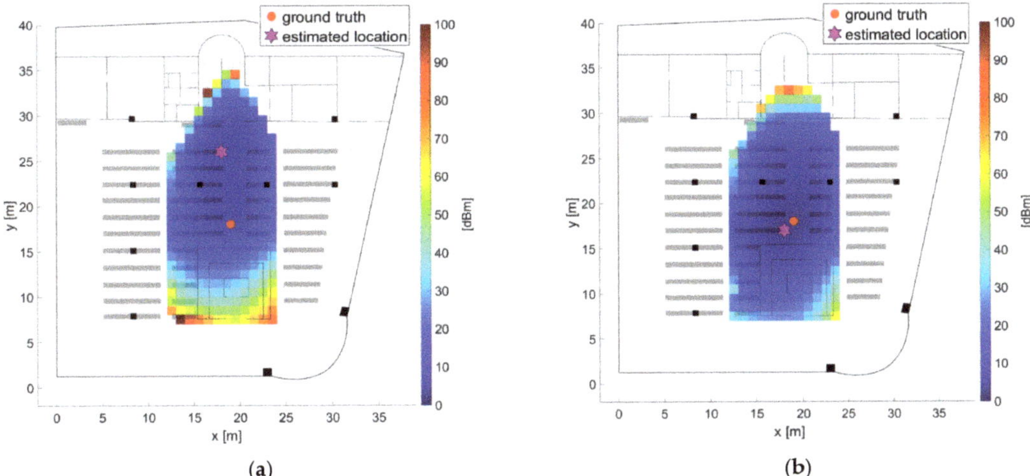

Figure 16. Two static positioning results on checkpoint 42 with the Samsung S3B with (**a**) worst and (**b**) overall best result.

Table 8. Deviations in [m] from the ground truth in dependence of the smartphone for the static measurements.

Smartphone	Orientation	Mean	Median	Standard Deviation
Nexus 5X	1	4.2	3.0	4.0
	2	3.1	2.2	2.6
OnePlus 5T	1	3.5	3.0	3.5
	2	3.1	2.0	4.1
Samsung S3A	1	1.8	1.0	2.3
	2	2.1	1.0	3.6
Samsung S3B	1	2.6	1.0	3.8
	2	2.9	1.0	4.4
Samsung S7	1	3.3	2.2	3.4
	2	2.8	1.4	3.1
Sony Z3	1	3.1	2.0	6.4
	2	2.1	1.0	4.2

6.3. Kinematic Positioning

In the case of the on-line kinematic measurements, the user walked along the trajectories back and forth with usual step speed of approximately $1\ \mathrm{ms}^{-1}$. In the following, the results of 12 measurement runs are presented. Because each smartphone requires a certain amount of time to perform a single Wi-Fi scan, i.e., its scan duration, a fingerprint could not be taken exactly at every checkpoint as the user pressed only an event button when passing by at a certain checkpoint and did not stop at this point. In order to determine the deviations at the checkpoints, the RSSI values had to be interpolated linearly. In addition to the deviations of the estimated positions from the ground truth, the kinematic measurements also determine the positions along the whole trajectory for each single scan. This allows that the walked trajectories can be reconstructed. Figures 17 and 18 therefore visualize the trajectories of two different measurement runs on the ground and second floor, respectively. Table 9 summarizes the deviations from the ground truth for the six different smartphones while walking back and forth. The deviations resulted in 2.7 m on average and a median of 1.4 m. As can be seen from Table 9, the largest mean deviation occurred with the Sony Z3 smartphone during the first measurement run with a value of 4.3 m. The reason for this large deviation is found in the long average scan duration of 4.1 s (compare with Table 4). Also the deviations of the Nexus 5X smartphone are larger which is also caused by the scan duration. As a result, the mesh points in the interpolation are lying further apart and their number is lower than as with short scan durations. Thus, the interpolation yield to a poorer approximation results in respective to the measured RSSI values. These results show that the time needed for a single Wi-Fi scan has a significant influence on the kinematic positioning results.

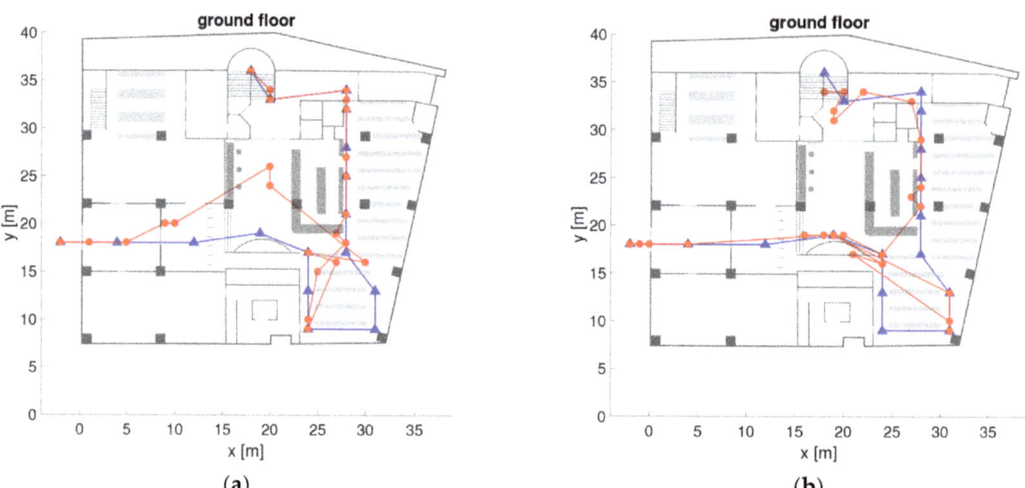

Figure 17. Trajectories of two measurement runs on the ground floor with estimated positions in red and ground truth in blue. (**a**) worst result; (**b**) good result.

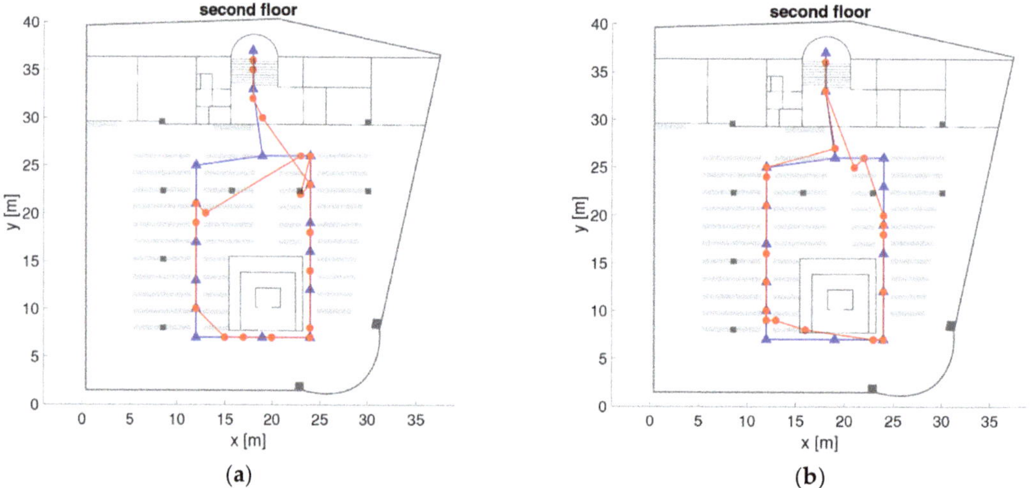

Figure 18. Trajectories of two measurement runs on the second floor with estimated positions in red and ground truth in blue. (**a**) worst result; (**b**) good result.

Table 9. Deviations in [m] from the ground truth in dependence of the smartphone for the kinematic measurement runs.

Smartphone	CP Start-End	Mean	Median	Standard Deviation
Nexus 5X	1-40	3.0	2.1	3.0
	40-1	2.9	2.2	2.7
OnePlus 5T	1-40	2.1	1.0	2.4
	40-1	1.9	1.0	2.7
Samsung S3A	1-40	1.6	1.0	2.3
	40-1	3.3	2.0	3.7
Samsung S3B	1-40	2.2	1.0	4.5
	40-1	2.9	1.0	5.0
Samsung S7	1-40	2.7	1.5	3.1
	40-1	2.0	1.0	2.4
Sony Z3	1-40	4.3	3.6	5.2
	40-1	3.9	3.0	3.9

6.4. Cramér-Rao Lower Bound

To investigate the deviations from the ground truth in more depth the Cramér-Rao Lower Bound (CRLB) was calculated. The CRLB is defined as the inverse of the Fisher information matrix (FIM) F [60–62]:

$$Cov(\theta) \geq F(\theta)^{-1} \tag{7}$$

The FIM F can be expressed as:

$$F(\theta) = \begin{bmatrix} F_{xx} & F_{xy} \\ F_{yx} & F_{yy} \end{bmatrix} \tag{8}$$

with:

$$F_{xx}(\theta) = \sum_{i=1}^{m} \rho \frac{(x_i - x_0)^2}{d_{i0}^4}$$

$$F_{xy}(\theta) = F_{yx}(\theta) = \sum_{i=1}^{m} \rho \frac{(x_i - x_0)(y_i - y_0)}{d_{i0}^4}$$

$$F_{yy}(\theta) = \sum_{i=1}^{m} \rho \frac{(y_i - y_0)^2}{d_{i0}^4}$$

and the channel constant ρ:

$$\rho = \left(\frac{10 \cdot n_p}{\sigma_i \cdot \ln 10} \right)^2 \tag{9}$$

where n_p is the path-loss exponent (typically between 2 and 4), σ_i is the standard deviation of the RSSI of AP_i, m is the number of APs and d_{i0} represents the true distance between AP_i and the unknown mobile device, which is numbered as 0.

Finally, the lower bound on Root Mean Square Error (RMSE) can be computed by,

$$RMSE \geq \sqrt{\text{trace}(F(\theta)^{-1})} \tag{10}$$

Figure 19 shows a visualization of the resulting Cramér-Rao Lower Bound (CRLB) on the RMSE for the ground and the second floor in the library. Low CRLB values visualized in dark blue indicate higher positioning accuracies during the on-line phase, while higher values in red mean lower accuracy. Especially on the ground floor, one can see two areas where the CRLB is 2 to 3 m (green-yellow areas), while in the other parts of the area it has only values of 0.5 to 1 m.

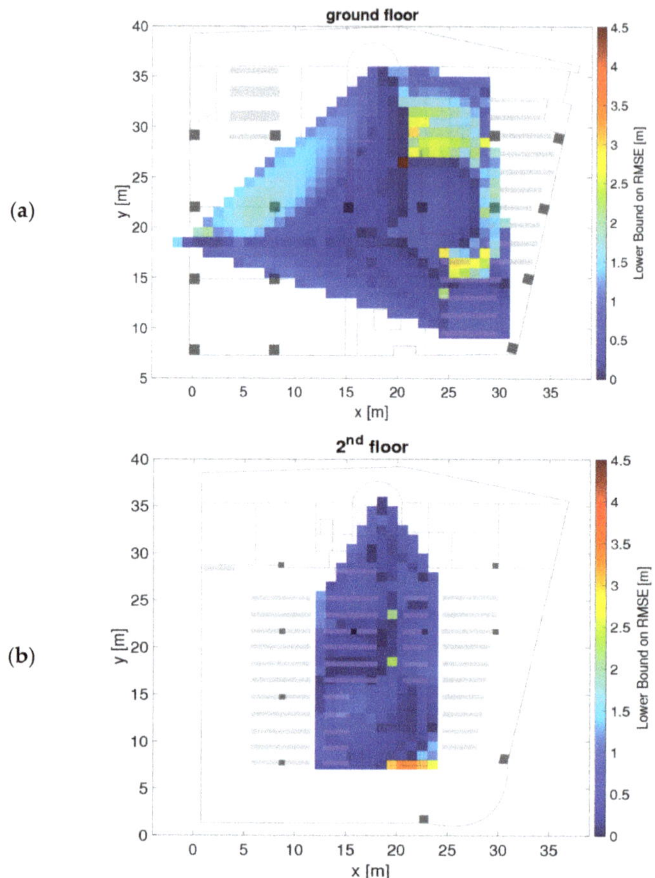

Figure 19. Visualization of the Cramér-Rao Lower Bound (CRLB) on the RMSE for (**a**) the ground and (**b**) the second floor.

6.5. Disscussion and Proposal for Performance Improvement

The results of the kinematic positioning tests indicate that the measured trajectories could be well reconstructed. Problems were seen only at the edges and in the corners on the ground floor and in the entrance area near the staircase on the second floor. If one compares the obtained results between the ground and second floor, however, differences in achievable positing accuracies can be seen. They are mainly caused by the existing building structures as the RSSI values do not vary significantly at the neighboring checkpoints on the ground floor. A significant difference on neighboring locations would facilitate a better matching success result. Although the second floor is a large reading room with many bookshelves, the resulting deviations on the checkpoints are smaller. The higher localization accuracies achieved could result from, on the one hand, the location of the bookshelves itself, which provide a significant variation of the RSSI values on the different checkpoints, and, on the other hand, due to the higher number of visible APs.

One major impact on the achievable positioning performance has not been considered so far. It relates to an optimization of the AP locations throughout the library. The current AP deployment enables only sufficient Wi-Fi communication services in most areas of the building. The APs are located in a rectangular shaped deployment at the same location on top of each other in every floor (apart from the ground floor). With AP rearrangement and additional deployment, it can be expected that higher positioning accuracies are achievable and a better service provided. Thus, future work will focus on this key point. It is proposed to deploy low-cost Raspberry Pi units serving as APs in addition. Retscher and Tatschl [63] have used Raspberry Pi units serving as APs and reference stations broadcasting as well as scanning and recording RSSI values at the same time. They introduced the Differential Wi-Fi (DWi-Fi) approach where reference stations as in Differential GNSS are deployed in the area of interest to derive correction parameters from the continuous sensed RSSI values of all involved APs and Raspberry Pi units (see also [1]). In Martínez-Gómez [64] Raspberry Pi units were employed as mobile devices. In our future research, it is planned to replace the smartphones by these devices in addition to the APs.

7. Path towards the Development of a Library Navigation and Information System

To assist students, staff and University visitors finding auditoriums and classrooms, offices and other rooms faster and easier, the positioning and navigation system can be combined with the in-house information system of TU Wien (TU Wien Information System & Services TISS) and with the e-learning platform TUWEL which is based on Moodle. Furthermore, additional application possibilities for location-related services are created. For instance, students could share their current location in order to be found faster by colleagues. A positioning system can also help to control and analyze people flows. These analyzes can later be a useful tool for, e.g., sustainable building development. In addition, short-term changes to the venue can be communicated more easily. The implementation of the presented positioning service at TU Wien can therefore lead to many new areas of application and thus contribute to an improvement of everyday life at the University. Especially for the library a navigation and information service is a very useful tool. To find a certain bookshelf it was seen in this study, however, that the integration with other technologies for positioning is required. Wi-Fi localization could be significantly improved if the new Round Trip Time (RTT) measurement protocol [65,66] is applied. In this case the double range between the transmitter, i.e., the AP, and the receiver, i.e., the smartphone, is derived by travel time measurement. Using RTT measurements ranges to the APs can be obtained with precisions on the decimeter level leading to higher localization accuracies than with common Wi-Fi fingerprinting [1,67]. The hardware of the APs, however, would need to be upgrade to be able to perform RTT measurements. Furthermore, currently not many smartphones on the market support these measurements. Another requirement would be to know the location of the AP precisely. If only the upgraded APs of the in-house Wi-Fi networks are used the location of the APs has to be surveyed once to obtain their 3D coordinates. The knowledge of the AP locations is not a requirement for location fingerprinting. Thus, a meaningful combination and integration of the RTT technology with fingerprinting is a promising solution. For further investigations the usage of low-cost Raspberry Pi units is foreseen. They should serve as APs and mobile devices enabling fingerprinting as well as RTT measurements.

Another improvement of localization performance in the kinematic positioning mode shall be achieved by the additional usage of the inertial sensors of the mobile devices. With smartphone accelerometers the distance travelled can be derived and with a gyroscope together with a magnetometer the direction of movement. Further developments are therefore focused on the integration of these sensors for continuous user localization.

Furthermore, other ways for the calibration of the RSSI recordings of different smartphone will be addressed in our future research. Apart from the calibration using a multivariate linear regression model the use of an in-motion calibration approach as in [68] shall be applied in order to cope with the inherent noise of Wi-Fi signals. Applying a window moving average filter to the raw RSSI recordings would lead to an improvement of the results.

A further task of investigation in this study is the integration with other technologies, such as Bluetooth LE beacons for areas with limited Wi-Fi coverage and serving as a backup solution as well as the usage of the RFID (Radio Frequency Identification) technology for book labeling and tracking. Thus, it is then possible to locate the correct book in the bookshelf itself and even detect if a book is taken out of the library without permission.

RFID can be easily used for book location and tracking as books can be labeled with very cheap passive tags.

At TU Wien, however, it is a requirement that the navigation and information system should not cause high additional costs for, e.g., for installation of new hardware. Furthermore, a wide variety of mobile devices should be capable to use the service. These two requirements were the main reasons why so far only Wi-Fi fingerprinting was considered in the first stage of this study.

8. Concluding Remarks and Outlook

The investigations in this study have shown that Wi-Fi fingerprinting can be used to achieve positioning accuracies on the meter level in the library building of TU Wien and that the direction taken is useful for the development of navigation and information services. It is expected that the positioning accuracies in the library can be increased by installation of additional APs under consideration of their deployment to provide a better distribution and geometry for localization. Since the APs on the upper floors of the library are all arranged in a rectangle deployment, the question can also be asked whether a rearrangement can improve positioning accuracy. The optimization of the geometry of the AP locations is especially a crucial requirement if new technologies, such as Wi-Fi RTT measurements, shall be employed for increasing the positioning accuracies and service performance. Additional deployment of hardware is foreseen in the future by using low-cost Raspberry Pi units broadcasting and receiving Wi-Fi signals.

To overcome a major disadvantage of location fingerprinting concerning the required labor-intensive system training, new approaches, such as the usage of crowdsourced RSSI data (see e.g., [69–72]) from all service users, will be employed. For crowdsourcing, users can provide their scanned Wi-Fi RSSI values to build-up and continuously update the fingerprinting database. As the comparison of the different measurement modes—static, stop-and-go and kinematic—in the off-line training phase has shown, the database creation in kinematic mode and the achievable positioning accuracies differ not much from the other two measurement modes. This means, that continuous system training can be carried out, which reduces the time required.

Author Contributions: Conceptualization. G.R.; methodology. G.R.; software. A.L.; validation. G.R. and A.L.; formal analysis. A.L.; investigation. A.L.; resources. A.L.; data curation. A.L.; writing—original draft preparation. G.R.; writing—review and editing. G.R. and A.L.; visualization. A.L.; supervision. G.R. All authors have read and agreed to the published version of the manuscript.

Funding: This research received no external funding.

Institutional Review Board Statement: Not applicable.

Informed Consent Statement: Not applicable.

Data Availability Statement: The data presented in this study are available on request from the corresponding author.

Conflicts of Interest: The authors declare no conflict of interest.

References

1. Retscher, G. Fundamental concepts and evolution of Wi-Fi user localization: An overview based on different case studies. *Sensors* **2020**, *20*, 5121. [CrossRef] [PubMed]
2. Chen, R.; Pei, L.; Liu, J.; Leppäkoski, H. WLAN and Bluetooth positioning in smart phones. In *Ubiquitous Positioning and Mobile Location-Based Services in Smart Phones*; IGI Global: Hershey, PA, USA, 2012; pp. 44–68.
3. Liu, H.; Darabi, H.; Banerjee, P.; Liu, J. Survey of wireless indoor positioning techniques and systems. *IEEE Trans. Syst. Man Cybern. Part C Appl. Rev.* **2007**, *37*, 1067–1080. [CrossRef]
4. Ali, W.H.; Kareem, A.A.; Jasim, M. Survey on wireless indoor positioning systems. *Cihan Univ. Erbil Sci. J.* **2019**, *3*, 42–47. [CrossRef]
5. Youssef, M.; Agrawala, A.; Shankar, A.U. WLAN location determination via clustering and probability distributions. In Proceedings of the First IEEE International Conference on Pervasive Computing and Communications (PerCom 2003), Fort Worth, TX, USA, 23–26 March 2003.

6. Sakpere, W.; Adeyeye-Oshin, M.; Mlitwa, N.B. A state-of-the-art survey of indoor positioning and navigation systems and technologies. *S. Afr. Comput. J.* **2017**, *29*, 145–197. [CrossRef]

7. Want, R.; Hopper, A.; Falcao, V.; Gibbons, J. The Active Badge location system. *ACM Trans. Inf. Syst.* **1992**, *10*, 91–102. [CrossRef]

8. Yi, K.Y.; Kim, D.Y.; Yi, K.M. Development of a Localization System based on VLC technique for an indoor environment. *J. Electr. Eng. Technol.* **2015**, *10*, 436–442. [CrossRef]

9. Akiyama, T.; Sugimoto, M.; Hashizume, H. Time-of-Arrival-based smartphone localization using Visible Light Communication. In Proceedings of the International Conference on Indoor Positioning and Indoor Navigation (IPIN 2017), Sapporo, Japan, 18–21 September 2017.

10. Jung, S.-Y.; Hann, S.; Park, C.-S. TDOA-based optical wireless indoor localization using LED ceiling lamps. *IEEE Trans. Consum. Electr.* **2011**, *57*, 1592–1597. [CrossRef]

11. Brena, R.F.; García-Vázquez, J.P.; Galván-Tejada, C.E.; Muñoz-Rodriguez, D.; Vargas-Rosales, C.; Fangmeyer, J. Evolution of indoor positioning technologies: A survey. *J. Sens.* **2017**. [CrossRef]

12. Moutinho, J.; Araújo, R.; Freitas, D. Indoor localization with audible sound—Towards practical implementation. *Pervasive Mobile Comput.* **2016**, *29*, 1–16. [CrossRef]

13. Li, J.; Han, G.; Zhu, C.; Sun, G. An indoor ultrasonic positioning system based on TOA for Internet of Things. *Mobile Inf. Syst.* **2016**, 1–10. [CrossRef]

14. Nakashima, Y.; Kaneto, R.; Babaguchi, N. Indoor positioning system using digital audio watermarking. *IEICE Trans. Inf. Syst.* **2011**, *94*, 2201–2211. [CrossRef]

15. Lopes, S.I.; Vieira, J.M.N.; Reis, J.; Albuquerque, D.F.; Carvalho, N.B. Accurate smartphone indoor positioning using a WSN infrastructure and non-invasive audio for TDoA estimation. *Pervasive Mobile Comput.* **2015**, *20*, 29–46. [CrossRef]

16. Woodman, O.; Harle, R. Concurrent scheduling in the Active Bat location system. In Proceedings of the 8th IEEE International Conference on Pervasive Computing and Communications Workshops (PERCOM Workshops 2010), Mannheim, Germany, 29 March–2 April 2010.

17. Minami, M.; Fukuju, Y.; Hirasawa, K.; Yokoyama, S.; Mizumachi, M.; Morikawa, H.; Aoyama, T. DOLPHIN: A practical approach for implementing a fully distributed indoor ultrasonic positioning system. In Proceedings of the Ubiquitous Computing UbiComp 2004, Nottingham, UK, 7–10 September 2004.

18. Priyantha, N.B.; Chakraborty, A.; Balakrishnan, H. The Cricket location-support system. In Proceedings of the 6th annual international conference on Mobile computing and networking (MobiCom 2000), Boston, MA, USA, 6–11 August 2000.

19. Batistic, L.; Tomic, M. Overview of indoor positioning system technologies. In Proceedings of the 41st International Convention on Information and Communication Technology, Electronics and Microelectronics (MIPRO 2018), Opatija, Croatia, 21–25 May 2018.

20. Cominelli, M.; Patras, P.; Gringoli, F. Dead on arrival: An empirical study of the Bluetooth 5.1 positioning system. In Proceedings of the 13th International Workshop on Wireless Network Testbeds, Experimental Evaluation & Characterization (WiNTECH '19), Los Cabos, Mexico, 25 October 2019.

21. Faragher, R.; Harle, R. Location fingerprinting with Bluetooth Low Energy Beacons. *IEEE J. Sel. Areas Commun.* **2015**, *33*, 2418–2428. [CrossRef]

22. Zhao, X.; Xiao, Z.; Markham, A.; Trigoni, N.; Ren, Y. Does BTLE measure up against WiFi? A comparison of indoor location performance. In Proceedings of the 20th European Wireless Conference, Barcelona, Spain, 14–16 May 2014; pp. 1–6.

23. Mainetti, L.; Patrono, L.; Sergi, I. A survey on indoor positioning systems. In Proceedings of the 22nd International Conference on Software, Telecommunications and Computer Networks (SoftCOM 2014), Split, Croatia, 17–19 September 2014.

24. Seco, F.; Jimenez, A.R. Smartphone-based cooperative indoor localization with RFID technology. *Sensors* **2018**, *18*, 266. [CrossRef] [PubMed]

25. Großwindhager, B.; Stocker, M.; Rath, M.; Boano, C.A.; Römer, K. SnapLoc: An ultra-fast UWB-based indoor localization system for an unlimited number of tags. In Proceedings of the 18th International Conference on Information Processing in Sensor Networks (IPSN '19), Montreal, QC, Canada, 16–18 April 2019.

26. Sadowski, S.; Spachos, P. RSSI-based indoor localization with the Internet of Things. *IEEE Access* **2018**, *6*, 30149–30161. [CrossRef]

27. Yeh, S.-C.; Hsu, W.-H.; Lin, W.-Y.; Wu, Y.-F. Study on an indoor positioning system using Earth's magnetic field. *IEEE Trans. Instrum. Meas.* **2020**, *69*, 865–872. [CrossRef]

28. Sheinker, A.; Ginzburg, B.; Salomonski, N.; Frumkis, L.; Kaplan, B.-Z.; Moldwin, M.B. A method for indoor navigation based on magnetic beacons using smartphones and tablets. *Measurement* **2016**, *81*, 197–209. [CrossRef]

29. Retscher, G. Indoor navigation. In *Encyclopedia of Geodesy, Earth Sciences Series*; Grafarend, E.W., Ed.; Springer: Cham, Switzerland, 2016. [CrossRef]

30. Gerstweiler, G.; Vonach, E.; Kaufmann, H. HyMoTrack: A mobile AR navigation system for complex indoor environments. *Sensors* **2016**, *16*, 17. [CrossRef] [PubMed]

31. Liu, J.; Chen, R.; Pei, L.; Guinness, R.; Kuusniemi, H. A hybrid smartphone indoor positioning solution for mobile LBS. *Sensors* **2012**, *12*, 17208–17233. [CrossRef]

32. Ettlinger, A.; Neuner, H.-B.; Burgess, T. Development of a Kalman Filter in the Gauss-Helmert model for reliability analysis in orientation determination with smartphone sensors. *Sensors* **2018**, *18*, 414. [CrossRef]

33. Retscher, G. Augmentation of indoor positioning systems with a barometric pressure sensor for direct altitude determination in a multi-storey building. *Cartogr. Geogr. Inf. Sci.* **2007**, *34*, 305–310. [CrossRef]

34. Ye, H.; Gu, T.; Tao, X.; Lu, J. Scalable floor localization using barometer on smartphone. *Wirel. Commun. Mob. Comput.* **2016**, *16*, 2557–2571. [CrossRef]
35. Werner, M.; Kessel, M.; Marouane, C. Indoor positioning using smartphone camera. In Proceedings of the International Conference on Indoor Positioning and Indoor Navigation (IPIN 2011), Guimaraes, Portugal, 21–23 September 2011.
36. Mautz, R. Indoor Positioning Technologies. Habilitation Thesis, ETH Zurich, Zurich, Switzerland, 2012.
37. Scaramuzza, D.; Fraundorfer, F. Visual odometry [Tutorial]. *IEEE Robot. Autom. Mag.* **2011**, *18*, 80–92. [CrossRef]
38. Ramezani, M.; Acharya, D.; Gu, F.; Khoshelham, K. Indoor positioning by visual-inertial odometry. In Proceedings of the ISPRS Annals of Photogrammetry, Remote Sensing and Spatial Information Sciences, Wuhan, China, 18–22 September 2017; Volume IV-2/W4.
39. Aqel, M.; Marhaban, M.H.; Saripan, M.I.; Ismail, N. Review of visual odometry: Types, approaches, challenges, and applications. *SpringerPlus* **2016**, *5*, 1897. [CrossRef]
40. Sahdev, R.; Chen, B.X.; Tsotsos, J.K. Indoor localization in dynamic human environments using visual odometry and global pose refinement. In Proceedings of the 15th Conference on Computer and Robot Vision (CRV 2018), Toronto, ON, Canada, 9–11 May 2018.
41. Retscher, G.; Tatschl, T. Indoor positioning with differential Wi-Fi lateration. *J. Appl. Geod.* **2017**, *11*, 249–269. [CrossRef]
42. Chen, X.; Kong, J.; Guo, Y.; Chen, X. An empirical study of indoor localization algorithms with densely deployed APs. In Proceedings of the Global Communications Conference GLOBECOM, Austin, TX, USA, 8–12 December 2014; pp. 517–522.
43. Schnabel, P. Elektronik-Kompendium. Available online: https://www.elektronik-kompendium.de/ (accessed on 1 November 2020). (in German).
44. RTR-GmbH. RTR—Frequenzbereiche. Available online: https://www.rtr.at/de/tk/FRQ_spectrum (accessed on 1 November 2020). (in German).
45. Retscher, G.; Leb, A. Influence of the RSSI scan duration of smartphones in kinematic Wi-Fi fingerprinting (paper 9743). In Proceedings of the FIG Working Week, Hanoi, Vietnam, 22–26 April 2019.
46. LeDoux, H.; Gold, C. An Effcient natural neighbour interpolation algorithm for geoscientific modelling. In *Developments in Spatial Data Handling*; Fisher, P.F., Ed.; Springer: Berlin/Heidelberg, Germany, 2005.
47. Lee, M.; Han, D. Voronoi tessellation based interpolation method for Wi-Fi radio map construction. *IEEE Commun. Lett.* **2012**, *16*, 404–407. [CrossRef]
48. Üreten, S.; Yongaçoğlu, A.; Petriu, E. A Comparison of interference cartography generation techniques in cognitive radio networks, In Proceedings of the 2012 IEEE International Conference on Communications (ICC), Ottawa, ON, Canada, 10–15 June 2012.
49. Kaemarungsi, K.; Krishnamurthy, P. Properties of indoor received signal strength for WLAN location fingerprinting. In Proceedings of the Mobile and Ubiquitous Systems: Networking and Services MOBIQUITOUS 2004, Boston, MA, USA, 22–26 August 2004; pp. 14–23.
50. Retscher, G.; Hofer, H. Wi-Fi location fingerprinting using an intelligent checkpoint sequence. *J. Appl. Geod.* **2017**, *11*, 197–205. [CrossRef]
51. Bahl, P.; Padmanabhan, V. RADAR: An in-building RF-based user location and tracking system. In Proceedings of the 19th Annual Joint Conference of the IEEE Computer and Communications Societies INFOCOM 2000, Tel-Aviv, Israel, 26–30 March 2000; Volume 2, pp. 775–784.
52. Honkavirta, V.; Perälä, T.; Ali-Löytty, S.; Piché, R. A comparative survey of WLAN location fingerprinting methods. In Proceedings of the IEEE 6thWorkshop on Positioning Navigation and Communication WPNC'09, Hannover, Germany, 19 March 2009; pp. 243–251.
53. Dawes, B.; Chin, K.-W. A comparison of deterministic and probabilistic methods for indoor localization. *J. Syst. Softw.* **2011**, *84*, 442–451. [CrossRef]
54. King, T.; Kopf, S.; Haenselmann, T.; Lubberger, C.; Effelsberger, W. COMPASS: A probabilistic indoor positioning system based on 802.11 and digital compasses. In Proceedings of the First ACM Workshop on Wireless Network Testbeds, Experimental Evaluation and Characterization (WiNTECH), Los Angeles, CA, USA, 29 September 2006.
55. Roos, T.; Myllymaki, P.; Tirri, H. A statistical modeling approach to location estimation. *IEEE Trans. Mob. Comput.* **2002**, *1*, 59–69. [CrossRef]
56. Gordon, N.; Salmond, D.; Smith, A. Novel approach to nonlinear/non-Gaussian Bayesian state estimation. *IEE Proc. F* **1993**, *140*, 107. [CrossRef]
57. Koch, K.-R. *Einführung in Die Bayes-Statistik*; Springer: Berlin/Heidelberg, Germany, 2000. (In German)
58. Yeung, W.M.; Zhou, J.; Ng, J.K.-Y. Enhanced fingerprint-based location estimation system in wireless LAN environment. In *Lecture Notes in Computer Science, Proceedings of the Emerging Directions in Embedded and Ubiquitous Computing Conference EUC 2007, Taipei, Taiwan, 17–20 December 2007*; Springer: Berlin/Heidelberg, Germany, 2007.
59. Moghtadaiee, V.; Dempster, A.G. Vector distance measure comparison in indoor location fingerprinting. In Proceedings of the International Global Navigation Satellite Systems IGNSS 2015 Conference, Gold Coast, Australia, 14–16 July 2015.
60. Patwari, N.; Ash, J.N.; Kyperountas, S.; Hero, A.O.; Moses, R.L.; Correal, N.S. Locating the nodes: Cooperative localization in wireless sensor networks. *IEEE Signal Process. Mag.* **2005**, *22*, 54–69. [CrossRef]
61. Laitinen, E.; Lohan, E. Access Point topology evaluation and optimization based on Cramér-Rao Lower Bound for WLAN indoor positioning. In Proceedings of the 2016 International Conference on Localization and GNSS (ICL-GNSS), Barcelona, Spain, 28–30 June 2016. [CrossRef]

62. Li, Q.; Li, W.; Sun, W.; Li, J.; Liu, Z. Cramér-Rao Bound analysis of Wi-Fi indoor localization using fingerprint and assistant nodes. In Proceedings of the 2017 IEEE 86th Vehicular Technology Conference (VTC-Fall), Toronto, ON, Canada, 24–27 September 2017. [CrossRef]

63. Retscher, G.; Tatschl, T. Differential Wi-Fi—A novel approach for Wi-Fi positioning ising lateration. In Proceedings of the FIG Working Week, Christchurch, New Zealand, 2–6 May 2016.

64. Martínez-Gómez, J.; Martínez, M.; Castillo-Cara, M.; Brea, V.M.; Orozco-Barbosa, L.; García, I. Spatial statistical analysis for the design of indoor particle-filter-based localization mechanisms. *Int. J. Distributed Sens. Netw.* **2016**, *12*, 8. [CrossRef]

65. Van Diggelen, F.; Want, R.; Wang, W. *How to achieve 1-m accuracy in Android*; GPS World: Cleveland, OH, USA, 3 July 2018.

66. Ibrahim, M.; Liu, H.; Jawahar, M.; Nguyen, V.; Gruteser, M.; Howard, R.; Bai, F. Verification: Accuracy evaluation of WiFi fine time measurements on an open platform. In Proceedings of the 24th Annual International Conference on Mobile Computing and Networking MobiCom '18, New Delhi, India, 29 October–2 November 2018.

67. Guo, G.; Chen, R.; Ye, F.; Peng, X.; Liu, Z.; Pan, Y. Indoor smartphone localization: A hybrid WiFi RTT-RSS ranging approach. *IEEE Access* **2019**, *7*, 176767–176781. [CrossRef]

68. del Horno, M.M.; García-Varea, I.; Orozco Barbosa, L. Calibration of Wi-Fi-based indoor tracking systems for Android-based smartphones. *Remote Sens.* **2019**, *11*, 1072. [CrossRef]

69. Kim, W.; Yang, S.; Gerla, M.; Lee, E.-K. Crowdsource based indoor localization by uncalibrated heterogeneous Wi-Fi devices. *Mob. Inf. Syst.* **2016**, 1–18. [CrossRef]

70. Rai, A.; Chintalapudi, K.K.; Padmanabhan, V.N.; Sen, R. Zee: Zero-effort crowdsourcing for indoor localization. In Proceedings of the 18th Annual International Conference on Mobile Computing and Networking Mobicom'12, Istanbul, Turkey, 22–26 August 2012; pp. 293–304.

71. Wu, C.; Yang, Z.; Liu, Y. Smartphones based crowdsourcing for indoor localization. *IEEE Trans. Mob. Comput.* **2015**, *14*, 444–457. [CrossRef]

72. Zhou, B.; Li, Q.; Mao, Q.; Tu, W.; Zhang, X.; Chen, L. ALIMC: Activity landmark-based indoor mapping via crowdsourcing. *IEEE Trans. Intell. Transp. Syst.* **2015**, *16*, 2774–2785. [CrossRef]

Article

Demonstration of Three-Dimensional Indoor Visible Light Positioning with Multiple Photodiodes and Reinforcement Learning

Zhuo Zhang [1,†], Huayang Chen [1,†], Weikang Zeng [1], Xinlong Cao [1], Xuezhi Hong [1,*] and Jiajia Chen [1,2,*]

1 Centre for Optical and Electromagnetic Research, South China Academy of Advanced Optoelectronics, South China Normal University, Guangzhou 510006, China; zhuo.zhang@coer-scnu.org (Z.Z.); huayang.chen@coer-scnu.org (H.C.); weikang.zeng@coer-scnu.org (W.Z.); xinlong.cao@coer-scnu.org (X.C.)
2 Department of Electrical Engineering, Chalmers University of Technology, Hörsalsvägen 9, 41296 Gothenburg, Sweden
* Correspondence: xuezhi.hong@coer-scnu.org (X.H.); jiajiac@chalmers.se (J.C.); Tel.: +86-2034725186 (X.H.); +46-317726034 (J.C.)
† These authors contributed equally to this work.

Received: 17 September 2020; Accepted: 10 November 2020; Published: 12 November 2020

Abstract: To provide high-quality location-based services in the era of the Internet of Things, visible light positioning (VLP) is considered a promising technology for indoor positioning. In this paper, we study a multi-photodiodes (multi-PDs) three-dimensional (3D) indoor VLP system enhanced by reinforcement learning (RL), which can realize accurate positioning in the 3D space without any off-line training. The basic 3D positioning model is introduced, where without height information of the receiver, the initial height value is first estimated by exploring its relationship with the received signal strength (RSS), and then, the coordinates of the other two dimensions (i.e., X and Y in the horizontal plane) are calculated via trilateration based on the RSS. Two different RL processes, namely RL_1 and RL_2, are devised to form two methods that further improve horizontal and vertical positioning accuracy, respectively. A combination of RL_1 and RL_2 as the third proposed method enhances the overall 3D positioning accuracy. The positioning performance of the four presented 3D positioning methods, including the basic model without RL (i.e., Benchmark) and three RL based methods that run on top of the basic model, is evaluated experimentally. Experimental results verify that obviously higher 3D positioning accuracy is achieved by implementing any proposed RL based methods compared with the benchmark. The best performance is obtained when using the third RL based method that runs RL_2 and RL_1 sequentially. For the testbed that emulates a typical office environment with a height difference between the receiver and the transmitter ranging from 140 cm to 200 cm, an average 3D positioning error of 2.6 cm is reached by the best RL method, demonstrating at least 20% improvement compared to the basic model without performing RL.

Keywords: reinforcement learning; 3D indoor positioning; visible light positioning

---

1. Introduction

The developments of location-based mobile services and the Internet of Things urgently need stable and precise indoor positioning technologies [1]. As the widely deployed global positioning system (GPS) has poor coverage and accuracy in the indoor environment, indoor positioning systems (IPS) that employ alternative radio frequency (RF) technologies (e.g., Bluetooth, RFID, iBeacon, Wi-Fi, and near-field communication [2–4]) have been investigated. However, the RF-based IPS (e.g., 1–3/5–15/0.1–0.3 m with Bluetooth/Wi-Fi/UWB, respectively [5]) can be largely affected by electromagnetic interference and multipath effect in a congested environment [6]. Compared with RF technologies, the visible light positioning (VLP) system has features with immunity to electromagnetic interference and high tolerance to multipath interference thanks to the domination of the LOS signal [7–11]. By simultaneously providing illumination and positioning services with the existing indoor lighting equipment (e.g., light-emitting diode (LED)), VLP with high positioning accuracy (e.g., in the order of centimeters [7]) is considered as one low cost and high energy efficiency solution for localization in the indoor environment.

Comparing with the two-dimensional (2D) positioning on a horizontal plane at a known height, three-dimensional (3D) positioning using the same setup is more challenging. In such a case, one needs to map the received signal to one more dimension (i.e., height) which increases the search space for the positioning process. To find the correct position in a larger searching space, the positioning algorithm and/or the hardware in 3D VLP systems are more complex than the 2D ones. From the hardware perspective, 3D VLP systems based on either a single photodiode (PD) or multiple PDs have been proposed. In the single-PD system, 3D positioning has been achieved by combining information from both the PD and the other hardware either at the receiver (e.g., accelerometer [12], rotatable platform [13]) or at the transmitter (e.g., steerable laser [14,15]), which is not necessarily simpler than the multiple-PD system from the system complexity perspective. 3D positioning based on a low complexity receiver with one PD only has also been proposed, which has additional requirements for the radiation patterns or geometric arrangement of LEDs to avoid ambiguity in height estimation [16,17]. 3D VLP systems using multiple PDs have also been proposed, in which the spatial or angular diversity of PDs are explored to estimate the 3D position of the receiver [18–20]. Though it needs more PDs at the receiver, it does not have any special requirement for the transmitter [16,17] and has shown the potential to reduce the number of LEDs for a simpler transmitter [13].

For the 3D positioning algorithm, trilateration and triangulation based methods are widely employed, in which the geometric relationship between the receiver and light sources (e.g., distance [19], incidence/irradiance angles [18,20]) is estimated from the received signal. One popular way to evolve from a 2D VLP algorithm to a 3D one is to conduct a brute-force search on several parallel 2D layers at various heights. After obtaining the horizontal positions on all candidate layers at a pre-defined height set, the estimated 3D position is determined as the one that most likely fulfills the constraint among the coordinates of the three dimensions [16,21,22]. To improve the efficiency in height estimation, a fast search method based on the golden section search (GSS) algorithm has been proposed which can significantly reduce the running time [16]. To further improve the positioning accuracy, machine learning (ML) techniques with outstanding nonlinear fitting capability have been introduced to the VLP systems. Supervised learning (SL) based VLP systems (e.g., neural network [23–25], random forest [26], and *K*-nearest neighbor [27]) have been proposed. However, the SL based VLP systems require sufficient training data to be prepared in advance, which increases the system complexity [27]. The performance of the SL positioning algorithms is also largely affected by the quality of the training. To avoid the above drawbacks, training-data-free ML techniques, such as reinforcement learning (RL), have been employed. Previous studies show that the application of RL in 2D VLP offers high and robust positioning accuracy [28,29]. Though the RL based 2D positioning algorithm has shown a higher tolerance to the error of a priori height information than the conventional one, the height of the object is still assumed to be known in advance under the 2D VLP framework. Moreover, in many applications with mobile devices, the exact height of the receiver is often unknown and could vary dynamically in a range much larger than the height error tolerance of the RL based 2D VLP algorithm.

In this paper, a 3D VLP system using multiple PDs and reinforcement learning is proposed which realizes high accuracy for 3D positioning without needs of data for off-line training. In the 3D VLP system, we first make a coarse estimation of the receiver height by exploring its relationship with the received signal strength (RSS) and then calculate the other two coordinates in the horizontal plane using trilateration [30]. To achieve high 3D positioning accuracy, three methods based on RL with different height update strategies are proposed. Experiments are carried out to evaluate the performance of the proposed methods under different receiver sizes. The results show that when the height difference between the receiver and the transmitter is within [140, 200]-cm, compared with the case without machine learning (i.e., the Benchmark), all three proposed RL based methods can improve 3D positioning accuracy robustly. Unlike our previous 2D VLP work [28,29] that only estimates the position on a horizontal plane and still requires the height information as an input, this paper is an extension, include three new major contributions: (i) methods for 3D positioning are investigated that output coordinates in all three dimensions without a priori information about any dimensions; (ii) two novel reinforcement learning processes are devised specifically for 3D VLP, which target accuracy enhancement in the horizontal plane and the vertical dimension, respectively, and a combination of them offers the highest 3D positioning accuracy; (iii) the effectiveness of the proposed RL based 3D positioning methods are demonstrated experimentally.

The remainder of the paper is organized as follows. The operation principles of different 3D positioning methods, including the basic model and three RL based methods (i.e., Method 1/2/3) are explained in Section 2. Section 3 shows the experimental setup for 3D VLP, compares the performance of different positioning methods, and analyze the impact of the receiver size. Finally, Section 4 draws conclusions.

2. Operation Principle

A multi-PD VLP system with M ($M \geq 3$) LEDs at the same height on the ceiling is considered in this study. Figure 1 shows the considered 3D VLP system setup and the signal processing flow, including the basic model without RL (later referred to as the benchmark), and three proposed methods that employ RL to improve positioning accuracy.

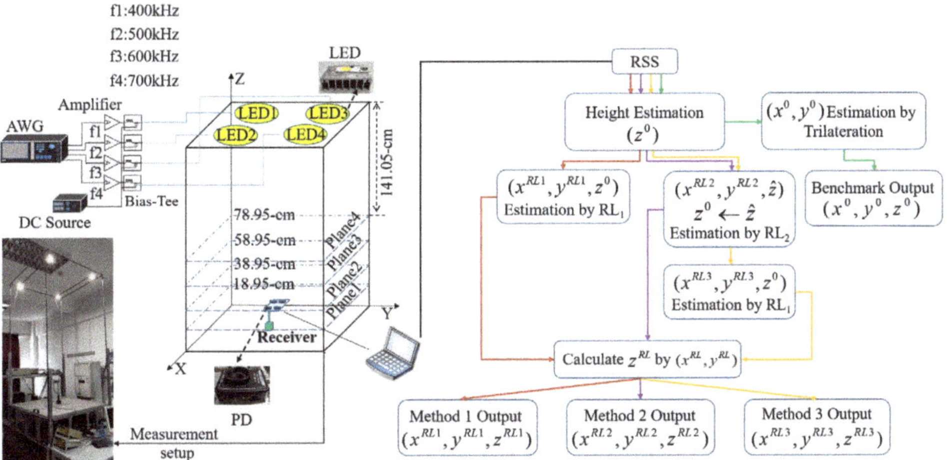

Figure 1. The 3D VLP system setup and the signal processing flow. The green, red, purple, and yellow lines represent the flow for the Benchmark, Method 1, Method 2, and Method 3, respectively. The inset shows the picture of our testbed.

The i-th ($i = 1, 2 \ldots M$) LED is located at $\left(L_i^x, L_i^y, L^z\right)$ and transmits a sinusoidal modulated signal with frequency f_i. At the receiver, N PDs are facing up at the same height, and the n-th ($n = 1, 2 \ldots N$) PD is located at (x_n, y_n, z). The received signal of the nth PD from all the LEDs is represented by

$s_n(t)$, whose power spectrum consists of M peak components at f_i ($i = 1, 2, \ldots, M$) [31]. After Fourier transformation, it can be expressed as [31]:

$$S_n^{f_i} = \frac{(P_i)^2(m+1)^2 A^2 \beta^2 h^{2(m+m')}}{4\pi^2 dn, i^{2(2+m+m')}} \tag{1}$$

in which P_i is the transmitted optical power of the ith LED, A is the PD area, β is the PD responsivity, $h = L^z - z$ is the height difference between the receiver and LEDs, m (m') is the Lambertian radiation pattern order of the LED (PD) and $d_{n,i}$ is the distance between the nth PD and the ith LED. Note that the irradiance angle and incidence angle are assumed to be the same in (1) as the PDs (LEDs) are facing up (down). The RSS of these components obtained by the N PDs from the M LEDs can be represented by a vector.

$$\boldsymbol{Rec} = \left[S_1^{f_1}, \ldots, S_1^{f_M}, \ldots, S_N^{f_1}, \ldots, S_N^{f_M} \right] \tag{2}$$

According to the location of the LEDs and PDs, we have:

$$\left(x_n - L_i^x \right)^2 + \left(y_n - L_i^y \right)^2 + h^2 = d_{n,i}^2 \tag{3}$$

2.1. Basic 3D Positioning Model

We first introduce a basic 3D positioning model, which is also referred to as benchmark later to show the accuracy improvement brought by the proposed reinforcement learning methods. Unlike the 2D VLP, the height of the receiver z, which equals to $L^z - h$, is unknown in the 3D VLP system and needs to be estimated. According to (1) and (3), the relationship between h and $S_n^{f_i}$ can be written as:

$$h \le dn, i = \left[\frac{C}{S_n^{f_i}} \left(\frac{h}{dn, i} \right)^{2(m+m')} \right]^{1/4} \le \left[\frac{C}{S_n^{f_i}} \right]^{1/4} \tag{4}$$

where $C = \frac{(P_i)^2(m+1)^2 A^2 \beta^2}{4\pi^2}$.

According to Equation (4), h is no more than the minimum value of $d_{n,i}$. We denote a coarse estimation of h as h_0, which equals d_{min} (i.e., the minimum value of the rightest term in Equation (4) among all possible combinations of N PDs and M LEDs) and is expressed as:

$$h0 = d_{min} = \text{mininum} \left[\frac{C}{S_n^{f_i}} \right]^{1/4} \forall n \in [1, N], \forall i \in [1, M] \tag{5}$$

With h_0, the 2D coordinates on the horizontal plane of the N PDs can be estimated by the conventional trilateration method using (1) and (3). Specifically, we estimate the nth PD's 2D coordinates by solving the following equations:

$$\begin{cases} 2x_n(L_b^x - L_a^x) + 2y_n(L_b^y - L_a^y) = d_{n,a}^2 - d_{n,b}^2 + (L_b^x)^2 - (L_a^x)^2 + (L_b^y)^2 - (L_a^y)^2 \\ 2x_n(L_c^x - L_a^x) + 2y_n(L_c^y - L_a^y) = d_{n,a}^2 - d_{n,c}^2 + (L_c^x)^2 - (L_a^x)^2 + (L_c^y)^2 - (L_a^y)^2 \end{cases} \tag{6}$$

where $a/b/c$ are the indexes of three different LEDs. As there are M LEDs on the ceiling, C_M^3 different pairs of equations can be established [30]. The output of trilateration is obtained by averaging these estimations to mitigate the impact of noise. Our positioning target is the coordinate of the center of the receiver. Assuming the PDs locate symmetrically at the corners of the receiver, the receiver position is obtained by averaging the estimated locations of the N PDs. The above 3D VLP system is referred to as the benchmark, whose output is $(x^0, y^0, z^0 = L^z - h_0)$ (see Benchmark Output in Figure 1).

2.2. Reinforcement Learning To Enhance 3D Positioning Accuracy

h_0 derived from (5) is a coarse estimation of the actual height h. The difference between h_0 and h may not be minor, particularly when the PD is not right below any LED. Since the estimation of the other two coordinates requires the height information as an input, the coarse estimation of h causes error propagation in the benchmark, which results in low positioning accuracy in all three dimensions. Inspired by our previous study [29] that RL can offer high tolerance to inaccurate h in the 2D VLP system, we propose to use RL to improve the positioning accuracy of the 3D VLP system.

The RL mechanism is shown in Figure 2, in which the *Agent* learns knowledge in the action-evaluation *Environment* and improves the *Action* by adapting to the *Environment* [32]. In the 3D VLP system, if the RSS and height are free from the impact of noise in RSS or height estimation error, we can get the exact 3D coordinates by using trilateration. Therefore, the *Environment* to be learned in the 3D VLP system is the error in RSS measurement and the height estimation (see the red box in Figure 2). In other words, the aim of RL is to learn and compensate for the above errors contained in the *Environment* to get a better estimation of the receiver position. As we have multiple PDs available at the receiver, the relative distances between PDs are fixed and can be used to assess the positioning error for reward calculation in RL. The relative distance error vector $\mathbf{E}_{dis}$ is used by the *Agent* to evaluate the *State* of *Environment*, which is defined as:

$$\mathbf{E}_{dis} = \left\{ dis_{(i,j)} - \hat{dis}_{(i,j)} \,\middle|\, i \neq j; i, j = 1, 2.., N \right\} \tag{7}$$

The $dis_{(i,j)}(\hat{dis}_{(i.j)})$ in Equation (7) denotes the real (calculated) distance between the i-th and j-th PDs. The $dis_{(1, 2)}$ of a four-PD receiver is shown in Figure 3 as an example. The *State* and *Reward* in the interaction between the *Agent* and *Environment* are defined as the maximum and average value of $\mathbf{E}_{dis}$, respectively:

$$State = \begin{cases} i, \text{if } \alpha_{i-1} < \max(\mathbf{E}_{dis}) \leq \alpha_i \text{ for } 1 \leq i < G \\ G, \text{if} \max(\mathbf{E}_{dis}) \geq \alpha_{G-1} \end{cases}, \tag{8}$$

$$Reward = \begin{cases} \frac{K-i}{K-1}*100, \text{ if } r_{i-1} < \text{average}(\mathbf{E}_{dis}) \leq r_i \text{ for } 1 \leq i < K \\ 0, \text{if average}(\mathbf{E}_{dis}) \geq r_{K-1} \end{cases}, \tag{9}$$

where $(\alpha_0, \alpha_1, \ldots, \alpha_{G-1})$ and $(r_0, r_1, \ldots, r_{K-1})$ are pre-determined constants based on accuracy requirements, G and K are the numbers of possible values for the *State* and *Reward*, respectively. The learning process in RL uses an action-evaluation strategy, where the consequences of actions (i.e., *Reward*) is used as the metric to help find the optimal action at a certain *State* of *Environment*. If the current *State* is not the target state (e.g., 1 in our study), the *Agent* takes an action to adjust the RSS and height coordinate.

There are different ways to conduct 3D positioning incorporating the RL. Pseudocode 1 shows the pseudocode for two methods with different height update strategies, namely RL_1 and RL_2. The RL_1 is used in Method 1 that adjusts the RSS without changing h except for the last action in learning (i.e., h is fixed to be h_0 when adjusting the RSS and only gets updated after the final RSS is obtained), while the RL_2 is used in Method 2 that adjusts RSS and h sequentially in each action. Specifically, in Method 2, (x_n^{new}, y_n^{new}) is obtained by using the updated RSS and $\hat{d}_{n,i} = (Ch^{2(m+m')}/S_n^{f_i})^{\frac{1}{2(2+m+m')}}$ based on trilateration in Equation (3), and then height difference is updated as $\hat{h}$ by averaging the N height differences between each LED and the receiver's plane, which can be expressed as:

$$\hat{h} = \frac{1}{N}\sum_{n=1}^{N} \sqrt{\left(Ch^{2(m+m')}/S_n^{f_i}\right)^{\frac{1}{2+m+m'}} - \left(x_n^{new} - L_i^x\right)^2 - \left(y_n^{new} - L_i^y\right)^2} \tag{10}$$

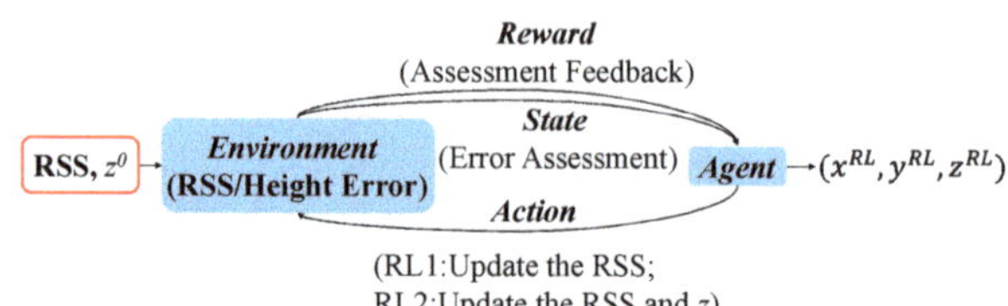

Figure 2. Schematic diagram of the reinforcement learning mechanism in the 3D VLP system.

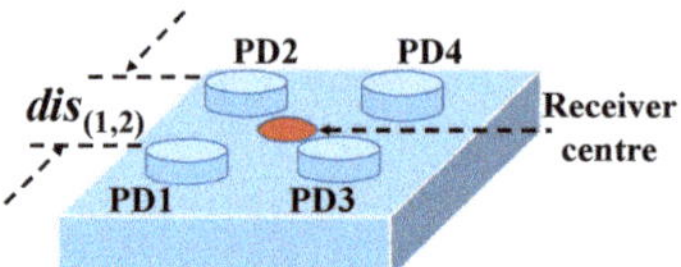

Figure 3. Receiver structure.

Pseudocode 1: Pseudocode for Method 1 and Method 2

---

1.   **Input:** the RSS vector **Rec**, the initial receiver height z^0
2.   **Output:** the refined RSS vector $\mathbf{Rec}^{RL}$, the 3D coordinates of the receiver (x^{RL}, y^{RL}, z^{RL})
3.   Calculate the coordinate of the nth PDs (x_n, y_n) with **Rec** and z^0 using trilateration
4.   Obtain $\mathbf{E}_{dis}$ using (7)
5.   Obtain the current *State* and *Reward*
6. $k = 0$
7.   **for** $k<$ the maximum number of iterations **do**
8. $k \leftarrow k+1$
9.   **for** *State* does not reach the target state and *Reward* is not less than the last one **do**

10.   Obtain $2M\times N$ new **Rec** for $2M\times N$ actions (i.e., increase/decrease one of the $M\times N$ received signal strengths by a pre-defined step in one action)
11.   **for** each new **Rec do**
12.   Calculate the new coordinates of the nth PDs
13.   Obtain $\mathbf{E}_{disr}$ using (7)
14.   Update *State* and *Reward*
15.   **end for**
16.   Choose the *Action* with maximum *Reward*
17.   Obtain $\mathbf{Rec}_{new}$ and update $\mathbf{Rec} \leftarrow \mathbf{Rec}_{new}$

           For Method 1 (Update the RSS)

18.   Calculate the new coordinates of the nth PDs (x_n^{new}, y_n^{new})
19.   Obtain $\hat{h}$ by (10) and update $z \leftarrow \hat{z} = L^z - \hat{h}$

           For Method 2 (Update the RSS and z)

20.   **end for**
21.   **end for**
22.   Obtain finally $\mathbf{Rec}^{RL}$
23.   Update z ($z \leftarrow z^0$ in RL1 and $z \leftarrow \hat{z}$ in RL2)
24.   Calculate the coordinate of the nth PDs (x_n^{RL}, y_n^{RL}) with $\mathbf{Rec}^{RL}$ and z
25.   Obtain h_{RL} by (10) and update $z^{RL} \leftarrow L^z - h_{RL}$
26.   Obtain the 3D coordinates of the receiver (x^{RL}, y^{RL}, z^{RL}) by averaging the coordinate of PDs

---

* z^{RL} corresponds to z^{RL1} and z^{RL2} for Method 1 and Method 2, respectively.

For the RSS adjustment in the RL_1/RL_2, each time one element of the RSS vector ***Rec*** is increased or decreased by step which is a minimum step to adjust the RSS values. After taking an action that modifies ***Rec*** (in RL_1 or RL_2) and h (in RL_2), the 3D coordinates of all PDs are obtained via trilateration and used to calculate its *Reward* based on a new $\mathbf{E}_{dis}$ according to Equation (9). The *Agent* chooses the *Action* with the maximum *Reward*, and update the *State* according to Equation (8).

Both methods continue the learning process until the target state or the maximum number of iterations. After learning, the estimated 3D coordinates of PDs after the last action in RL are saved. The receiver's 3D coordinates (i.e., $(x^{RL1}, y^{RL1}, z^{RL1}=L^z-h_{RL1})$ in Method 1 and $(x^{RL2}, y^{RL2}, z^{RL2}=L^z-h_{RL2})$ in Method 2) are obtained by averaging the coordinates of PDs and used as the final outputs (see Method 1 Output and Method 2 Output in Figure 1).

It is worth noting that the two methods concentrate on positioning accuracy improvement in the horizontal plane and height, respectively. In the RL_1, the learning process only puts the efforts to optimize the X and Y coordinates in the horizontal plane, while the RL_2 does one-step refinement for both the height and RSS in each action. It is also shown in the results (see Section 3), the two methods cannot achieve positioning accuracy improvement in all three dimensions simultaneously. Therefore, we combine the RL_1 and the RL_2, which is referred to as Method 3. Since our previous research in [29] shows that reinforcement learning can tolerate the inaccuracy of h to some extent, in Method 3 we use the RL_2 to update h and RSS, which are followed by the RL_1 to update the X and Y coordinates. Finally, the height estimation is refined according to Equation (10), and $(x^{RL3}, y^{RL3}, z^{RL3}=L^z-h_{RL3})$ is obtained (see Method 3 Output in Figure 1). The pseudocode for Method 3 is shown in Pseudocode 2.

Pseudocode 2: Pseudocode for Method 3

---

1. Input: the RSS vector ***Rec***
2. Output: Coordinate of the receiver $\left(x^{RL3}, y^{RL3}, z^{RL3}\right)$.
3. Estimate h_0 with (5) and $z^0=L^z-h_0$
4. Run RL_2 to obtain Rec^{RL} and $\hat{z}$
5. Update $Rec \leftarrow Rec^{RL}, z^0 \leftarrow \hat{z}$
6. Run RL_1 to obtain the 3D coordinate of the receiver $\left(x^{RL3}, y^{RL3}, z^0\right)$
7. Refine height z^{RL3}
8. Obtain the final 3D coordinates of the receiver $\left(x^{RL3}, y^{RL3}, z^{RL3}\right)$

To better illustrate the RL processes in different 3D VLP methods, Table 1 summarizes the features of the three proposed methods. The RL-based 2D VLP method (i.e., PWRL in [29]) is also listed for comparison.

Table 1. Summary of different RL-based VLP methods.

Method RL Element	3D VLP Method 1	3D VLP Method 2	3D VLP Method 3	2D VLP PWRL [29]
Input	(1) Measured RSS and (2) height estimated based on the basic 3D positioning model			(1) Measured RSS and (2) exact height
Environment	Errors in RSS measurement and height estimation			RSS error
Action	RSS adjustment under an estimated height(RL_1)	RSS and height adjustments (RL_2)	RSS adjustment (in both RL_1 and RL_2) and height adjustment (in RL_2)	Only RSS adjustment, where height is known.
State	Determined by the relative distance error with (8)			
Reward	Determined by the relative distance error with (9)			

3. Experiment Investigation

3.1. Experimental Setup

The performance of the proposed 3D VLP methods is investigated experimentally. Figure 1 shows the experimental setup. There are four LEDs (Cree CXA2435) on the ceiling with coordinates of (24.2, 19.8, 218.9), (83.5, 19.7, 218.9), (22.7, 78.1, 218.9), (82.6, 77.8, 218.9) in centimeter (cm), respectively. Four sinusoidal signals of frequency (400/500/600/700 kHz) from four signal generators are amplified and then combined with direct current (DC) signals via Bias-Tees (ZFBT-4R2GW+) to drive the four LEDs, respectively. As shown in Figure 3, the receiver consists of four PDs (PDA100A2) on the four corners. To ensure that the signal from all four LEDs can be received by the PD (field of view: ~60°) in the 120 cm × 120 cm area, the height difference between the PD and the LED of our test space should be larger than 71 cm. To investigate the impact of receiver size on the performance of the proposed 3D VLP methods, the distance between adjacent PDs is adjusted (i.e., $dis_{(1,2)}$ = 10/20/30/40 cm). In order to get ground truth locations of PDs and LEDs, we divide the area of a solid aluminium plate into many 10 cm × 10 cm grids with a ruler/tape measure which has the resolution of 1 mm and use the lower left side as the origin. The PD is mounted with an optical mounting post on a base which is moved on the grid to change the 2D coordinates on horizontal planes (see Figure 4a). The height of the PD is adjusted by changing the length of the optical mounting post on the base, and is measured manually with a ruler. The horizontal and height coordinates of LEDs are determined by finding their projections on the solid aluminium plate and their distance to this plate with the help of a plumb bob (see Figure 4b,c). To lower the measurement error, the averaged value of multiple measurements is used as ground truth locations. We take measurements at four test planes of different heights with 20 cm spacing, whose Z coordinates are 18.95/38.95/58.95/78.95 cm, corresponding to 199.95/ 179.95/159.95/139.95 cm for h, respectively. The height difference between the receiver and the ceiling in the testbed is about [140, 200]-cm, which emulates the cases of positioning a hand-held device in a typical office environment. Note that the tilt of a hand-held device could severely affect the positioning accuracy as Equation (1) no longer holds. As the average elbow height for a mixed male/female human population is 104.14 cm when he/she stands up [33], this offers about ± 30 cm margin for a room with a ceiling height of 270 cm. In case the height difference is larger, a stronger light source is needed to guarantee a reasonable signal-to-noise ratio (SNR) for high accuracy positioning [30]. For each test plane, four PDs are adjusted to the same height, and samples are taken at 49 uniformly distributed locations in the 120 cm × 120 cm area. The RSS at the receiver is measured using a spectrum analyzer (8593E, Agilent, Elgin, IL, USA) with a sweep time of 30 ms and averaged over 10 measurements. For example, the measured RSS in the center of Plane 4 are 0.354-µW, 0.292-µW, 0.309-µW, 0.319-µW for the sine wave signals from the four LEDs, respectively. For a practical receiver of small form factor, discrete Fourier transform of the temporal samples from an analog-to-digital converter can be conducted to measure the signal strength at different frequencies. The detailed parameters of the experimental setup are listed in Table 2.

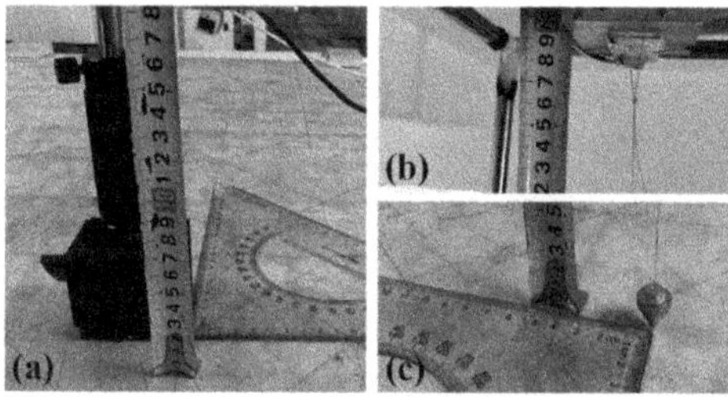

Figure 4. Measurements of the ground truth locations of (a) PD, (b) and (c) LED.

Table 2. Experimental parameters.

Parameter	Value
Space size(length × width × height)	120 × 120 × 220 (cm)
Coordinates of LED1/LED2/LED3/LED4	(24.2, 19.8, 218.9)/ (83.5, 19.7, 218.9)/ (22.7, 78.1, 218.9)/ (82.6, 77.8, 218.9) (cm)
f1/f2/f3/f4	400/500/600/700 (kHz)
LED voltage	18.0 (V)
LED current	0.32 (A)
Lambertian order of LED (*m*)	1.78
Lambertian order of PD (*m'*)	3.56
Distance between PD1 and PD2 ($dis_{(1,2)}$)	10/20/30/40 cm
Heights of Plane 1/2/3/4	18.95/38.95/58.95/78.95 (cm)
Height difference between receiver 1/2/3/4 to LEDs (*h*)	199.95/179.95/159.95/139.95 (cm)

To balance running time and positioning accuracy, we set $_\Delta step$ to 0.1 μw to adjust *Rec* in each action during the learning process. $K = G = 5$, $(\alpha_0, \alpha_1, \alpha_2, \alpha_3, \alpha_4) = (0, 0.2, 0.5, 1, 2)$ in cm, and $(r_0, r_1, r_2, r_3, r_4) = (0, 0.05, 0.125, 0.25, 0.5)$ in cm. In general, the accuracy performance is improved when the number of iterations increases and exhibits a trend of convergence when the number of iterations exceeds a certain value. The number of iterations shall not be too small to achieve the state of convergence. On the other hand, since the processing time and computational complexity of the algorithm increase with a larger number of iterations, the number of iterations shall not be too large. Therefore, the maximum allowable number of iterations in RL based methods is set to 1000 empirically in this experiment to balance the complexity and positioning accuracy.

3.2. Performance Evaluation

We run Method 1/2/3 off-line with MatLab (MathWorks, Natick, MA, USA) on a desktop computer (i5 processor @2.29 GHz (Intel, Santa Clara, CA, USA) with 16 GB RAM) and the measured average processing time is 0.96/0.44/0.69-s, respectively. Figure 5 shows the spatial distribution of 3D positioning error for four different positioning methods (i.e., the benchmark and methods 1/2/3) when $dis_{(1,2)}$ equals to 40 cm. The 3D/2D positioning errors are the Euclidean distance between the real coordinates and the calculated coordinates of the receiver in the 3D/2D space, respectively. To illustrate the 3D positioning accuracy intuitively, we take the actual position of the sampling point as the center of the sphere and the 3D positioning error as the radius of the sphere. The radius r_{sphere} is defined as:

$$r_{sphere} = \sqrt{(x - x_{real})^2 + (y - y_{real})^2 + (z - z_{real})^2}, \tag{11}$$

where (x, y, z) denotes the output of the positioning algorithms and $(x_{real}, y_{real}, z_{real})$ is the real coordinate of the receiver. The non-uniform distribution of errors is observed, which is the interplay of there location-dependent factors: (a) SNR which is higher at the center of test plane, (b) inaccurate a priori information about the VLP system (e.g., m and m' in (1)) that may cause significant overestimation or underestimation of the distance between PD and LED, and (c) the error in approximating the actual height difference with h_0 in Equation (5) which varies for different incidence/irradiance angles of the PD-LED pair used in the calculation of h_0.

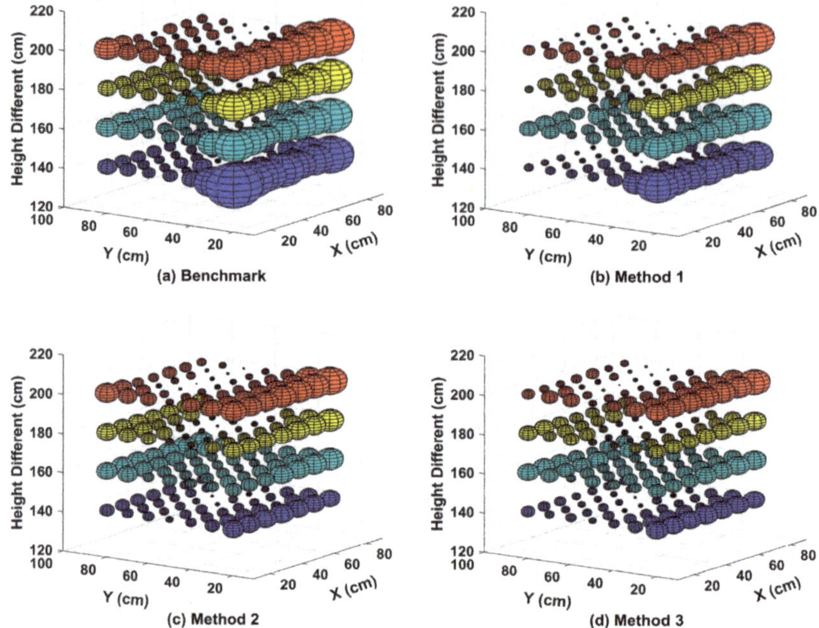

Figure 5. Spatial distribution of the 3D positioning error at different heights in the case of $dis_{(1,2)} = 40$ cm for the (**a**) Benchmark, (**b**) Method 1, (**c**) Method 2, and (**d**) Method 3.

At the edges of test planes where the SNR is lower, the 3D positioning error is larger than that in the central of test planes with higher SNR. If the overestimation of the distance between the PD and LED happens (e.g., a result due to factor (b)), the approximation error in Equation (5) will be larger. For example, we find that the Lambertian model for LED1/LED2 with the parameters in Table 2 causes overestimation of the distance between LED and PD. This leads to significant larger positioning errors in the region with smaller Y which uses LED1/LED2 to calculate h_0. In general, all three RL based methods achieve higher 3D positioning accuracy than the benchmark in the test planes. Method 2 can reduce the error of some points to very small (e.g., test points on the left half of Figure 5c). However, the positioning error with Method 1 is more uniformly distributed in some planes (e.g., $h = 139.95/159.95$ cm in Figure 5b,c). Regardless of the height of the receiver's plane, Method 3 offers the best performance among the four methods.

To further analyze the impact of RL on positioning errors in different dimensions, Figure 6a–c give the cumulative distribution function (CDF) of height/2D/3D error, respectively. Here, 2D error represents the error in the horizontal plane. Plane 2 (i.e., h = 179.95 cm) is used as an example in Figure 6. As shown in Figure 6a–c, all three RL based methods can reduce the height/3D positioning error. For the height dimension, the improvement in Method 3 is most significant, which can reduce the height error from ~5.4 cm to ~3.5 cm for 90% of the test points. Thanks to the additional height update procedure, Method 2 outperforms Method 1 in terms of height estimation accuracy. As shown in Figure 6b, Methods 1 and 3 perform similarly and reduce the 2D positioning error significantly when compared with the Benchmark. For Method 2, though more points are having lower positioning error when compared with the Benchmark, the number of points with larger positioning error also increases. For example, the ratios of points with 2D positioning error of ≤1.76 cm (≥3.0 cm) are 71% and 57% (25% and 17%) for Method 2 and the Benchmark, respectively. This is consistent with the enhanced non-uniformity by Method 2 shown in Figure 5c. In Figure 6c, the 3D positioning error of

90% test points with the Benchmark is less than ~5.4 cm, which can be reduced to less than ~4.9 cm, ~4.0 cm, and ~3.6 cm by Methods 1–3, respectively. As Method 3 exhibits superior performance in both the height dimension and the XY plane, it offers the best 3D positioning performance among the four tested algorithms.

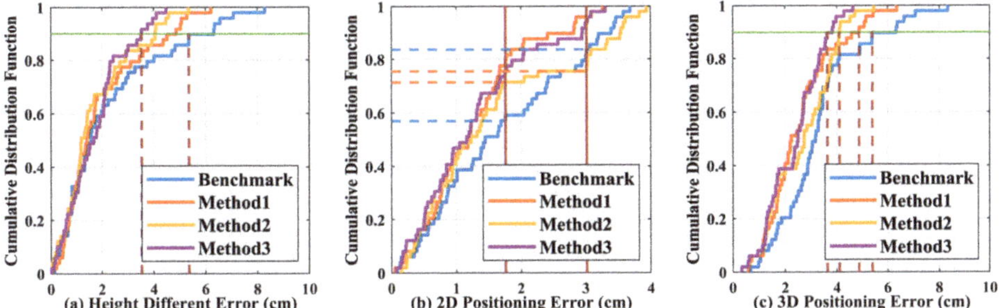

Figure 6. The cumulative distribution function of (**a**) height, (**b**) 2D, and (**c**) 3D positioning errors at Plane 2 in the case of $dis_{(1,2)}$ = 40 cm.

Figure 6 implies that Method 2 outperforms Method 1 in the height dimension, while Method 1 outperforms Method 2 in the horizontal plane. Method 3 inherits the advantages of Method 1 and Method 2, performing the best in all dimensions. The performance superiority of the different RL based methods at different dimensions can be attributed to their unique learning mechanisms (see Figure 2 and Pseudocodes 1 and 2). The RL_1 focuses on optimization of the 2D positioning error, and updates the height estimation only at the end of the learning process, while the RL_2 updates the height estimation in each action, which improves the height estimation accuracy but no further optimization in the horizontal plane. For Method 3, it first uses RL_2 to get a better estimation of the height and then uses RL_1 to optimize the rest two coordinates (see Pseudocode 2). For the CDF, the tested error is a continuous random variable. As we keep each measured point as an individual test, there are always some steps in the CDF curves. As shown in Figure 6, we always give the upper bound of test errors for the proposed RL based algorithms (i.e., Methods 1–3) but the lower bound of test errors for the Benchmark. It is a conservative way to show the benefits brought by the RL. More test points might help to estimate more accurate improvement but would not make the concluding results not true.

Figure 7a shows the mean 3D positioning error obtained by the Benchmark and Method 3 for $dis_{(1,2)}$ = 10/20/30/40 cm at different heights. 80% confidence intervals of the positioning error are also given in Figure 7a (i.e., the vertical bars). The improvement of 3D positioning accuracy with RL is obvious. The upper bounds of Method 3 are even smaller than the lower bounds of the Benchmark. Under different distances between adjacent PDs, Method 3 obtains a mean 3D positioning error below 3.2 cm regardless of the size of the receiver. The results also indicate that the performance of the two methods varies randomly in small ranges with respect to the height of test plane. Figure 7b shows the mean 3D positioning error obtained by the Benchmark and Method 3 for $dis_{(1,2)}$ = 10/20/30/40 cm in the entire test space. The average 3D positioning errors with different receiver sizes are within [2.51, 2.69] cm and [3.15, 4.02] cm for Method 3 and the Benchmark, respectively, revealing an obvious reduction of the average 3D positioning error by at least 20%. Moreover, it also clearly indicates that the positioning performance is more stable (i.e., less variation of positioning errors) when the RL is implemented.

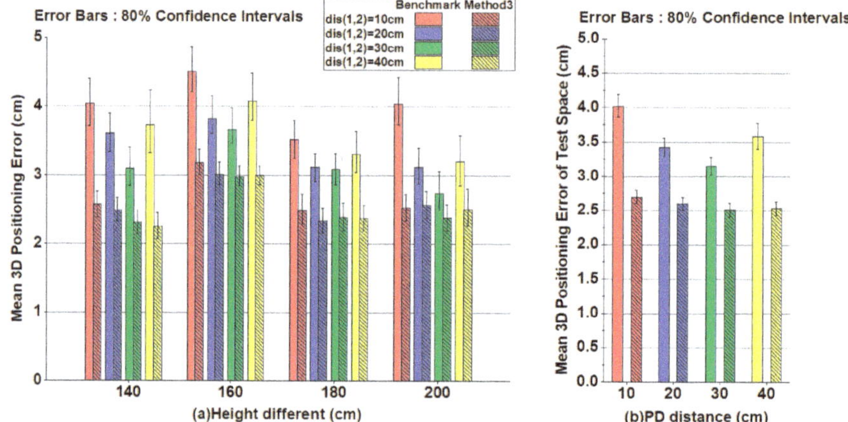

Figure 7. (**a**) Mean 3D positioning error at different heights for $dis_{(1,2)}$ = 10/20/30/40 cm with Benchmark/Method 3. (**b**) Mean 3D positioning error in the test space for $dis_{(1,2)}$ = 10/20/30/40 cm with Benchmark/Method 3.

4. Conclusions

A 3D indoor VLP system with reinforcement learning to enhance the positioning accuracy is proposed and experimentally investigated. The three proposed RL based methods share the *Agent-Environment* interaction framework with properly defined *State/Action/Reward*, but employ different height update strategies. The experimental results show that thanks to the learning process, all three RL based positioning methods outperform the Benchmark in terms of 3D positioning accuracy. The results also verify that Method 1 (Method 2) with RL_1 (RL_2) offers a significant improvement in the horizontal plane (height dimension) over the Benchmark. By combining RL_1 and RL_2, Method 3 offers the highest positioning accuracy not only in the 3D space but also in the height dimension and the horizontal plane, respectively. For the test planes with height difference from 140 cm to 200 cm, the mean 3D positioning error has been significantly improved (>20%) by Method 3 compared with the Benchmark. Moreover, the RL also reduces the variation of the 3D position error compared to the Benchmark with receivers of different sizes.

Author Contributions: Conceptualization, X.H. and J.C.; Data curation, W.Z. and X.C.; Funding acquisition, X.H. and J.C.; Writing—original draft, Z.Z. and H.C.; Writing—review & editing, X.H. and J.C. All authors have read and agreed to the published version of the manuscript.

Funding: This work is supported by the Swedish Foundation for Strategic Research, the Swedish Research Council, STINT joint China-Sweden mobility program, the National Natural Science Foundation of China (NSFC) (61605047, 61671212, 61550110240) and Golden Seed Project of South China Normal University (20HDKC04).

Conflicts of Interest: The authors declare no conflict of interest.

References

1. Ling, R.W.C.; Gupta, A.; Vashistha, A.; Sharma, M.; Law, C.L. High precision UWB-IR indoor positioning system for IoT applications. In Proceedings of the 2018 IEEE 4th World Forum on Internet of Things (WF-IoT), Singapore, 5–8 February 2018; pp. 135–139.
2. Sthapit, P.; Gang, H.S.; Pyun, J.Y. Bluetooth based indoor positioning using machine learning algorithms. In Proceedings of the 2018 IEEE International Conference on Consumer Electronics—Asia (ICCE-Asia), Jeju, Korea, 24–26 June 2018; pp. 206–212.

3. Andrushchak, V.; Maksymyuk, T.; Klymash, M.; Ageyev, D. Development of the iBeacon's positioning algorithm for indoor scenarios. In Proceedings of the 2018 International Scientific-Practical Conference Problems of Infocommunications, Science and Technology (PIC S&T), Kharkiv, Ukraine, 9–12 October 2018; pp. 741–744.

4. Tang, S.; Tok, B.; Hanneghan, M. Passive indoor positioning system (PIPS) using near field communication (NFC) technology. In Proceedings of the 2015 International Conference on Developments of E-Systems Engineering (DeSE), Dubai, UAE, 12–15 December 2015; pp. 150–155.

5. Ultra-Wideband Technology for Indoor Positioning-by Infsoft. Available online: https://www.infsoft.com/technology/positioning-technologies/ultra-wideband (accessed on 30 October 2020).

6. Soner, B.; Ergen, S.C. Vehicular Visible Light Positioning with a Single Receiver. In Proceedings of the 2019 PIMRC, Istanbul, Turkey, 8–11 September 2019; p. 8904408.

7. Rahman, A.B.M.M.; Li, T.; Wang, Y. Recent advances in indoor localization via visible light: A survey. *Sensors* **2020**, *20*, 1382. [CrossRef] [PubMed]

8. Gu, W.; Aminikashani, M.; Kavehrad, M. Indoor visible light positioning system with multipath reflection analysis. In Proceedings of the 2016 IEEE International Conference on Consumer Electronics (ICCE), Las Vegas, NV, USA, 9–11 January 2016; pp. 89–92.

9. Wang, Q.; Luo, H.; Men, A.; Zhao, F.; Gao, X. Light positioning: A high-accuracy visible light indoor positioning system based on attitude identification and propagation model. *Int. J. Distrib. Sens. Netw.* **2018**, *14*, 1550147718758263. [CrossRef]

10. Komine, T.; Nakagawa, M. Fundamental analysis for visible-light communication system using LED lights. *IEEE Trans. Consum. Electron.* **2004**, *50*, 100–107. [CrossRef]

11. Lee, K.; Park, H.; Barry, J.R. Indoor channel characteristics for visible light communications. *IEEE Commun. Lett.* **2011**, *15*, 217–219. [CrossRef]

12. Yasir, M.; Ho, S.W.; Vellambi, B.N. Indoor positioning system using visible light and accelerometer. *J. Lightwave Technol.* **2014**, *32*, 3306–3316. [CrossRef]

13. Li, Q.L.; Wang, J.Y.; Huang, T.; Wang, Y. Three-dimensional indoor visible light positioning system with a single transmitter and a single tilted receiver. *Opt. Eng.* **2016**, *55*, 106103. [CrossRef]

14. Lam, E.W.; Little, T.D.C. Indoor 3D Location with Low-Cost LiFi Components. In Proceedings of the 2019 Global LIFI Congress (GLC), Paris, France, 12–13 June 2019; pp. 1–6.

15. Lam, E.W.; Little, T.D.C. Resolving height uncertainty in indoor visible light positioning using a steerable laser. In Proceedings of the 2018 IEEE International Conference on Communications Workshops, KansasCity, MO, USA, 20–24 May 2018; pp. 1–6.

16. Plets, D.; Almadani, Y.; Bastiaens, S.; Ljza, M.; Martens, L.; Joseph, W. Efficient 3D trilateration algorithm for visible light positioning. *J. Opt.* **2019**, *21*, 05LT01. [CrossRef]

17. Plets, D.; Bastiaens, S.; Ijza, M.; Almadani, Y.; Martens, L.; Reas, W.; Stevens, N.; Joseph, W. Three-dimensional visible light positioning: An Experimental Assessment of the Importance of the LEDs' Locations. In Proceedings of the 2019 International Conference on Indoor Positioning and Indoor Navigation (IPIN), Pisa, Italy, 30 September–3 October 2019; pp. 1–6.

18. Yang, S.H.; Kim, H.S.; Son, Y.H.; Han, S.K. Three-dimensional visible light indoor localization using AOA and RSS with multiple optical receivers. *J. Lightwave Technol.* **2014**, *32*, 2480–2485. [CrossRef]

19. Xu, Y.; Zhao, J.; Shi, J.; Chi, N. Reversed three-dimensional visible light indoor positioning utilizing annular receivers with multi-photodiodes. *Sensors* **2016**, *16*, 1245. [CrossRef] [PubMed]

20. Yasir, M.; Ho, S.W.; Vellambi, B.N. Indoor position tracking using multiple optical receivers. *J. Light. Technol.* **2016**, *34*, 1166–1176. [CrossRef]

21. Han, S.K.; Jeong, E.M.; Kim, D.R.; Son, Y.H.; Yang, S.H.; Kim, H.S. Indoor three-dimensional location estimation based on LED visible light communication. *Electron. Lett.* **2013**, *49*, 54–56.

22. Gu, W.; Kavehard, M.; Aminikashani, M. Three-dimensional indoor light positioning algorithm based on nonlinear estimation. In Proceedings of the SPIE OPTO, San Francisco, CA, USA, 13–18 February 2016; p. 97720V.

23. Liu, P.; Mao, T.; Ma, K.; Chen, J.; Wang, Z. Three-dimensional visible light positioning using regression neural network. In Proceedings of the 2019 15th International Wireless Communications and Mobile Computing Conference (IWCMC), Tangier, Morocco, 24–28 June 2019; pp. 156–160.

24. Alonso-González, I.; Sánchez-Rodríguez, D.; Ley-Bosch, C.; Quintana-Suárez, M. Discrete indoor three-dimensional localization system based on neural networks using visible light communication. *Sensors* **2018**, *18*, 1040. [CrossRef] [PubMed]
25. Zhang, S.; Du, P.; Chen, C.; Zhong, W.D. 3D indoor visible light positioning system using RSS ratio with neural network. In Proceedings of the 2018 23rd Opto-Electronics and Communications Conference (OECC), Jeju, Korea, 2–6 July 2018; pp. 1–2.
26. Jiang, J.; Guan, W.; Chen, Z.; Chen, Y. Indoor high-precision three-dimensional positioning algorithm based on visible light communication and fingerprinting using K-means and random forest. *Opt. Eng.* **2019**, *58*, 1. [CrossRef]
27. Xu, S.; Chen, C.C.; Wu, Y.; Wang, X.; Wei, F. Adaptive residual weighted K-nearest neighbor fingerprint positioning algorithm based on visible light communication. *Sensors* **2020**, *20*, 4432. [CrossRef] [PubMed]
28. Zhang, Z.; Chen, H.; Hong, X.; Chen, J. Accuracy enhancement of indoor visible light positioning using point-wise reinforcement learning. In Proceedings of the 2019 Optical Fiber Communications Conference and Exposition (OFC), San Diego, CA, USA, 3–7 March 2019.
29. Zhang, Z.; Zhu, Y.; Zhu, W.; Chen, H.; Hong, X.; Chen, J. Iterative point-wise reinforcement learning for highly accurate indoor visible light positioning. *Opt. Express* **2019**, *27*, 22161–22172. [CrossRef] [PubMed]
30. Zhang, W.; Chowdhury, M.I.S.; Kavehrad, M. Asynchronous indoor positioning system based on visible light communications. *Opt. Eng.* **2014**, *53*, 045105. [CrossRef]
31. Guo, X.; Shao, S.; Ansari, N.; Khreishah, A. Indoor localization using visible light via fusion of multiple classifiers. *IEEE Photonics J.* **2017**, *9*, 1–16. [CrossRef]
32. Milioris, D. Efficient indoor localization via reinforcement learning. In Proceedings of the 2019 IEEE International Conference on Acoustics, Speech and Signal Processing (ICASSP), Brighton, UK, 12–17 May 2019; pp. 8350–8354.
33. Anthropometry | Sustainable Ergonomics Systems—Ergoweb. Available online: https://ergoweb.com/anthropometry/ (accessed on 30 October 2020).

Publisher's Note: MDPI stays neutral with regard to jurisdictional claims in published maps and institutional affiliations.

 *sensors*

Article

Indoor Localization Based on Infrared Angle of Arrival Sensor Network

Damir Arbula and Sandi Ljubic *

University of Rijeka, Faculty of Engineering, 51000 Rijeka, Croatia; damir.arbula@riteh.hr
* Correspondence: sandi.ljubic@riteh.hr

Received: 8 October 2020; Accepted: 2 November 2020; Published: 4 November 2020

Abstract: Accurate, inexpensive, and reliable real-time indoor localization holds the key to the full potential of the context-aware applications and location-based Internet of Things (IoT) services. State-of-the-art indoor localization systems are coping with the complex non-line-of-sight (NLOS) signal propagation which hinders the use of proven multiangulation and multilateration methods, as well as with prohibitive installation costs, computational demands, and energy requirements. In this paper, we present a novel sensor utilizing low-range infrared (IR) signal in the line-of-sight (LOS) context providing high precision angle-of-arrival (AoA) estimation. The proposed sensor is used in the pragmatic solution to the localization problem that avoids NLOS propagation issues by exploiting the powerful concept of the wireless sensor network (WSN). To demonstrate the proposed solution, we applied it in the challenging context of the supermarket cart navigation. In this specific use case, a proof-of-concept navigation system was implemented with the following components: IR-AoA sensor prototype and the corresponding WSN used for cart localization, server-side application programming interface (API), and client application suite consisting of smartphone and smartwatch applications. The localization performance of the proposed solution was assessed in, altogether, four evaluation procedures, including both empirical and simulation settings. The evaluation outcomes are ranging from centimeter-level accuracy achieved in static-1D context up to 1 m mean localization error obtained for a mobile cart moving at 140 cm/s in a 2D setup. These results show that, for the supermarket context, appropriate localization accuracy can be achieved, along with the real-time navigation support, using readily available IR technology with inexpensive hardware components.

Keywords: infrared sensor; angle of arrival; indoor localization; wireless sensor networks; navigation

---

1. Introduction

In recent years, we have been witnessing a rapid increase in the availability of commercial indoor localization solutions. This is not a surprise as smartphone users are already accustomed to outdoor location-based services. Precise indoor localization is the only significant technological obstacle to extend these services to the area where many users spend most of their time. Indoor localization is, therefore, the Holy Grail problem in ubiquitous computing, context-aware applications, and, specifically, location-based Internet of Things (IoT) services.

The outdoor localization problem is solved by Global Positioning System (GPS), a satellite-based navigation system consisting of a network of 24 satellites placed into orbit. The basis of the outdoor localization method is the combination of the line-of-sight (LOS) radio propagation from the satellite transmitter to the receiver and the fact that it can be predicted and even real-time calibrated using information from the referent stations on the ground. Distance from the satellites with known locations can be precisely estimated based on the time it takes for the signal to reach the receiver; therefore, the receiver can be positioned using the multilateration method.

Unlike in the open spaces, signal propagation indoors is affected by complex interactions with a large number of fixed and moving obstacles through reflection, refraction, and scattering. Hence, it is very difficult or even not possible to predict the signal path and to correlate distance from the receiver with the propagation time, signal strength, or any other signal parameter. As noted in the following subsection, where we give an overview of the commercial systems and research solutions to the indoor localization problem, non-line-of-sight (NLOS) and multipath propagation is the main problem that various authors try to solve or even avoid. For example, Belloni et al. [1] target their application to obstacle-free open indoor spaces since the angle-of-arrival (AoA) method is very sensitive to multipath propagation, resulting in poor localization accuracy. On the other hand, the LOS signal propagation is characterized by the fact that the intensity of the LOS component of the signal is significantly higher than other components. Thus, the signal propagation path between transmitter and receiver can be modeled with a straight line, allowing for precise and accurate transmitter location estimation.

Indoor positioning systems (IPS) face an interesting technical challenge due to the wide variety of promising sensor technologies that can be applied, each one with different pros and cons. Brena et al. [2] provided a helpful systematization of the respective field, by introducing a comprehensive review of the literature that involves the technological perspective of IPS evolution, a classification scheme for different technological approaches, and a presentation of the existing research trends. In the related survey, authors concluded that "there is not yet an overall satisfying solution for the IPS problem", and, while addressing the specific problem of locating merchandise in retail stores, they argued that "not a single technology or combination of technologies is both feasible and satisfying". Shang et al. [3] have also presented a detailed survey in this field, however, they focused on a review of the improvement schemes for indoor mobile location estimation. Among many methods and techniques for enhancing location estimation, they analyzed the possibility of fusing spatial context. Namely, they tackled graph-based motion models of an indoor space, an instance of which we utilized in our research.

The indoor localization field is broad and there are many different solutions but their performance is very difficult to evaluate without the proper context. Therefore, in the following overview, we specifically focus on systems that target or can be used in our showcase application: supermarket aisle-level localization, which itself has been given a lot of attention, specifically cart localization within high shelves surrounded corridors.

1.1. Related Work: Indoor Localization Methods and Solutions

In this subsection, we present current state-of-the-art indoor localization systems, both in the research and development phase, as well as those that are already commercially available. Presented systems are roughly divided into two groups, based on the type of signal used, and one additional group consisting of infrastructure-free systems that require no signal (Table 1).

Table 1. Taxonomy of the existing indoor localization methods.

Infrastructure	RF	WiFi, Bluetooth Low Energy (BLE), Ultra-wideband (UWB), Radio-frequency identification (RFID)
	Light	Visible Light Positioning (VLP), Infrared (IR)
Infrastructure-free		Magnetic, Sensor fusion, OCR

Infrastructure based methods are using either radio frequency (RF) signals or modulated visible light sources to be able to estimate the position and, as such, require specific infrastructure to be installed on-site. Infrastructure, in this sense, consists of highly available WiFi access points (AP) or specialized equipment that has to be installed with specific localization intent, such as radio-frequency identification (RFID) scanners, Bluetooth Low Energy (BLE) beacons, Ultra-wideband (UWB) beacons, or modulated light sources.

1.1.1. Systems Using RF Signal

WiFi is the dominating wireless technology for indoor data transfer and, as such, the WiFi signal is the ideal candidate to exploit for localization purposes. WiFi localization systems are predominantly using Received Signal Strength (RSS) as the principal method for the estimation of the distance between WiFi AP and mobile nodes. Standard indoor NLOS multipath propagation decreases the correlation between distance and RSS to the point where, using estimated distances, only basic proximity-based localization is possible. On the other hand, short range and the abundance of different available WiFi APs are enabling a pattern matching method or fingerprinting to become the de-facto standard localization technique for indoor location-based services on consumer devices [4]. However, this method requires a site survey fingerprint in advance and localization performance is highly sensitive to changes in the environment, i.e., in a supermarket with moving people. The most common algorithm utilized for WiFi fingerprinting is weighted K-nearest neighbors (WKNN), which calculates K-nearest neighboring points to a mobile user. Typical problems associated with the WKNN involve a difference in observed AP sets during offline and online stages, and a possibility for some of the K neighbors to be physically far from the user. Enhancements of the default WKNN have thus been proposed, that change the number of considered neighbors dynamically—either by using RSS-based filtering [5], or a more sophisticated clustering algorithm [6]. Issues and requirements of the WiFi-based localization systems are furthermore alleviated by the incorporation of inertial motion unit (IMU) data readily available on modern smartphone devices and by using the Simultaneous Localization and Mapping (SLAM) technique. Systems based on this technique, such as Apple WiFiSLAM [7], achieve localization accuracy around 2 m. Yang and Shao [8] obtained even more promising results by using multiple antennas on WiFi APs and the combination of distance and AoA estimation along with the capability of filtering NLOS measurements. The authors report localization error from 2.2 m up to 0.5 m by using one or several WiFi access points, respectively. WiDeo system, introduced in Reference [9], represents one of the most encouraging efforts in providing WiFi-based indoor motion tracking. It utilizes specially developed WiFi AP with antenna array, as well as backscatter analysis, i.e., composite reflected signal examination wherein the amplitude, time-of-flight (ToF), and AoA parameters are all estimated. The WiDeo thus provides a possibility to trace subject motions without the need for any accompanying device, with reported median localization accuracy of 0.8 m and motion tracking accuracy of 7 cm. Since WiDeo's accuracy is in line with the localization accuracy of the solution proposed in this paper, an appropriate comparison is given in the Discussion section, highlighting the pros-and-cons for using the related systems in the target supermarket environment.

Systems using Bluetooth Low Energy (BLE) technology are commercialized under different brand names, such as iBeacon by Apple or Eddystone by Google. They consist of a number of beacons with known positions publishing their ID and mobile nodes that are localized through estimated distances from beacons using the multilateration method. This method is somewhat similar to the WiFi RSS method, but a low range warrants smaller cells and lower distance estimation errors. Just like with WiFi, the advantage of this method is its availability on all present-day smartphones, while, at the same time, deployed devices are cheaper, smaller, more portable, and energy-efficient. Nevertheless, it is hard to achieve sub-meter precision, i.e., Faragher and Harle [10] report tracking accuracies of <2.6 m in 95% of the time with a density of one beacon in 30 m^2. Currently, these systems are mostly used for proximity-based localization and point-of-interest services.

Furthermore, BLE and WiFi can be combined. Kriz et al. [11] report sub-meter localization error median in a 52 × 43 m office building equipped with 4 WiFi access points and 17 BLE beacons. The downside of their method is a relatively long scan and measure delay taking from 6 to 10 s to reach stated localization accuracy. There are many commercial systems present relying on WiFi and BLE, such as AisleLabs, and those that are combining BLE, WiFi, and SLAM methods, such as indoo.rs. The latter combination has reported accuracy from 2 to 5 m, depending on the density and placement of beacons.

Ultra-wideband (UWB) is an RF technology for a short-range, high bandwidth communication with a high temporal resolution, resulting in centimeter-level accuracy [12]. Localization systems based on UWB are mostly using time-of-arrival (ToA) and time-difference-of-arrival (TDoA) of RF signals to estimate the distance between the transmitter and the receiver. To be able to use ToA methods they need to perform precise time synchronization of anchor nodes, somewhat similar to GPS satellites. Although there are techniques that mitigate this synchronization challenge [13], this requirement further complicates the overall system. Commercial UWB systems, such as Sewio, utilize many anchor nodes in the LOS range of mobile nodes they are tracking. Although the accuracy of UWB systems is very competitive, the limited range of the anchor nodes in the supermarket configuration, along with their specialized hardware design, and consequently high price, results in prohibitive installation costs.

Radio-frequency identification (RFID) is fairly mature and available technology mainly used for object tagging and identification, but there are also many examples of using RFID in indoor localization scenarios. One of the first usages was the LANDMARC system [14] that consisted of a small number of RFID readers with a high range and a large number of active RFID tags divided into two sets: landmark tags with known locations and mobile tags with unknown locations. Landmark tags were used for continuous calibration providing partial resistance to changes in the environment, thus enabling more accurate mobile tag location estimation. Reported accuracy is from 1 to 2 m, but authors did note several important issues, such as long scan time (7.5 s interval between readings) and inconsistent emitting signal strengths of RFID tags. Ryoo and Das [15] utilized RFID to enable supermarket cart localization. They report a median of localization error limited to 5 cm with a 90-percentile error of 15 cm. This method is using carts equipped with passive RFID tags, while RFID readers are installed directly above the aisle on a fixed height 2 m above the cart. The localization algorithm is based on a distance estimation between the reader and the tag. Distance is estimated using $\Delta\phi/\Delta f$ slope obtained from phase response measurements through different interrogation channels. The problematic aspect of this method is long measurement time, around 400 ms for each channel. Since there is a minimum of 5 channels, it adds up to 2 s during which the cart has to remain stationary in order to estimate distance and location. Other drawbacks of this system include the high cost of multiple RFID readers, each with multiple antenna setup and appropriate cabling.

1.1.2. Systems Using Light Sources

Another signal source that can be exploited for localization purposes is light, either infrared or visible. With light-emitting diode (LED) technology becoming the new standard in ambient illumination there are numerous possibilities to harness its properties, such as Visible Light Communication (VLC) and Visible Light Positioning (VLP) [16]. The basic principle of VLP operation is that each light source serves as a beacon whose modulated radiation can be captured by a light sensor, usually a front-facing smartphone camera. Radiation from each light source is modulated (e.g., by fixed frequency or by transmitting Manchester-encoded data), and can be uniquely identified by a mobile node. The identification of multiple light sources in an image allows the positioning of a smartphone using AoA. There are many examples of VLC and VLP systems both in research and development and in the commercial phase. Kuo et al. [17] present the Luxapose system which, in a laboratory environment, achieves decimeter-level localization accuracy using a high resolution 33 MP smartphone camera and 5 LED beacons. On the other hand, Qiu et al. [18] are using simple and inexpensive external light sensors in a 4.7 m × 8.6 m indoor environment with 12 modulated LEDs, attaining sub-meter precision. Their approach requires a data collection phase similar to the WiFi fingerprinting method. Among commercially available systems, we can highlight those from companies, like ByteLight, GE Lighting, and Philips [19]. Typical drawbacks of VLP systems are considerable computational requirements for real-time image processing and location estimation and high energy demands, as well as high initial installation cost.

Infrared (IR) communication technology is widely adopted, inexpensive and readily available. IR signals are used in many different applications ranging from consumer remote controls to data transfer (IrDA). One of the first indoor localization systems Active Badge [20] was using IR signals. This system was intended for personnel tracking using a set of tags each emitting IR signal with a unique code every 15 s. Signals are picked up by Badge Sensors installed at various rooms inside the building providing room-level accuracy. Badge Sensors were powered and connected to a network using a special 4 wire system using telephone twisted-pairs cable and RS232 data-transfer format.

In more recent research [21], IR beacons are detected using a low-resolution CCD camera fitted with an IR filter on a mobile robot, a method similar to the already mentioned VLP solution [17]. Although the setup seems simple, and only a small number of LEDs in a field-of-view (FoV) is required to achieve decimeter-level precision, the problem is that the beacon signal is not identifiable; thus, their positions are hardcoded. To be able to identify beacons, a large number of modulated LED sources needs to be installed and powered full time. This requires an adequate energy source, either through separate cabling or battery, both options being rather expensive.

The problem of multipath (MP) propagation is the most prominent among the indoor positioning systems based on optical signals. Namely, the receiver in such systems usually senses the line of sight component of the signal, as well as other MP components, due to light reflections and refractions in the indoor environment. Since the received signal components can vary in power strength and phase, the localization accuracy of the underlying system can be significantly reduced.

A model of IR signal reflections on any kind of surface material is proposed in Reference [22] that can be applied to characterize the multipath behavior of optical signals in applications, such as indoor positioning and VLC communications. The respective model is derived according to the experimental measurements on three different materials (terrazzo, foam board, and plasterboard). In Reference [23], authors propose a model to determine the multipath effect in indoor environments when the shape and characteristics of the environment (e.g., reflection features of the materials) are known a priori. The related model can be applied for indoor positioning, irrespectively, of both the underlying system and the utilized measurement type (e.g., RSS, phase of arrival (PoA), differential phase of arrival (DPoA)). For example, when analyzing the MP effect in AoA-based systems, wherein the signal phase information is not relevant, it is necessary to know the signal strength reaching the detector from each element in the environment after a certain number of rebounds. The mentioned model comes with an algorithm that calculates the signal strength in the MP scenario. In recent research [24], a Position Sensitive Device (PSD) sensor was used for experimental testing of MP effects in IR-based indoor positioning. The positioning has been calculated using AoA and PoA techniques, and the errors caused by the MP have been analyzed. The obtained results showed that the MP effects for AoA, unlike for PoA, have little impact on the indoor positioning accuracy.

1.1.3. Infrastructure-Free Systems

Unlike systems that are based on specific installed infrastructure providing RF or light signals and allowing estimation of distance, angle, and consequently position, some systems demand no specific equipment to estimate indoor location.

The first system of that kind uses the fact that the Earth's magnetic field is distorted by structural steel elements in a building and that this distorted field has a certain temporally stable signature that can be mapped. Related methods are somewhat similar to the WiFi fingerprinting, the key difference being that the Earth's magnetic field is stable and undisturbed even by large moving metal objects (i.e., elevator cabin) on distances above 1 m from the magnetic sensor. Chung et al. [25] report accuracy within 1.64 m for 90% of the time using a simple RMS-based nearest neighbor searching algorithm for the localization. On the other hand, they also note that the chance of error increases with the size of the fingerprint map and propose a hybrid solution with WiFi fingerprinting that can complement repeating magnetic signatures and set upper bound on localization error for larger maps. This method is further investigated by Shu et al. [26], along with the more sophisticated augmented particle filter (APF) localization algorithm and IMU-based tracking used to help in the proper timing of the magnetic field measurements. The interesting fact in the context of this paper is that the authors describe the supermarket environment as the most challenging one (others being office building and underground parking garage). Their experimental results verify that description since they report that their system achieves 90 percentile localization accuracy of 8 m in the supermarket environment using the magnetic field alone. Finally, it is worth noticing that although the infrastructure is not required, this method requires mapping the magnetic field which can take significant effort and time. Representative commercial implementation of the magnetic field sensing is one by the brand IndoorAtlas.

Another innovative system that requires no infrastructure is Google Tango Project. Three key features of the Tango Project system are (1) motion tracking, (2) area learning, and (3) depth perception. Tango can be used both to map indoor spaces and to estimate location within by using a standard gyroscope, accelerometers along with the wide-angle camera and depth techniques, such as Structured Light, Time of Flight, and Stereo Vision. The project is still in its research and development phase, and Tango-enabled devices are becoming available on the market only recently. Tango-enabled devices can be used in indoor localization and navigation context, and, currently, besides the unavailability of the hardware, the main obstacle is the power required for computation.

Finally, it is worth mentioning the Monocular localization system [27] that can be used as a complement to Google Tango. This system uses real-time video camera-based optical character recognition (OCR) and building floor plan with a mapping of prominent signs locations, such as store logos above entrances. Using this information, it is possible to estimate location relative to a detected visual cue.

Based on the research review presented above, a comparison of indoor localization systems' main characteristics is summarized in Table 2. Along with the characteristics of different localization methods (typical accuracy, installation costs estimate, energy consumptions, and main drawbacks), we highlight the representative commercial solutions that can be considered as readily available for applying in the supermarket context. By outlining the commercial examples, we point out the fact that some of the largest (and the most influential) companies, such as Apple, Philips, and Google, recognize the importance of indoor localization, and actively contribute in the respective field.

Table 2. Comparison of related indoor localization systems.

Method	Commercial Example	Typical Accuracy	Install. Costs	Energy Cons.	Main Drawbacks
VLP	ByteLight, Philips	50 cm	high	high	high computational requirements (real-time image processing) and installation costs
BLE	iBeacon (Apple)	>2 m	medium	low	low signal range, hard to achieve sub-meter precision
UWB	Sewio	30 cm	high	low	the need for precise time synchronization of anchor nodes, low range, specialized high-priced hardware design
WiFi	WiFiSLAM (Apple)	1–2 m	low	medium	site survey fingerprinting, high sensitivity to changes in the environment
Magnetic	IndoorAtlas	1–2 m	no	medium	magnetic field mapping, error increases with the size of the fingerprinting map
Sensor fusion	Project Tango	N/A	no	high	R&D phase, limited availability
IR AoA	proposed solution	3 cm (static 1D) 20–50 cm (mobile 1D) 40 cm (static 2D) <1 m (mobile 2D)	low	low	low range, IR collision

1.2. The Overview of the Proposed Solution

In this paper, we present a novel infrared (IR) sensor and AoA estimation algorithm relying on low range LOS signal propagation. The sensor is furthermore applied in a novel localization method based on tracking mobile IR transmitters. In order to provide LOS signal sensing throughout the environment and gather measurements from mobile nodes, we exploit the powerful concept of the wireless sensor network (WSN). We believe that this combination has the potential to overcome some of the issues within current state-of-the-art indoor localization systems.

Although the proposed method can be utilized in many different applications, in this paper, we tackle the specific aisle-level cart navigation use case. The environment in this use case is characterized by many narrow corridors, high or moving obstacles, such as shelves or customers. Many current systems fall short in this kind of environment, specifically because of the unpredictable and changing signal propagation. In this context, our showcase system is using WSN nodes which are equipped with IR AoA sensors and distributed above the aisles. WSN measures a signal from the infrared transmitters installed on the carts and delivers those measurements to the localization server, thus enabling real-time cart localization. The key advantages of this system are inexpensive installation and maintenance, and competitive localization precision as demonstrated in conducted experiments.

The related research efforts most often focus exclusively on the design of a specific sensor with an attempt to enhance indoor localization accuracy but without further utilization within a system that would assist the end-user to navigate in the target indoor environment. In other words, the related work often lacks the well-rounded solution built upon the underlying localization technology. In this sense, our contributions are based on the development of all modules required for indoor navigation and their successful integration into the proof-of-concept system. The system targets the supermarket navigation context and involves the following:

- Novel IR AoA sensor, made of inexpensive off-the-shelf components, enabling AoA estimation with an error around $1°$,
- Wireless sensor network, based on the proposed IR AoA sensor, which provides infrastructural support for real-time navigation,
- Localization strategy/method/algorithm, utilizing the proposed WSN and a spatial context (aisle graph), with suitable localization accuracy,
- Supermarket navigation model based on shelves graph and aisles graph,
- Server, API, and client applications suite, demonstrating both the features and the look-and-feel of the proposed system.

2. Materials and Methods

2.1. Angle-of-Arrival Sensor

As with every conventional outdoor navigation system, the integral component of the indoor navigation system is the one used for mobile node localization. The proposed method consists of measuring the strength of the IR signal on the IR phototransistors placed on a specifically constructed sensor and the estimation of the angle-of-arrival of the IR signal from the measurement data. This localization method achieves high accuracy with simple low-cost hardware, while requiring LOS between the IR transmitter and IR sensor. Therefore, the key technical properties are novel sensor design and angle-of-arrival estimation algorithm.

In our research, we opted for IR-based technology, with two main goals in mind: (1) to propose a LOS-based sensor design that would be inexpensive to produce—using cheap off-the-shelf components, and (2) to develop the proof-of-concept solution, utilizing the network of such sensors, that would be fittingly accurate in the target (supermarket) context.

When it comes to the already available sensors that provide AoA measurements, we can outline devices used in the Cricket Compass System [28] working with ultrasound signals, different antenna array systems [29], and rotating laser systems [30]. Furthermore, a PSD (Position Sensitive Device) sensor has been successfully utilized for indoor positioning using AoA techniques, with the obtained localization error below 1 cm [31]. QADA (Quadrant Photodiode Angular Diversity Aperture) sensor also showed to be a part of the promising angle-based localization apparatus, as it was used in an IR indoor positioning system that provides the absolute error of 0.9° in the estimation of the polar angle, and 12 cm of absolute localization error [32].

However, all these mentioned solutions are neither simple nor low-priced nor fully adequate for straightforward installation in the supermarket venue. For example, some of them require appropriate cabling, which we wanted to avoid from the very beginning. Regarding the sensor costs, we can outline PSD and QADA devices that were used in the abovementioned research (Hamamatsu S5991-01 PSD, and QADA receiver QP50-6-18u-TO8) and which hold a price level of USD 180–200 and USD 120, respectively. Following our main idea, we opted for a novel design of a much more affordable AoA sensor.

2.1.1. Design

The inspiration for the creation of a new type of sensor was drawn from the research by Song et al. [33], in which authors introduce a new type of digital camera. The camera has a size of 1 cm in diameter and contains a total of 180 micro-lens oriented in different directions [34].

The key idea of the proposed design is to utilize an array of IR phototransistors placed on the circular rim and directed outwards to detect the angle of arrival (AoA) of the incoming infrared signal. AoA could be estimated using measured data and known specific radiant sensitivity of IR phototransistors.

Initial advantages of this method are the usage of small and inexpensive off-the-shelf components, such as IR phototransistors, along with the readily available dedicated ATtiny45 microcontroller and the 16-channel multiplexer, as well as the fact that the principle of the operation of the hardware part is rather simple. The sensor is controlled via a one-wire protocol that is used to select the appropriate channel on the multiplexer, allowing the measurement of selected IR phototransistor output by the host node AD converter.

The first-generation prototype was our preliminary design of the novel target IR AoA sensor, which showed to be only a "debugging" step in the process of building the final, i.e., the second-generation prototype. Namely, the first-generation prototype used a simple design with through-hole components and was able to estimate AoA with an average error of around 10°. We found this error to be quite large, so we did not consider the related prototype for further work on the localization system. Instead, we tackled the unfitting size and imprecise positioning of IR phototransistors (in the first-generation prototype) by introducing surface mounted (SMT) components and pick-and-place automated assembly. By doing so, we developed the second-generation prototype (shown in Figure 1) with the lower AoA estimation error. We were able to further reduce this error, up to 1°, by utilizing a specific calibration procedure and the corresponding estimation algorithm (described in the following subsections). The second-generation prototype was thus used in our WSN-based localization solution and related experiments.

Figure 1. Infrared (IR) angle-of-arrival (AoA) sensor prototype.

2.1.2. Calibration

The main source of estimation error proved to be the relative radiant sensitivity of each SMT IR phototransistor, which is unique (Figure 2) and significantly different from the ideal characteristic specified in the datasheet [35]. This fact presented an issue since radiant sensitivity characteristic is the principal parameter for enabling AoA estimation.

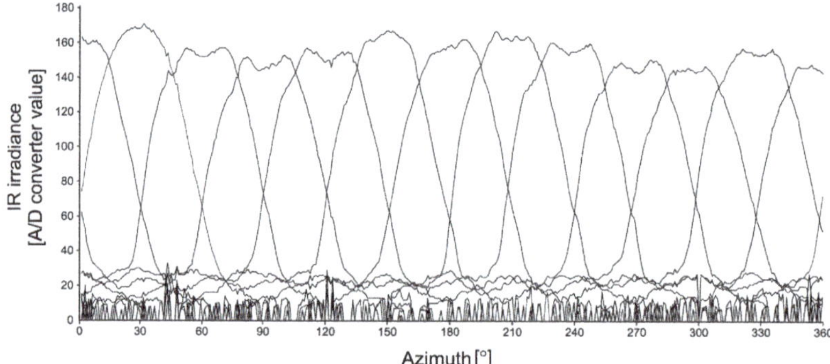

Figure 2. AoA phototransistors' IR irradiance measurements with varying angle of arrival of the incoming IR signal. Each curve corresponds to one of 12 phototransistors on the AoA sensor. Maximal values for each phototransistor are achieved when the transmitter is positioned near the phototransistor axis, i.e., directly in front of the phototransistor. IR irradiance is measured as a voltage drop on resistors serially connected to phototransistors. The phototransistor collector current and the corresponding voltage drop are proportional to the measured irradiance. The voltage is displayed as a 10-bit A/D converter readout.

The solution to the estimation error induced by the unique radiant sensitivity of each phototransistor was to implement an automated calibration platform (Figure 3) and to use it to record true radiant sensitivities for all transistors like the one shown in Figure 2. The central controlling part of the platform is the calibration server implemented as a RaspberryPi computer running the iPython Notebook kernel. Controlling software communicates with sensor nodes via a connected JeeLink sensor node and controls the rotation of the stepper motor via general-purpose input/output (GPIO) connectors and the power amplifier. The test platform is fully manageable from a personal computer using the iPython Notebook client, i.e., any Internet browser.

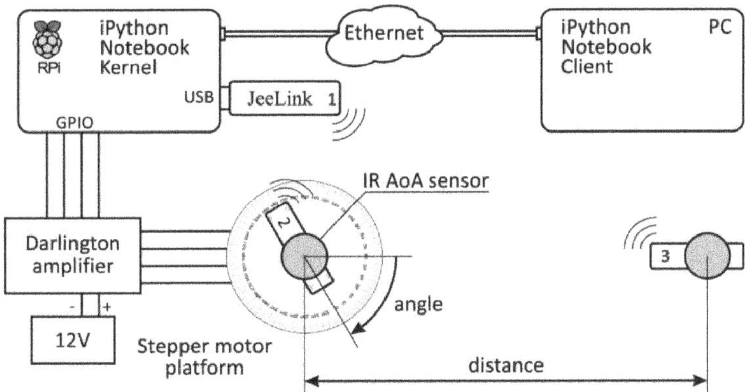

Figure 3. AoA sensor calibration system [36].

The sensor network part of the platform consists of three nodes:

- The JeeLink node, in the scheme labeled as node 1, serves as a gateway: to send commands to nodes 2 and 3, and to receive measured data from node 2.
- Node 2 is mounted on a rotating platform and attached to the sensor being calibrated.
- Node 3 is an infrared transmitter, i.e., it serves as a controlled IR radiation source with known distance and AoA relative to node 2.

For each step of the stepper motor, node 1, attached to the calibration server, triggers a flash of the IR diodes on the node 3 and, at the same time, measurement is taken by each of the 12 IR phototransistors on the sensor attached to the node 2. Measured data is then transmitted back to node 1 and stored on the client computer.

After the measurement has been made, the stepper motor rotates one half-step, or in this case for 0.9° in the clockwise direction, and the procedure repeats for all 360°, i.e., for 400 half-steps. The obtained data consists of a true AoA taken from the known position of the stepper motor and 12×400 IR measurements taken from 12 phototransistors for each of 400 different AoA.

Unlike the typical IR transmission (e.g., with remote control device), wherein the signal is modulated in order to separate it from the ambient light, in our case only a DC signal is used. We did not consider modulating the emitted signal because, in our solution, AoA estimation relies exclusively on the relative strength of the signal received on the sensor's phototransistors.

Possible MP effects were not taken into consideration during the calibration procedure. Namely, we did not experience any unexpected issues in this matter, as long as the sensor or the transmitter was not too close (few centimeters) to some reflective object. In all other cases, the LOS component showed to be a predominant part of the received signal, and, as such, it is de-facto exclusively used for AoA estimation. Our calibration platform was placed in the center of the room, away from the walls and other obstacles, so we can fairly assume that there was no significant MP effect on the sensor calibration.

2.1.3. Estimation Algorithm

To be able to estimate the angle of arrival, the first observation that needs to be made is that the phototransistor readout presents the sum of the IR irradiance from the transmitting node and the ambient. To filter out ambient radiation, the sensor needs to make a measurement at the moment in which all IR transmitters in the range are off. In the following text, all phototransistor readouts are considered to be already filtered, i.e., subtracted by measured ambient radiation.

The second observation can be made from real (Figure 2) and nominal (Figure 4b) phototransistors array radiant sensitivity characteristics: for any given angle, only three phototransistors have their output above the level that can be used for the AoA estimation. Therefore, the estimation algorithm selects three phototransistors with the highest output. Since those three phototransistors are always successive and the middle one has maximal output with the following two being its counterclockwise and clockwise neighbors, their outputs are marked as v_m, v_{ccw}, and v_{cw}, respectively.

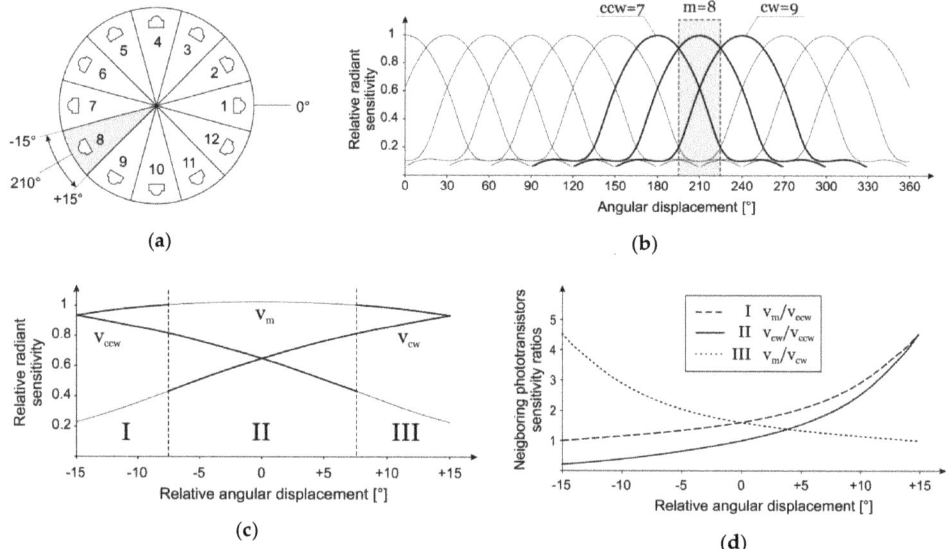

Figure 4. (**a**) AoA sensor layout; (**b–d**) diagrams used in the estimation algorithm, obtained from nominal phototransistor sensitivity as defined in the datasheet.

The third observation is that absolute values of the signal irradiance readout cannot be used as is, since they are dependent on the unknown distance between sensor and transmitter. The upside is the fact that the ratio of their values does not depend on distance; thus, the following three values can be used: v_m/v_{ccw}, v_m/v_{cw}, and v_{cw}/v_{ccw}.

Once phototransistors indices m, cw, and ccw are selected, further AoA estimation is performed in the range $[-15°, 15°]$, relative to the orientation of phototransistor with the maximal value v_m. After measurement data are obtained, the estimation is performed by selecting one ratio and matching its value with calibration data. The selection of the ratio that is used in the estimation depends on the segment of the given range in which AoA is being estimated, and this itself can be estimated using the default v_{cw}/v_{ccw} ratio.

The criteria for the selection of the ratio (from three possible ratios v_m/v_{cw}, v_m/v_{ccw}, and v_{cw}/v_{ccw}) is twofold: (1) expected absolute values that are used in the ratio need to be as high as possible (Figure 4c), and (2) the rate of change, i.e., the absolute derivative of ratios, should be as high as possible, as well (Figure 4d). Using these criteria, the estimation range is divided into three segments, marked as I, II, and III and ratios used are v_m/v_{ccw}, v_{cw}/v_{ccw}, and v_m/v_{cw}, respectively.

Finally, calibration measurement data is used to calculate all three ratios for any given angle. Strictly speaking, for each of 12 different sections (Figure 4a), i.e., for each triplet of neighboring sensors (m, ccw, cw), calculated ratios from measured values are fitted with polynomials and only coefficients of corresponding polynomials are stored for the estimation. Since all polynomials are monotone (Figure 4d) in the segment they are used, the estimation algorithm itself is reduced to a simple binary search.

In order to systematically visualize the required steps in the proposed AoA estimation algorithm, we additionally present the corresponding flowchart (Figure 5).

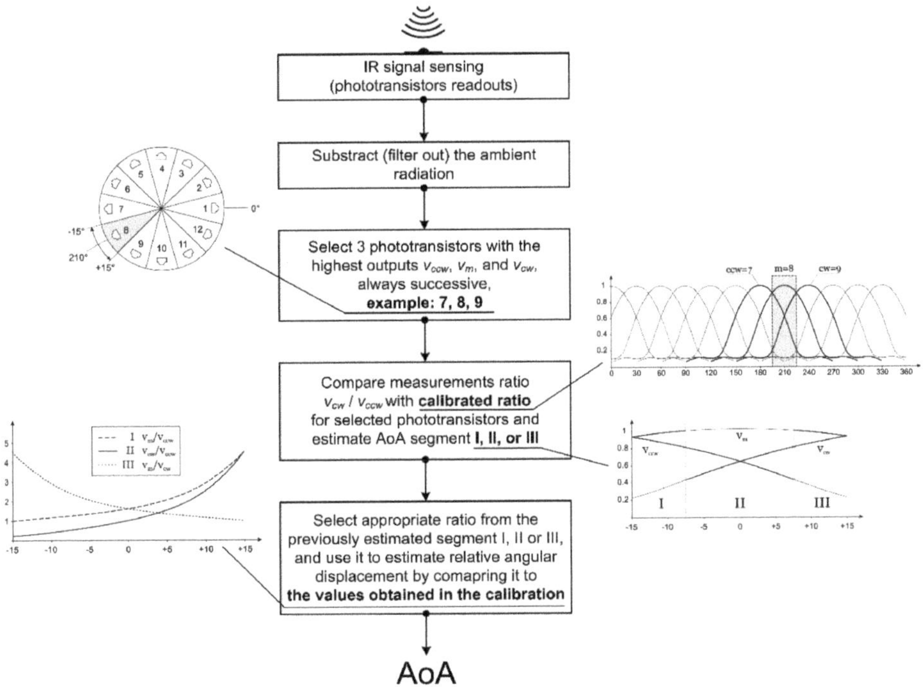

Figure 5. The flowchart of the proposed AoA estimation algorithm.

Custom calibration and the described algorithm were used in the evaluation of 16 different AoA sensors. As shown in Figure 6, this approach reduced the standard deviation of the estimation error to around 1°. The main drawback of the IR AoA sensor is its reduced range: We have been able to capture IR signals and estimate the AoA with a distance between transmitter and sensor up to 4 m. On the other hand, since the algorithm for the AoA estimation utilizes the ratio of the detected IR levels on different phototransistors, the range could be further increased simply by using more powerful IR transmitters and AD converters with a variable reference voltage.

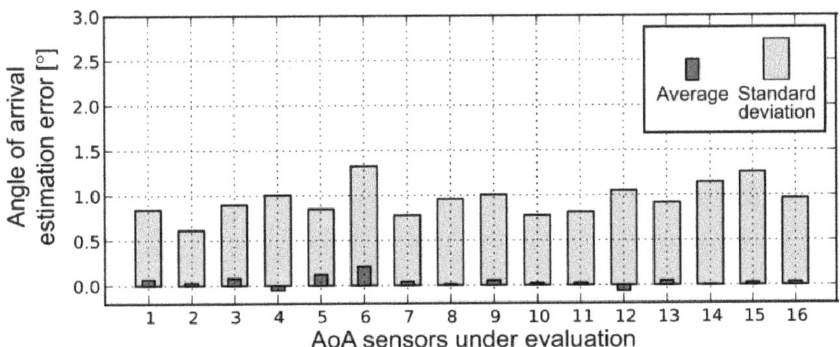

Figure 6. Angle of arrival estimation error for 16 different sensors.

2.2. Showcase Application: Supermarket Navigation

Our showcase application assumes enriching in-situ shopping experience by providing cart localization and navigation services within supermarkets aisles. Supporting this context is motivated by current customer demands, as well as by recent trends that suggest increasing involvement of modern technologies within the shopping process, i.e., smart shopping [37]. According to a 2016 study by ECE [38], a German project manager for shopping centers, end users want to utilize digital services in brick-and-mortar stores. The study showed that, among people younger than 40, every third person needs some sort of in-store guidance system. Billa, one of the most famous Austrian supermarket chains, has a mobile application that utilizes Bluetooth Beacon technology for providing location-based promotions in 11 selected shopping venues. French multinational retailer Carrefour, as well as Target, the second-largest discount store retailer in the United States, both invested in VLP-based indoor navigation prototype solutions, to assist their customers in finding what they are looking for more easily [39,40].

A supermarket is a typical example of a dynamic indoor space with dense obstacles. As noted in the Related Work subsection, there is a large number of approaches for solving indoor localization problem in such a context. Most of them are tackling a hard problem with complicated multipath signal propagation. Those approaches are either using different heuristic methods to overcome the lack of closed-form solution for NLOS propagation or trying to ensure simple LOS propagation by dense dissemination of beacons, demanding complicated and high-cost installation.

The key aspect of the proposed system, code-named Navindo, is to integrate readily available and inexpensive technology, such as the AoA IR sensor described in the previous section, with state-of-the-art low-power communication technology provided by modern wireless sensor nodes. In general, Navindo demonstrates how to overcome drawbacks of any system in need of LOS propagation using inexpensive, autonomous, and easily deployed wirelessly connected nodes. The benefit is twofold: (1) the system is accurate since there is an abundance of the high-quality LOS signal throughout the environment, and (2) the system is simple and inexpensive to install and maintain. Navindo consists of 3 components, as shown in Figure 7.

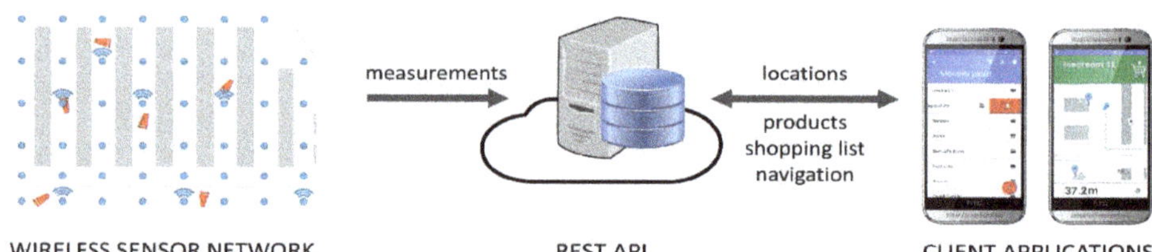

Figure 7. Navindo indoor navigation system. (1) Wireless sensor network with nodes deployed at fixed locations (i.e., above aisles in the supermarket) and simple IR transmitters (tags) on mobile objects that are being located—carts. (2) Application programming interface (API) that provides support for managing both the location data and the information about the target navigation area. (3) Client mobile applications used for accessing, managing, and visualizing location data.

2.2.1. Wireless Sensor Network

The key piece of the proposed solution is a wireless sensor network that extends the usable region for high-precision LOS mobile node localization based on the IR AoA sensor. The main purpose of the wireless sensor network is to provide infrastructural support for real-time navigation in the supermarket context. Having in mind the typical organization of the brick-and-mortar shopping venue, consisting of shelves and corridors, a straightforward WSN topology is assumed which relies on placing sensors above the carts' movement area. Hence, nodes equipped with IR AoA sensor are positioned directly above aisles, every few meters, preferably clipped onto light sources that are usually hanging low enough to evenly illuminate products on the shelves. Each node with its IR AoA sensor can track the IR signal along the aisle with the standard deviation of error below $1°$, as described in the previous section.

Additionally, each cart is equipped with a simple IR transmitter and, currently, its sole function is to transmit its ID number allowing both identification and localization from the sensor nodes side. The benefits of leaving RF transceiver out of mobile nodes are twofold: (1) cart is kept as simple as possible, and (2) the number of RF transmitting nodes is reduced, consequently reducing packet collisions and decreasing latency and throughput of the WSN.

Deployment of the previously described system is simple and straightforward; thus, the cost of the related installation is expected to be low. Since the density of the WSN directly determines the price of the system installation, the overall budget can be predicted based on the required number of WSN nodes and the expected number of the IR transmitters (carts). In the following text, we present empirical results that suggest that one sensor node per every 3 m of the corridor represents a suitable density. Furthermore, we introduce the possibility to reduce system operating costs even more, by proposing energy harvesting scenarios.

Testbed. In our testbed, we used commercially available JeeNode nodes based on Atmel ATmega328p microcontroller and RFM12B radio module. The radio module is using an 868 MHz ISM frequency band, with a data rate of 49.261 kb/s, and an indoor range from 10 to 20 m. Mobile nodes (carts) are equipped with simple IR diodes and the radio module turned off. Sensing nodes, positioned above corridors, are equipped with the proposed IR AoA sensors. A single gateway node was equipped with the EtherCard extension module based on the ENC28J60 chip and connected to the Internet using a wired local area network.

WSN communication protocol is based on the JeeLib library [41] with an address space of 250 groups with 30 regular nodes each. The media access part of the protocol implements a simple CRC-16 algorithm for error detection but with no collision avoidance. Thus, in the proof-of-concept laboratory setup, consisting of no more than 10 nodes, we used a simple centralized algorithm based on iterative queries. In this algorithm, the gateway node was also a master node, i.e., the one issuing all queries, waiting for response messages, and forwarding the corresponding data. This node was sequentially querying other nodes for fresh AoA measurements: The response message payload was a pair of values consisting of the ID of the transmitter and the AoA measurement. After each query, the gateway node simply forwarded the measurements to the localization server API.

In this setup, sensor nodes are continuously listening for IR signals and storing all measurements until they receive a query from the master node. This way all RF collisions were avoided with an obvious issue of poor scaling of the network, but that was an inherent drawback of the selected platform.

Our future work plan includes upgrading the platform to the custom Texas Instruments CC2538 or CC2650 SoC based node. The plan is to implement an algorithm on top of the Contiki operating system and use its protocol stack consisting of state-of-the-art protocols, such as IEEE 802.4.15, 6LoWPan, RPL (Routing Protocol for Low-Power and Lossy Networks), and CoAP (Constrained Application Protocol), thus enabling large address space that can host a much larger network, multi-hop routing, low power operation, etc.

Measurement protocol. Every measurement is initiated by the IR transmission from the node placed on the mobile cart, after which sensing WSN nodes in its range detect IR signal and estimate AoA, thus enabling estimation of the cart location as shown in Figure 8. To unambiguously identify the transmitting cart, every IR transmission is prefixed with the cart ID encoded with a slightly customized NEC IR protocol. This protocol is the de-facto standard and, as such, used by many consumer electronics, mostly for remote control. In our setup, the ID is an 8-bit number allowing identification of 256 different carts. The duration of the IR transmission is the sum of the modulated ID prefix and continuous IR signal used for AoA measurement. Using the NEC protocol for encoding ID prefix takes 40.5 ms, and the measurement signal takes an additional 25 ms, resulting in a total transmission duration of 65.5 ms. After the detection of the IR signal, the sensing node decodes the mobile node ID and then, from a measurement of the incoming IR signal on 12 IR transistors, estimates the angle of arrival.

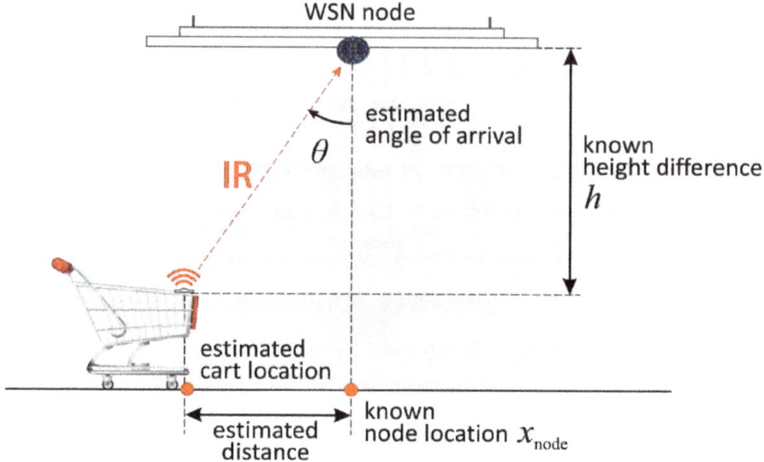

Figure 8. Cart location can be estimated in 1D (along the aisle) using AoA measurement in combination with the prior knowledge of the wireless sensor network (WSN) node location and height difference between node and IR transmitter. In this setup, the AoA sensor is rotated in order to estimate the angle of arrival in the x–z plane. The location of the cart is calculated using an estimated angle, a priori known AoA sensor position, and simple trigonometry relation: $x_{cart} = x_{node} + h \cdot \tan(\theta)$.

IR AoA sensor, in its current form, can detect and measure AoA of just one IR transmitter at a time. The number of transmitters in the low sensor range is relatively small and transmission duration takes only 65.5 ms, with the potential to be further reduced. Therefore, to handle multiple transmitters, we opted to use Time Division Multiplexing (TDM). Since we wanted to keep transmitters simple and offline, i.e., with no RF communication, we did not introduce additional synchronization overhead required for TDM. Consequently, the collision of IR packets from nearby carts presented a potential issue. Each cart is usually in the range of only 2 to 3 different sensor nodes, so we chose to solve the collision problem using simple transmission delay randomization. In our prototype system, IR signals from the carts were transmitted with a randomly chosen delay between transmits, ranging from 0.5 s to 1.5 s. Collision probability depends on the number of carts in the range of the sensor.

An additional advantage of using IR signals is directed radiation (towards sensors installed above corridors) and relatively small signal range. This property significantly reduces the collision probability, thus avoiding the need for time synchronization of the transmitters, given the small duty cycle required for transmitting. Thus, in our analysis, we considered 1 to 6 moving carts in the range of one AoA sensor. The resulting average times between cart location updates are presented in Figure 9.

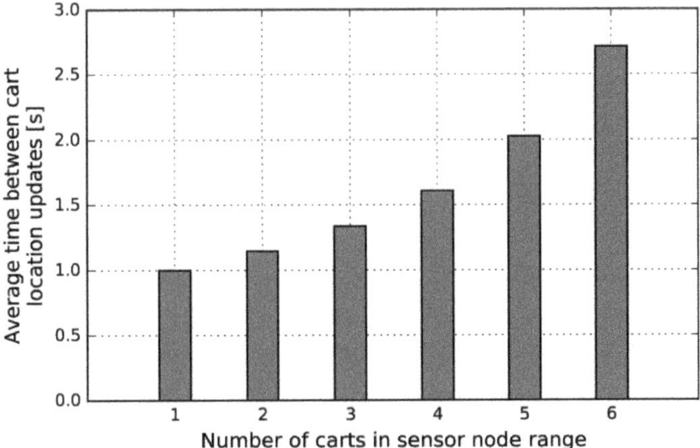

Figure 9. The average time between cart location updates depending on IR package collision probability or, more precisely, on the number of carts in the sensor node range. With a standard deployment density of 1 sensor every 3 m, there is a high probability of multiple sensors in the IR range of the mobile node, further reducing latency.

We did not consider FDMA (Frequency-Division Multiple Access) or CDMA (Code-Division Multiple Access) instead of TDM, simply because we think that the average time between cart location updates, as provided with TDM, is appropriate for the target context. Specifically, we find the "in-aisle crowding scenario", wherein more than 4 shopping carts are visible to only a single AoA sensor, less likely to happen. Hence, for a typical WSN topology (1 sensor per every 2–3 m) and a typical shopping scenario, we expect cart location updates to take place every 1–1.5 s. We find this frequency to be suitable for the target localization service, i.e., for real-time navigation support within supermarket corridors.

Localization algorithm. To be able to estimate the location of the mobile node, using just AoA measurement, the position (location and orientation) of sensing nodes must be known at installation time. Thus, the localization algorithm can map each node i to the following tuple (x_i, y_i, h, φ_i), that is, the location of the node in 3D: x_i, y_i, height h (usually the same for all sensors), and the orientation of the sensor in xy plane, φ_i. This is a reasonable demand which introduces additional benefits: simplifying topology control, designating node roles, and setting transmission parameters, such as timings and signal strengths.

Another prerequisite for the localization, and later for the navigation, is the aisles graph, seeing that all estimated locations are subsequently mapped onto its edges. The aisles graph represents a navigation layer of the supermarket map. Generally, WSN nodes are positioned on the edges of the aisles graph and are oriented in the direction of the edge they reside on, as can be seen in Figure 10.

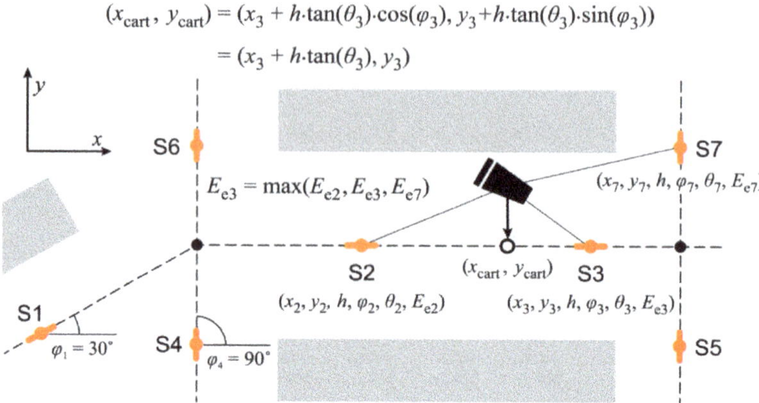

$$(x_{cart}, y_{cart}) = (x_3 + h \cdot \tan(\theta_3) \cdot \cos(\varphi_3),\ y_3 + h \cdot \tan(\theta_3) \cdot \sin(\varphi_3))$$

$$= (x_3 + h \cdot \tan(\theta_3),\ y_3)$$

Figure 10. Sensing nodes S1–S7 are placed on the edges of the aisles graph (dashed line). After the IR transmission from the cart took place, nodes S2, S3, and S7 performed measurement and AoA estimation. Measured irradiance on the node S3 was the highest so the cart was localized using estimated AoA θ_3 and the position of the sensing node S3 in the Equation (1). The orientation of the node S3 is $\varphi_3 = 0°$; thus, the estimated location is $(x_3 + h \cdot \tan(\theta_3),\ y_3)$.

As described previously, after each IR transmission the AoA measurement and estimation are initiated on every WSN node in the range. If the measurement was successful, cart ID, estimated AoA, and the maximum measured irradiance level are all sent to the gateway node as soon as possible. On the reception of measurements, the gateway simply forwards them to the localization server using REST API. The localization algorithm itself is executed on the server. Input for this algorithm is the set of measurements initiated by the same IR transmission. Measurement of the sensing node i is defined by the tuple (i, θ_i, E_{ei}), that is, ID of the sensing node, estimated AoA θ_i, and maximum measured irradiance E_{ei}.

The localization algorithm is consisting of the following three steps:

1. Select measurement → from all measurements in the set, pick the one with the highest maximum measured irradiance E_{ei}. This step is based on the simple heuristic assuming that the highest irradiance measurement correlates with the lowest distance between the transmitter and the sensor, and, more importantly, with the lowest geometric dilution of precision (GDOP).

2. Estimate location from selected measurement → selected measurement, along with the position of the corresponding sensing node, is used in the simple equation to estimate cart location:

$$(x_{cart}, y_{cart}) = (x_i + d \cdot \cos(\varphi_i),\ y_i + d \cdot \sin(\varphi_i)) \tag{1}$$

where d is the Euclidean distance of the estimated projection of the cart position onto the line passing through the sensing node in the direction of its orientation φ_i:

$$d = h \cdot \tan(\theta_i) \tag{2}$$

3. Estimate the location on the aisles graph → find the nearest point on the aisles graph edge from the estimated location. This step is usually straightforward since the aisles graph itself is constructed according to the positions of the sensors; thus, the distance of the estimated location from the graph tends to be zero. As will be described later, this mapping of the location to the aisles graph edges is important for the shortest path navigation to the products on the shelves.

Using the described algorithm, the real trajectory of the mobile node is estimated as the sequence of locations placed on the edges of the aisles graph. Each estimated location is the result of the successfully received IR transmission as presented in Figure 11.

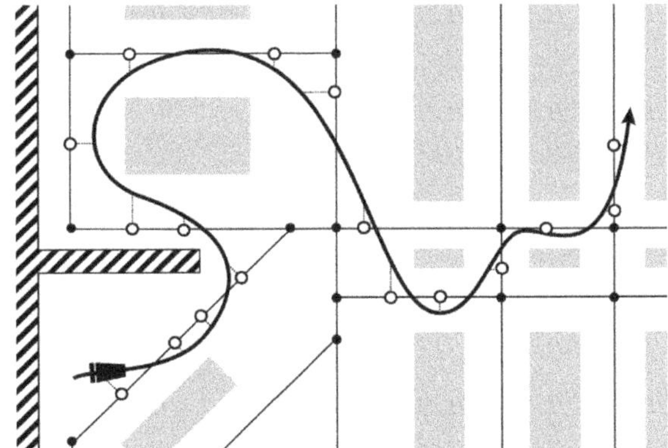

Figure 11. Localization strategy based on aisles graph. Vertices of the aisles graph are marked with •
(black dot), and estimated locations are marked with ○ (white dot). WSN nodes are not visible. All
estimated locations reside on the aisles graph edges. The cart is localized after each IR transmission.

As stated previously, in the presented localization solution, the position (both the location and
the orientation) of sensing nodes must be known at installation time. Consequently, while setting the
WSN topology at the target indoor environment, sensors should be installed precisely, following the
specified distance and in the direction of the aisle-graph edges. If a particular sensor is not oriented
correctly (with a certain angular deviation from the aisle-graph edge), then the associated error will
propagate in absolute amount and thus affect the location estimation correspondingly.

The localization accuracy of the WSN described above, which utilizes novel IR AoA introduced in
this paper, is thoroughly tackled in the Results section. However, for the sake of completeness, we continue
this part of the paper by presenting the remaining parts of the Navindo indoor navigation system.

2.2.2. Server and API

The part that connects all Navindo components in one system is the server and its front-facing
application programming interface (API). It is utilized both as permanent storage of measurement
data gathered from the wireless sensor network and for the implementation of the business logic for
all client applications. The chosen software stack includes Debian Linux OS, PostgreSQL database,
Nginx web server, and Gunicorn application server. API was implemented using Python programming
language, Django, and Django Rest Framework package. It implements several functions, such as:

- WSN measurements retrieval and storage,
- WSN node layout management,
- Cart location estimation and update,
- Product locations management,
- Store layout management,
- User signup and login,
- Token-based authentication,
- User cart registration,
- Shopping lists management,
- Shopping list based shortest path navigation (directions).

On the server, each supermarket is modeled using two graphs: the aisles graph and the shelves graph
(Figure 12). Carts and WSN nodes are located on the edges of the aisles graph, and products are located
on the edges of the shelves graph. Each product can be mapped from the shelves graph to the nearest
point(s) on the aisles graph. This way all navigation algorithms, such as the shortest path and the traveling

salesman, are performed on the aisles graph, while locations of all objects in the 2D coordinate system are preserved. For the simulation purposes, an existing retailer's webshop with more than 20 thousand items was automatically scraped, and obtained items were algorithmically distributed on the shelves graph edges.

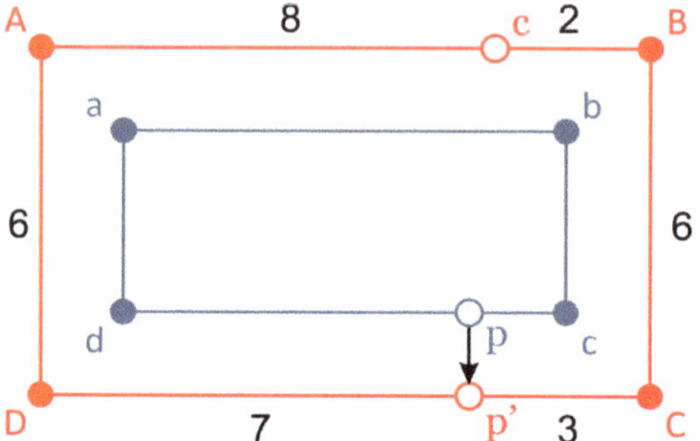

Figure 12. Shelves and aisles graphs: $\{a, b, c, d\}$ is the set of vertices of the shelves graph with the product location p on the edge (c, d) and $\{A, B, C, D\}$ is the set of vertices of the aisles graph with cart location c on the edge (A, B). The product location is mapped to the aisles graph as p' and the shortest path from cart to product is (c, B, C, p') with length 11.

2.2.3. Client Applications

In our proof-of-concept solution, we developed a client mobile application for Android devices that encompasses shopping list utilities and indoor navigation services (Figure 13). After the initial registration with the Navindo system, the user is allowed to manage shopping lists by making use of the product database.

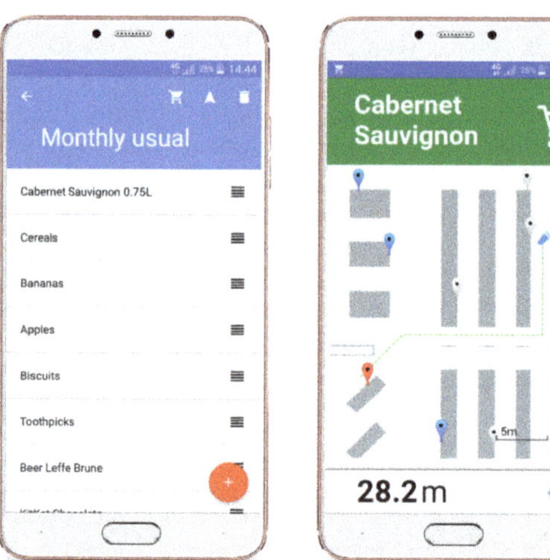

Figure 13. Mobile application screenshots. Shopping list editor screen on the left and the navigation screen on the right. The dotted line on the navigation screen represents the shortest path on the aisles graph from the current estimated cart location to the location of the next product in the shopping list. The more detailed view of the corresponding aisles graph is presented in Figure 11.

Once the user enters a supermarket and gets a shopping cart, navigation services can be enabled. In the mobile application, this is done by cart registration activity wherein the user is required to enter the cart ID (information provided on the cart itself). Navindo system provides the user with two basic options for routing within a store: (1) using shortest paths to all products from the list in the predefined order and (2) reordering the shopping list automatically to generate the global shortest route for picking all products (an instance of Traveling Salesman solution).

Navigation activity is designed in a way to resemble Google Maps user experience (Figure 13). As the user is moving inside the store, the position of the shopping cart is updating in real-time. If a smartphone running the Navindo application is equipped with the compass sensor, the current orientation of the shopping cart will be visualized on the map, as well. To utilize this option in full scale, the smartphone device should be mounted on the trolley, for which simple holders can be used (similar to car phone mounts).

Once all products from the shopping lists are collected, the mobile application routes the user to the counters zone, thus completing the navigation assist. Localization context can therefore be deactivated by unregistering the related cart ID.

To further augment location-aware shopping, as well as to demonstrate extra benefits of ubiquitous computing in the supermarket settings, we also developed an accompanying Navindo smartwatch application. According to the inherent limitations of the smartwatch I/O capabilities, only a subset of smartphone application functions is provided. The related use cases are shown in Figure 14. Samsung Gear 2 watch, running Tizen OS, was used in our proof-of-concept solution.

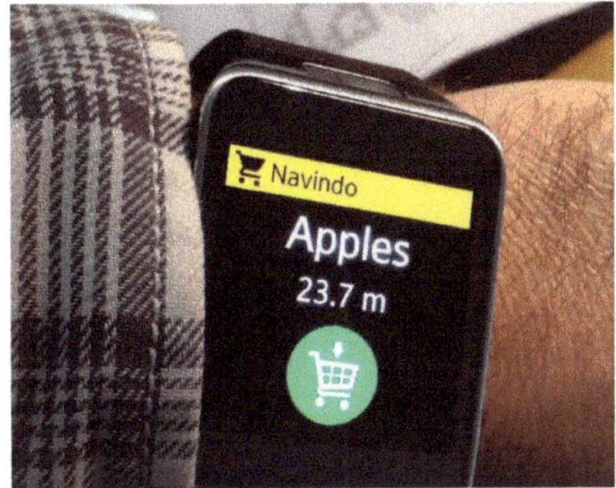

Figure 14. Smartwatch application.

2.2.4. Prospective IoT Services

Along with the on-site customers, using mobile applications on their gadgets (smartphones and smartwatches), retail management represents another user group that can substantially benefit from the Navindo system. Since the information of both the products and the gathered locations are available in real-time, extensive data analytics can be performed and subsequently used in order to boost efficiency and sales numbers. According to the collected location-based information, a detailed insight can be provided for the following cases: customers movement routes within a store, time of dwelling in the particular zones (heat maps), crowding scenarios, products purchase frequencies, customer feedback to promotional offers, a correlation between product locations and purchase volume, etc. Such analysis could further lead to strategic decisions about promotional offers, personalized marketing, shopping gamification (coupons and prizes), product placement and exposure, store layout optimization [42], human resources planning, and supermarket activities in general.

3. Results

In this section, we present the testing procedures and the results concerning the localization performance of the proposed AoA-sensor-based WSN solution. We investigated the localization accuracy of the WSN by performing four different evaluations as follows:

- **E1:** Static-1D (empirical, laboratory settings),
- **E2:** Mobile-1D (empirical, laboratory settings),
- **E3:** Static/Mobile-2D (empirical, laboratory settings),
- **E4:** Large-scale Mobile-2D (simulation).

In the first experiment (E1), we examined localization accuracy considering, altogether, four possible sensor deployment patterns. The effect of the WSN density was assessed using a proof-of-concept setup with sensing nodes positioned every 2 or 3 m, along the corridor (Figure 15). The height of the sensing nodes above the IR transmitter level was set to 2 m and 3 m. We opted for a 2 × 2 experimental design, pragmatically considering two discrete values for both sensor spacing and sensor height, having in mind the current signal range of the proposed IR AoA sensor. For every combination of sensor spacing d and height h, the cart was moved down the corridor and localized every 10 cm. The reference trajectory was a straight 8 m line along the aisle, and the cart was kept static in each measurement point (thus the Static-1D label). Resulting localization errors are presented in Figure 16, revealing the centimeter-level accuracy of the proposed solution for the corresponding setup.

Figure 15. E1-Testbed: numbered nodes (1, 2, 3) are equipped with an IR AoA sensor, and on the handle of the cart is an IR transmitter. Its design is similar to the sensing node since the IR AoA sensor has IR diodes on the opposite side of the IR phototransistors.

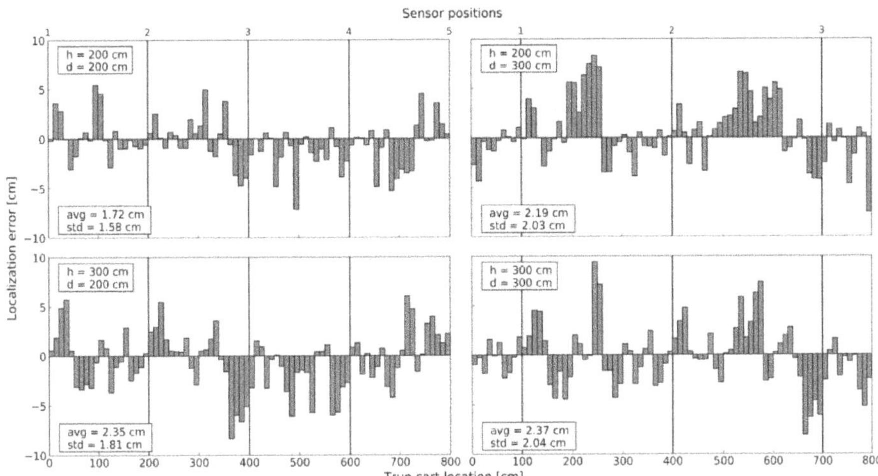

Figure 16. Localization error experimentally measured in the 8-m-long corridor for every 10 cm. Distribution of nodes with IR AoA sensor *d* was one in every 2 and every 3 m, left and right column, respectively. The height of the sensing nodes above the IR transmitter *h* was set to 2 and 3 m, top and bottom row, respectively. As can be seen from the presented results, in all scenarios, localization error did not exceed 10 cm.

According to the obtained results, we can reasonably recommend WSN density with 2–3 m sensor spacing, as well as sensor placement onto the lighting infrastructure. It is assumed that supermarket lights are placed above the aisles, 2–3 m above the IR transmitter level, which is typically a case in order to evenly illuminate all shelves. In the described experiment E1 all measurements were performed with the static IR transmitter. However, in the real-world scenario, the transmitter is mobile and, due to the described transmission delay, the real-time localization error depends on movement speed. To evaluate the localization error of the mobile transmitter we performed the second experiment (E2).

In the experiment E2, we used an IR transmitter attached to the Pioneer AT-3 mobile robot platform. The robot was set on a straight 8-m-long trajectory, representing the cart movement along the aisle, with constant velocities of 70 cm/s (maximum robot speed) and 35 cm/s. Three AoA sensors were distributed along the trajectory every 3 m and were positioned 3 m above the IR transmitter on the robot. Obtained localization errors are presented in Figure 17.

As expected, the error is dependent on movement speed since the time delay between two consecutive location updates is kept in the constant range. Location update is performed immediately after the IR transmission event, reducing the localization error to the level obtained in the static context. We find these results to be suitable for supermarket navigation, considering the usual movement speed of customers and the context in which fine-grained localization is needed only when the customer is moving slowly.

Having in mind that the cart's true position can be outside of the aisles graph, the goal of the third experiment (E3) was to evaluate the localization accuracy of the proposed solution in a more realistic Mobile-2D context. This time we set the robot (Figure 18a) on a 7.7-m-long path within the specially designed topology wherein 6 sensors were set in a 2D mesh (Figure 18b). This topology determines the corresponding aisles graph on which the robot can be localized by our solution. All sensors were positioned 3 m above the IR transmitter level. In order to make the experiment setup similar to the supermarket scenario, the given route was positioned between the shelf mock-ups made of cardboard boxes. Altogether, six "shelves", each 220 cm high, were thus used to simulate the target context (Figure 18c,d). Although there were three people in the laboratory during the experiment (two experiment administrators and a robot operator), we did not formalize any intentional user movement in that space. Hence, the MP effects were not specifically analyzed.

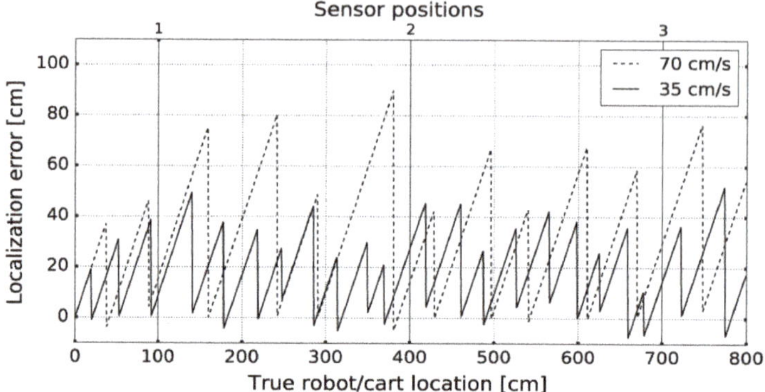

Figure 17. Localization error of the mobile transmitter for two different velocities experimentally measured in the 8-m-long corridor. The height of the sensing nodes was set to 3 m above the transmitter. The transmission delay is uniformly distributed between 0.5 s and 1.5 s. It can be seen that error is significantly and rapidly decreasing after the IR transmission events. Although varying, localization error showed to be bounded below 50 cm for 35 cm/s and below 90 cm for 70 cm/s. The localization error in this context represents a displacement of the estimated location from the real location in 1D (i.e., the displacement on a robot trajectory line); thus, it can be negative.

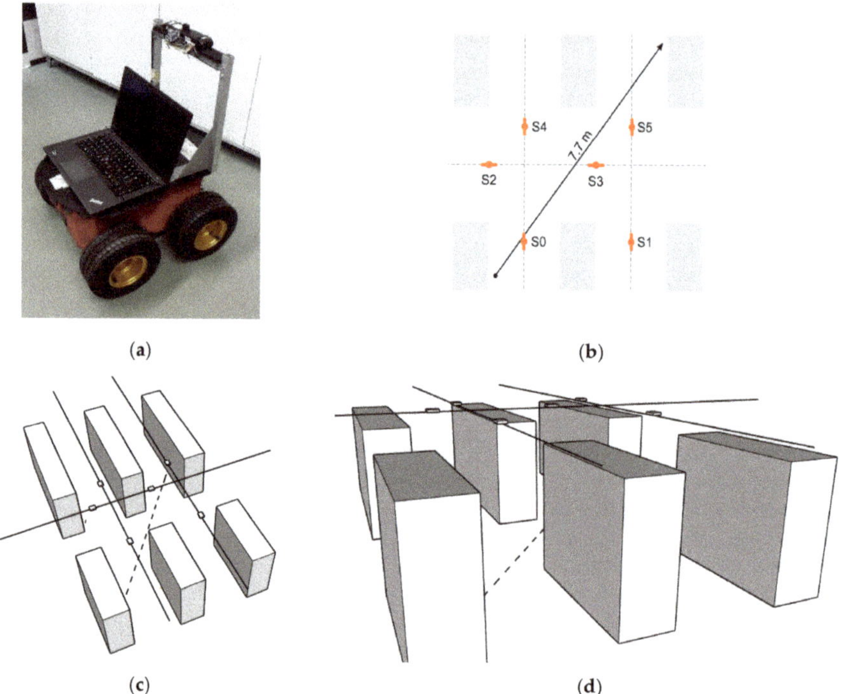

Figure 18. (**a**) Pioneer AT-3 mobile robot platform with IR transmitter attached to its handle; (**b**) Static/Mobile-2D experiment setup (WSN topology and movement trajectory used in E3); (**c**,**d**) 3D models of the E3 setup: six AoA sensors are placed above the shelf mock-ups made of cardboard boxes.

Given that the used AT-3 robot can easily maintain a constant speed on the straight line, we opted for the linear trajectory once again, which allowed simple calculation of fine-grained ground truth 2D locations in real-time. The robot traveled along the same trajectory with constant velocities set as in the Experiment E2 (70 cm/s, 35 cm/s). Additional static localization, like in the E1, was performed with a 10 cm resolution on the same path. Localization errors obtained in the given E3 setup are presented in Figure 19.

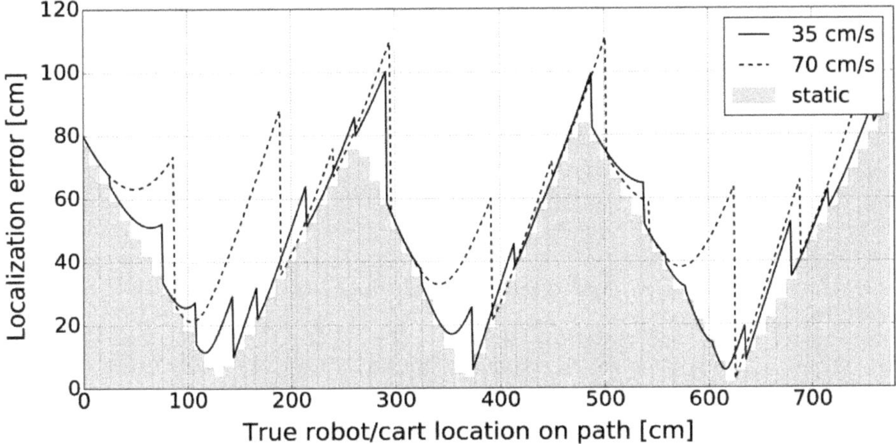

Figure 19. Localization error obtained in E3. Considering the low Static-1D error from the E1 experiment, we can conclude that the Static-2D localization error mainly originates from the distance between the cart true location and the aisles graph and, to a lesser extent, from the AoA measurement error. The Mobile-2D localization error additionally includes a component related to the movement speed and the IR transmit time delays as examined in the experiment E2.

The localization error in E3, as opposed to E1 and E2, contains an additional component (distance from the aisles graph; thus, the localization accuracy in the 2D mobile context is expectedly lower.

Finally, to assess the general localization error of the proposed solution, we carried out a simulation (E4) based on a large-scale movement trajectory (Figure 20).

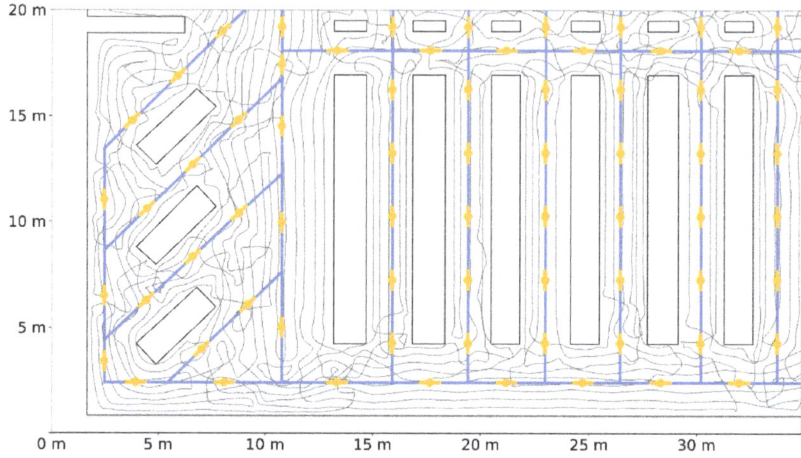

Figure 20. Large-scale Mobile-2D simulation: a part of the supermarket topology with IR AoA sensors, aisles graph, and cart trajectory.

For an 80 × 38 m supermarket layout, we designed a WSN topology with exactly 301 IR AoA sensors that were simulated according to the real characteristics of the 16 prototype sensors. We distributed simulated sensors along the virtual corridors so that every location in the supermarket has at least one sensor in the IR transmitter range. The corresponding aisles graph was constructed according to the store layout and the designed WSN topology. In the described setup, we inserted a large-scale cart trajectory that uniformly covers available movement space for the given layout. This 5000-m-long trajectory was obtained using the ergodicity-based coverage algorithm [43], targeting the homogeneous area coverage of the simulated cart locations. We simulated shopping cart movement along the given trajectory using three speeds: 35 cm/s, 70 cm/s, and 140 cm/s. Ground truth locations from simulated trajectory were compared to estimated locations provided by WSN, and localization error was thus inspected. Table 3 summarizes the results of the E4 evaluation.

Table 3. Localization errors obtained in the simulation. Along with this paper, we provide a Supplementary Video File which thoroughly demonstrates error calculation in E4.

Cart Speed	Mean Error [cm]	STD [cm]
35 cm/s	63.4	39.8
70 cm/s	73.6	40.2
140 cm/s	99.5	50.5

In all empirical evaluations (E1, E2, and E3), we measured the ground truth information manually. Sophisticated equipment was not at our disposal, and all the measurements were obtained using a simple laser pointer and a laser distance measuring device. The same gadgets were used for setting the experiment scenarios (sensor placement in E1, E2, and E3, as well as shelf mock-ups layout in E3). For static localization purposes, ground truth measurements were taken at the exact positions for which the system provided the corresponding estimations. On the other hand, for the mobile context, we manually measured only the starting and ending point of the robot movement trajectory, while the other ground truth locations were calculated (according to the constant speed of the robot). Although we did not formally determined the accuracy and precision of the ground truth data, we can assume, given the magnitude of the localization error, that these factors do not significantly affect it. Hence, we consider the effects of ground truth precision and accuracy small enough to be neglected concerning the obtained localization error.

We find obtained localization performance suitable for the supermarket navigation context. Namely, in a scenario where a customer is looking for a specific product on the shelf, and/or wants to get visual feedback about the cart location within a dense supermarket map, localization on the aisles graph will provide adequate information. Put differently, in a typical supermarket layout with 2–3 m wide corridors, localization on the aisle centerline, along with the cart orientation visualization, should be sufficient for user-friendly blue-dot navigation. Localization error of the proposed solution generally depends on the indoor venue layout (e.g., wider corridors), as well as on the movement trajectory, but appropriate WSN topology can be utilized to mitigate this error.

Due to the simple design and favorable position of nodes, there is an additional opportunity of using energy harvesting as the main power source, thus limiting or even avoiding network energy maintenance costs. Stationary nodes could be positioned onto the lights which would open up the possibility of using photovoltaic cells to power them. Simple IR transmitters placed on carts only need to transmit when moving, so they could be powered using energy from the cart motion (i.e., wheels rotational energy).

4. Discussion

As shown within the simulation (experiment E4), which considers the effect of the proposed localization method in a larger space (supermarket level) with a topology that includes corridors and high shelves, for the obtained localization error level (0.6–1 m, depending on the cart speed) it is sufficient to provide 301 AoA sensors on a gross area of 3040 m^2. We consider such a WSN density (~1 sensor per 10 m^2) to be a suitable solution, given the trade-off between the cost estimation of the related infrastructure and the provided localization accuracy in the target context. Specifically, knowing the production costs of both the proposed IR AoA sensor (~USD 13) and the proposed IR transmitter (~USD 9), we can assume the total cost for an 80 × 38 m retail venue with exactly 100 shopping carts: USD 4.813. We find this amount to be a rational investment for supermarket management, seeing that the shopping experience could be considerably enhanced via localization and navigation services. A more densely deployed sensors would allow even better localization accuracy (as shown in experiment E1), however, according to the already mentioned cost-accuracy trade-off, the idea is to use a reasonable number of sensors within a given indoor environment. As stated at the end of the previous section, the infrastructure maintenance costs could be furthermore reduced by utilizing energy harvesting (on both the sensor and the transmitter side), which is part of our future work plans.

Regarding the deployment aspects, since the proposed solution localizes the user on the aisle graph (i.e., in the corridors between the shelves with products), the WSN topology design is rather straightforward. Namely, the proposed sensors should be placed above all the aisles in which users are expected to be walking through (and not above the shelves), with a distance of 2 m to 3 m along the corridors (as demonstrated in conducted experiments). Since no cabling is needed for powering and connecting the WSN infrastructure, we assume supermarket lights and cart handles to be a pragmatic choice for placing the IR AoA sensors and IR transmitters, respectively. In most cases, the lighting infrastructure in a supermarket corresponds to the expected WSN topology, as lights are usually placed above the aisles, in a way to evenly illuminate all shelves.

Seeing that the proposed solution, based on the novel IR AoA sensor, enables cart localization with the corresponding error between 0.6 m and 1 m (in the mobile 2D scenario), the question of comparability with other mature solutions with similar accuracy (e.g., RFID-based and BLE-based) arises. When it comes to the RFID-based localization in the supermarket scenario, we are usually considering carts equipped with RFID tags, and RFID readers deployed at the venue according to the given density. As demonstrated in Reference [15], the problematic aspect of the related RFID-based method can be a long measurement time, and, consequently, the need for the cart to remain stationary (up to 2 s) in order to estimate its current location. Furthermore, if a passive RFID system is selected as the backbone for the localization solution, then one must consider the high cost of multiple RFID readers. For example, high-performance RFID readers, such as the *Impinj Speedway RAIN RFID Reader* (used for baggage tracking, retail inventory management, etc.) reaches a price of over USD 1.000 per single unit. On the other hand, BLE beacons seem to be the most competitive hardware in terms of cost estimation, seeing that the average price for a single beacon is around USD 25 (it depends on the manufacturer, the transmission range, and the form factor of the beacon). However, typical localization accuracy within the BLE-based systems is around 2 m, which makes them more suitable for less precise services, such as proximity-based localization and point-of-interest detection.

One can argue that 2 m localization error, achievable via mature RFID or BLE solutions, can represent an adequate accuracy level for cart localization in supermarkets. We agree that such accuracy can be considered satisfactory, but, at the same time, we think that the end-user should, as far as possible, be provided with highly usable localization services, here including seamless and precise navigation support. Hence, one of our goals was to provide a direction-finding interface, similar to that from the well-known navigation applications, such as Google Maps. In this sense, the proposed system supports real-time localization (and, consequently, navigation) wherein the localization error changes dynamically, depending on the cart movement speed and the distance from the aisle graph. We can assume that a larger localization error will not affect (negatively, to a greater extent) the usability of the navigation service when the user searches for the target shelf at a higher speed. However, once the customer reduces the cart speed coming in front of the shelf, where the target product should be found, then a lower localization error can considerably affect (positively) the overall user experience. If the user, while fine-searching for a target product, stops in the middle of the corridor (quite a possible scenario), then the proposed solution can localize the corresponding cart at the decimeter level (as demonstrated in the E1 experiment). In addition, when considering RFID and BLE solutions, we have to take into account the fact that shelves for them are practically "transparent". This means that 2 m error can imply location estimation in a corridor adjacent to the ground truth, effectively making the respective error fairly larger for the end-user. Conversely, in our LOS system, the shelves represent the obstacles for the IR signal, and, following the aisle-graph model, the cart will be localized in the right corridor.

Regarding the comparison of the proposed IR AoA sensor with the existing IR sensors for localization (some of them are tackled in Section 1.1.2), we think that the results of such a comparison would be difficult to generalize for the target setting. Namely, our localization method depends on the specific spatial context, i.e., the mobile transmitter can be localized on the aisle graph exclusively, which inherently contributes to the localization error. However, we are willing to retain that component of the total localization error, believing that the aisle-graph model represents a suitable trade-off for the supermarket environment.

The showcase application of the proposed indoor localization principle is providing a cart navigation service within a supermarket venue on the aisle level. Mobile applications for smartphone and smartwatch devices are developed, confirming the utility of location-based data provided by the underlying WSN in real-time. The proposed system has a better cost-benefit ratio when compared to competing solutions, considering the tradeoff between required installation and maintenance expenses and achieved localization precision.

To the best of our knowledge, the WiDeo system [9] represents the best competing approach in terms of both the achieved accuracy and expected installation costs. The object being traced using WiDeo does not have to be accompanied by any supplementary device, which is a noteworthy advantage among existing localization solutions. However, according to the reported experiments, WiDeo can trace only five independent concurrent motions without worsening its accuracy which is, according to the authors, "sufficient for a home environment, but not for work environments where a far greater amount of motion is expected". Moreover, it must be noted that WiDeo utilizes a method wherein motion tracing accuracy significantly outperforms (absolute) localization accuracy (80 cm level). Last but not least, the current version of the WiDeo implementation does not support localization in real-time.

The supermarket context, specifically tackled in this paper, is a typical example of a dynamic indoor space with dense obstacles and numerous moving objects (humans and shopping carts). In such a target context, real-time localization is of utmost importance for providing navigation service. Independent objects, e.g., shopping carts, not only have to be concurrently localized, but they also have to be unambiguously identified. For example, localization and identification should seamlessly work when target objects' positions are very close (by means of both the estimated AoA and distance), or in cases wherein moving trajectories interfere and continuously overlap.

According to the abovementioned, the proposed Navindo indoor navigation system seems to be a suitable solution for a supermarket domain, given that it utilizes WSN topology and IR AoA sensors developed with such context in mind. The showcase application proved that the proposed localization solution can be easily deployed in order to provide accurate aisle level navigation in real-time.

The proposed IR-based localization principle is completely orthogonal to any RF-based solution, meaning that related approaches can be combined to boost localization performance for a given setup. The advantages of the proposed method are entirely complementary to the shortcomings of the RF localization systems.

5. Conclusions

In this paper, a supermarket navigation system, which relies on a novel IR AoA sensor, WSN-based localization infrastructure, and graph-based motion model, is introduced and described. The system is based on the LOS propagation of the IR signals, and a localization algorithm that uses measurements and AoA estimation provided by the IR AoA sensor. A proof-of-concept implementation demonstrated how inexpensive, autonomous, and easily deployable wireless nodes can be utilized to provide suitable localization accuracy for the target context. Several factors can have an impact on the localization error of the proposed solution, e.g., AoA measurement error, applied WSN topology (sensor density) for a given store layout, the relation between movement speed and IR transmit time delays, and movement trajectory (distance from the aisles graph). Altogether, four evaluation procedures were performed to investigate localization performance. The accuracy of estimated location was firstly observed in a 1D static context for different WSN densities, according to the given number of utilized AoA sensors and varied distance between WSN nodes and IR transmitters. The effect of moving speed on the localization accuracy in 1D and 2D setups was evaluated, as well, both empirically and via simulation. Since the proposed solution estimates cart location on the aisles graph exclusively, different movement trajectories were put under test: the straight 1D path along the sensor line, the straight 2D path beneath the sensor mesh, and the large-scale tortuous trajectory within the simulation environment. All obtained results, ranging from centimeter-level accuracy (Static-1D) to 1 m mean localization error (Mobile-2D simulation), are presented and discussed in detail.

Our future work plan consists of addressing detected limitations in the current version of the system and exploring potentials for further system improvement. This especially holds for a thorough investigation of the possibilities for increasing the IR signal range and analysis of the multipath propagation effects, as well as for evaluating the system in the real-world scenario, i.e., out of the laboratory context. As noted previously, we intend to upgrade the system platform by utilizing a state-of-the-art protocol suite with an enhanced hardware base that should support larger WSN topologies, as well as the implementation of power control through energy harvesting.

Supplementary Materials: The following are available online at http://www.mdpi.com/1424-8220/20/21/6278/s1, Video S1: E4-simulation.mp4.

Author Contributions: Conceptualization, D.A.; methodology, D.A. and S.L.; software, D.A. and S.L.; validation, D.A. and S.L.; formal analysis, D.A. and S.L.; investigation, D.A. and S.L.; resources, D.A. and S.L.; data curation, D.A. and S.L.; writing—original draft preparation, D.A. and S.L.; writing—review and editing, D.A. and S.L.; visualization, D.A. and S.L.; supervision, D.A. and S.L.; project administration, D.A. and S.L.; funding acquisition, *none*. All authors have read and agreed to the published version of the manuscript.

Funding: This research received no external funding.

Conflicts of Interest: The authors declare no conflict of interest.

References

1. Belloni, F.; Ranki, V.; Kainulainen, A.; Richter, A. Angle-based indoor positioning system for open indoor environments. In Proceedings of the 2009 6th Workshop on Positioning, Navigation and Communication, Hannover, Germany, 19 March 2009; pp. 261–265. [CrossRef]
2. Brena, R.F.; García-Vázquez, J.P.; Galván-Tejada, C.E.; Muñoz-Rodriguez, D.; Vargas-Rosales, C.; Fangmeyer, J. Evolution of Indoor Positioning Technologies: A Survey. *J. Sens.* **2017**, *2017*, 1–21. [CrossRef]
3. Shang, J.; Hu, X.; Gu, F.; Wang, D.; Yu, S. Improvement Schemes for Indoor Mobile Location Estimation: A Survey. *Math. Probl. Eng.* **2015**, *2015*, 1–32. [CrossRef]
4. Sen, S.; Choudhury, R.R.; Radunovic, B.; Minka, T. Precise indoor localization using PHY layer information. In Proceedings of the 10th ACM Workshop on Hot Topics in Networks—HotNets '11, Cambridge, MA, USA, 14–15 November 2011; Association for Computing Machinery (ACM): New York, NY, USA, 2011; pp. 1–6.
5. Shin, B.; Lee, J.H.; Lee, T.; Kim, H.S. Enhanced weighted K-nearest neighbor algorithm for indoor Wi-Fi positioning systems. In Proceedings of the 8th International Conference on Computing Technology and Information Management (NCM and ICNIT), Seoul, Korea, 24–26 April 2012; pp. 574–577.
6. Hu, X.; Shang, J.; Gu, F.; Han, Q. Improving Wi-Fi Indoor Positioning via AP Sets Similarity and Semi-Supervised Affinity Propagation Clustering. *Int. J. Distrib. Sens. Netw.* **2015**, *11*. [CrossRef]
7. Huang, J.; Millman, D.; Quigley, M.; Stavens, D.; Thrun, S.; Aggarwal, A. Efficient, generalized indoor WiFi GraphSLAM. In Proceedings of the 2011 IEEE International Conference on Robotics and Automation, Shanghai, China, 9–13 May 2011; IEEE: Piscataway, NJ, USA, 2011; pp. 1038–1043.
8. Yang, C.; Shao, H.-R. WiFi-based indoor positioning. *IEEE Commun. Mag.* **2015**, *53*, 150–157. [CrossRef]
9. Joshi, K.; Bharadia, D.; Kotaru, M.; Katti, S. WiDeo: Fine-grained device-free motion tracing using RF backscatter. In Proceedings of the 12th USENIX Conference on Networked Systems Design and Implementation (NSDI'15), Oakland, CA, USA, 4–6 May 2015; USENIX Association: Berkeley, CA, USA, 2015; pp. 189–204.
10. Faragher, R.; Harle, R. Location Fingerprinting With Bluetooth Low Energy Beacons. *IEEE J. Sel. Areas Commun.* **2015**, *33*, 2418–2428. [CrossRef]
11. Kriz, P.; Maly, F.; Kozel, T. Improving Indoor Localization Using Bluetooth Low Energy Beacons. *Mob. Inf. Syst.* **2016**, *2016*, 1–11. [CrossRef]
12. Gezici, S.; Tian, Z.; Giannakis, G.B.; Kobayashi, H.; Molisch, A.F.; Poor, H.V.; Sahinoglu, Z. Localization via ultra-wideband radios: A look at positioning aspects for future sensor networks. *IEEE Signal Process. Mag.* **2005**, *22*, 70–84. [CrossRef]
13. Song, L.; Zou, H.; Zhang, T. A Low Complexity Asynchronous UWB TDOA Localization Method. *Int. J. Distrib. Sens. Netw.* **2015**, *2015*, 1–9. [CrossRef]
14. Ni, L.M.; Liu, Y.; Lau, Y.C.; Patil, A.P. LANDMARC: Indoor Location Sensing Using Active RFID. *Wirel. Netw.* **2004**, *10*, 701–710. [CrossRef]
15. Ryoo, J.; Das, S.R. Phase-based Ranging of RFID Tags with Applications to Shopping Cart Localization. In Proceedings of the 18th ACM International Conference on Modeling, Analysis and Simulation of Wireless and Mobile Systems-MSWiM '15, Cancun, Mexico, 2–6 November 2015; Association for Computing Machinery (ACM): New York, NY, USA, 2015; pp. 245–249.
16. Armstrong, J.; Sekercioglu, Y.A.; Neild, A. Visible light positioning: A roadmap for international standardization. *IEEE Commun. Mag.* **2013**, *51*, 68–73. [CrossRef]
17. Kuo, Y.-S.; Pannuto, P.; Hsiao, K.-J.; Dutta, P. Luxapose. In Proceedings of the 20th Annual International Conference on Mobile Computing and Networking-MobiCom '14, Maui, HI, USA, 7–11 September 2014; Association for Computing Machinery (ACM): New York, NY, USA, 2014; pp. 447–458.
18. Qiu, K.; Zhang, F.; Liu, M. Visible Light Communication-based indoor localization using Gaussian Process. In Proceedings of the 2015 IEEE/RSJ International Conference on Intelligent Robots and Systems (IROS), Hamburg, Germany, 28 September–2 October 2015; pp. 3125–3130. [CrossRef]

19. Wright, M. Philips Lighting Deploys LED-based Indoor Positioning in Carrefour Hypermarket. In *LEDs Magazine*. 21 May 2015. Available online: https://www.ledsmagazine.com/architectural-lighting/retail-hospitality/article/16696804/philips-lighting-deploys-ledbased-indoor-positioning-in-carrefour-hypermarket (accessed on 31 October 2020).

20. Want, R.; Hopper, A.; Falcão, V.; Gibbons, J. The active badge location system. *ACM Trans. Inf. Syst.* **1992**, *10*, 91–102. [CrossRef]

21. Hijikata, S.; Terabayashi, K.; Umeda, K. A simple indoor self-localization system using infrared LEDs. In Proceedings of the 2009 Sixth International Conference on Networked Sensing Systems (INSS) 2009, Pittsburgh, PA, USA, 17–19 June 2009; Volume 17, pp. 1–7. [CrossRef]

22. De-La-Llana-Calvo, Á.; Galilea, J.L.L.; Gardel, A.; Rodríguez-Navarro, D.; Bravo, I.; Tsirigotis, G.; Iglesias-Miguel, J. Modeling Infrared Signal Reflections to Characterize Indoor Multipath Propagation. *Sensors* **2017**, *17*, 847. [CrossRef] [PubMed]

23. De-La-Llana-Calvo, Á.; Galilea, J.L.L.; Gardel, A.; Rodríguez-Navarro, D.; Bravo, I.; Tsirigotis, G.; Iglesias-Miguel, J. Modeling the Effect of Optical Signal Multipath. *Sensors* **2017**, *17*, 2038. [CrossRef] [PubMed]

24. De-La-Llana-Calvo, Á.; Galilea, J.L.L.; Gardel, A.; Rodríguez-Navarro, D.; Bravo, I.; Zapata, F.E. Characterization of Multipath Effects in Indoor Positioning Systems by AoA and PoA Based on Optical Signals. *Sensors* **2019**, *19*, 917. [CrossRef] [PubMed]

25. Chung, J.; Donahoe, M.; Schmandt, C.; Kim, I.-J.; Razavai, P.; Wiseman, M. Indoor location sensing using geo-magnetism. In Proceedings of the 9th International Conference on Mobile systems, Applications, and Services (MobiSys'11), Washington, DC, USA, 28 June–1 July 2011; pp. 141–154.

26. Shu, Y.; Bo, C.; Shen, G.; Zhao, C.; Li, L.; Zhao, F. Magicol: Indoor Localization Using Pervasive Magnetic Field and Opportunistic WiFi Sensing. *IEEE J. Sel. Areas Commun.* **2015**, *33*, 1443–1457. [CrossRef]

27. Wang, S.; Fidler, S.; Urtasun, R. Lost Shopping! Monocular Localization in Large Indoor Spaces. In Proceedings of the 2015 IEEE International Conference on Computer Vision (ICCV), Santiago, Chile, 7–13 December 2015; IEEE: Piscataway, NJ, USA, 2015; pp. 2695–2703.

28. Priyantha, N.B. The Cricket Indoor Location System. Ph.D. Thesis, Massachusetts Institute of Technology, Cambridge, MA, USA, 2005.

29. Tragas, P.; Kalis, A. A Printed ESPAR Antenna for Mobile Computing Applications. In Proceedings of the 12th European Wireless Conference—Enabling Technologies for Wireless Multimedia Communications, Athens, Greece, 2–5 April 2006; pp. 1–4.

30. Nasipuri, A.; El Najjar, R. Experimental Evaluation of an Angle Based Indoor Localization System. In Proceedings of the 2006 4th International Symposium on Modeling and Optimization in Mobile, Ad Hoc and Wireless Networks, Boston, MA, USA, 3–7 April 2006; IEEE: Piscataway, NJ, USA, 2006; pp. 1–8.

31. Rodríguez-Navarro, D.; Galilea, J.L.L.; De-La-Llana-Calvo, Á.; Bravo, I.; Gardel, A.; Tsirigotis, G.; Iglesias-Miguel, J. Indoor Positioning System Based on a PSD Detector, Precise Positioning of Agents in Motion Using AoA Techniques. *Sensors* **2017**, *17*, 2124. [CrossRef] [PubMed]

32. Aparicio-Esteve, E.; Hernandez, A.; Urena, J.; Villadangos, J.M.; Ciudad, F. Estimation of the Polar Angle in a 3D Infrared Indoor Positioning System based on a QADA receiver. In Proceedings of the 2019 International Conference on Indoor Positioning and Indoor Navigation (IPIN), Pisa, Italy, 30 September–3 October 2019; pp. 1–8. [CrossRef]

33. Song, Y.M.; Xie, Y.; Malyarchuk, V.; Xiao, J.; Jung, I.; Choi, K.-J.; Liu, Z.; Park, H.; Lu, C.; Kim, R.-H.; et al. Digital cameras with designs inspired by the arthropod eye. *Nat. Cell Biol.* **2013**, *497*, 95–99. [CrossRef] [PubMed]

34. Hsu, J. Insect-Eye Camera Offers Wide-Angle Vision for Tiny Drones. In *IEEE Spectrum*. 1 May 2013. Available online: https://spectrum.ieee.org/robotics/robotics-hardware/insecteye-camera-offers-wideangle-vision-for-tiny-drones (accessed on 31 October 2020).

35. Vishay Semiconductors. VEMT2023SLX01 IR phototransistor, NPN, SMD, Technical Data Sheet. 2013. Available online: https://www.vishay.com/docs/84166/vemt2023slx01.pdf (accessed on 31 October 2020).

36. Arbula, D. Distributed algorithm for node localization in anchor free wireless sensor network. Ph.D. Thesis, Faculty of Electrical Engineering and Computing, University of Zagreb, Zagreb, Croatia, 2014.

37. Li, R.; Song, T.; Capurso, N.; Yu, J.; Couture, J.; Cheng, X. IoT Applications on Secure Smart Shopping System. *IEEE Internet Things J.* **2017**, *4*, 1945–1954. [CrossRef]

38. ECE Market Research. *At Your Service: Die Bedeutung von Services in Shopping-Centern.* Technical Report. 2016. Available online: https://www.ece.com/fileadmin/media/E1_Presse/Publikationen/ECE_At_Your_Service_deu.pdf (accessed on 31 October 2020).

39. Halper, M. Carrefour Guides Shoppers to in-store Discounts via the Ceiling Lights. In *LUX Review*. 21 May 2015. Available online: www.luxreview.com/2015/05/21/carrefour-guides-shoppers-to-in-store-discounts-via-the-ceiling-lights (accessed on 31 October 2020).

40. Halper, M. Target's IoT Trial Expands to 100 Stores. In *LUX Review*. 17 November 2015. Available online: https://www.luxreview.com/2015/11/17/target-s-spy-lights-trial-expands-to-100-stores/ (accessed on 31 October 2020).

41. Wippler, J.-C. JeeLib Library. Available online: https://github.com/jeelabs/jeelib (accessed on 31 October 2020).

42. Hwangbo, H.; Kim, J.; Lee, Z.; Kim, S. Store layout optimization using indoor positioning system. *Int. J. Distrib. Sens. Netw.* **2017**, *13*. [CrossRef]

43. Ivic, S.; Crnkovic, B.; Mezic, I. Ergodicity-Based Cooperative Multiagent Area Coverage via a Potential Field. *IEEE Trans. Cybern.* **2016**, *47*, 1983–1993. [CrossRef] [PubMed]

Letter

A Recursive Algorithm for Indoor Positioning Using Pulse-Echo Ultrasonic Signals

Salvatore A. Pullano [1], **Maria Giovanna Bianco** [1], **Davide C. Critello** [1], **Michele Menniti** [1], **Antonio La Gatta** [2,*,†] **and Antonino S. Fiorillo** [1,*,†]

1   Department of Health Sciences, University "Magna Græcia" of Catanzaro, Viale Europa, 88100 Catanzaro, Italy; pullano@unicz.it (S.A.P.); mg.bianco@unicz.it (M.G.B.); critello@unicz.it (D.C.C.); menniti@unicz.it (M.M.)
2   LRO—London Research Organization, 207 Regent Street, London W1B3HH, UK
*   Correspondence: lagatta.a@sarf.ch (A.L.G.); nino@unicz.it (A.S.F.)
†   The authors equally contributed to the manuscript.

Received: 15 August 2020; Accepted: 2 September 2020; Published: 4 September 2020

Abstract: Low frequency ultrasounds in air are widely used for real-time applications in short-range communication systems and environmental monitoring, in both structured and unstructured environments. One of the parameters widely evaluated in pulse-echo ultrasonic measurements is the time of flight (TOF), which can be evaluated with an increased accuracy and complexity by using different techniques. Hereafter, a nonstandard cross-correlation method is investigated for TOF estimations. The procedure, based on the use of template signals, was implemented to improve the accuracy of recursive TOF evaluations. Tests have been carried out through a couple of 60 kHz custom-designed polyvinylidene fluoride (PVDF) hemicylindrical ultrasonic transducers. The experimental results were then compared with the standard threshold and cross-correlation techniques for method validation and characterization. An average improvement of 30% and 19%, in terms of standard error (SE), was observed. Moreover, the experimental results evidenced an enhancement in repeatability of about 10% in the use of a recursive positioning system.

Keywords: ultrasonic transducers; time of flight estimation; pulse-echo technique; ferroelectric films; piezopolymer

1. Introduction

Over the years, ultrasonic technology has been applied in variegated fields ranging from underwater acoustics [1,2], medical imaging [3] and biomedical devices [4,5]. Apart from the above, indoor localization systems have reached a widespread consensus as they are inexpensive, space-saving and less prone to interference due to environmental light or heat sources [6–9]. In-air ultrasounds were amply investigated to retrieve information about unstructured environments in 3D tracking and motion detection [6,7,10–15]. Although most technologies (infrared radiation, radio frequency, artificial vision) are currently developed and commercialized, systems based on ultrasounds can be realized with simple hardware [10,16], combining multiple coplanar transmitters [17] or in association with multiple receivers [18], easily achieving a sub-mm resolution.

However, the performances of the 3D ultrasonic positioning system can be significantly improved by working on hybrid technologies or a novel algorithm [12,19].

Conversely, in-air ultrasounds, which usually range from 30 to 120 kHz, are poorly suitable in the case of long distances and for the most sophisticated fine-grained local positioning systems (LPSs), because of the signal wavelength, and the wide lobe of irradiation of the available transducers [14,20]. Recent literature reports different attempts to overcome the limitation of commercially available transducers in terms of bandwidth, sensitivity and directivity, by introducing novel geometries and by the optimization of acoustic wave propagation [7,21–26]. LPSs are usually realized with a combination of multiple transmitters/receivers, properly positioned around the target area. The emitted signal and the received echo provide different basic information, such as the receiving object distance, through a time of flight (TOF) estimation or other information about target characteristics, as in the case of bio-inspired echolocation systems [27,28]. The simplest and most common way to detect TOF echo signals is the threshold method, in which the detection occurs when a signal crosses a predetermined threshold [29–31]. It is generally characterized by a lower accuracy introduced by the sampling frequency, low signal-to-noise ratio (SNR), and difficulty in setting an optimized threshold. The introduced delay is generally nonconstant, resulting in a variable offset error. Another widely used technique involves the cross-correlation function to estimate the TOF of pulse-echo signals by varying the time observation point [31–33]. The latter is also exploited in natural bio-sonar, in a neural approach for calculating the temporal correlation between pulse and echo [29,34,35]. Other approaches exploit artificial intelligence techniques as a probabilistic algorithm, artificial neural networks, k-nearest neighbor or support vector machine to evaluate the position of an object and improve automatized learning [36,37].

Although for most applications they provide a sufficient level of accuracy, they are inherently sensitive to SNRs, distortion and other factors such as fluctuations of sound velocity and the proximity of other objects. In this paper, a modified cross-correlation technique, based on pulse-echo analysis, is investigated for a recursive TOF evaluation. The transmitted pulse and the echo are generated by curved polyvinylidene fluoride (PVDF) transducers previously investigated for robotic applications, characterized by a low quality factor and high coupling in air [7,28]. The technique is based on a recursive cross-correlation analysis and the use of a template signal as a reference. The TOF is evaluated with respect to a calibrated echo signal, resulting in an improved accuracy and repeatability during continuous target monitoring. The proposed approach is directed to the development of a new algorithm which, together with the advancements in sensors technologies, can provide improvements in real-time driver monitoring and behavior, especially if integrated with complementary technologies (e.g., alcohol monitoring, fatigue recognition systems).

2. Materials and Methods

2.1. Ultrasound Sensors

The application of ultrasonic sensors in determining the x, y, z coordinates of an object in a working space (e.g., cockpit, robot space) can be used complementarily with optical systems or alone as a valid alternative to optical methods with a reduced sensitivity to noise, dust, lighting conditions, etc. [38]. In SONAR (Sound NAvigation and Ranging) systems, the resolution can be correlated with the spectral content of the received signals. The radial resolution in a sonar system is a function of the bandwidth, whereas the azimuth resolution is a function of the system opening [39,40]. In air, the time of flight of ultrasonic waves at different frequencies can be considered almost the same; thus, the resolution is limited by the data acquisition and processing. The propagation medium introduces an attenuation which depends on different factors like beam dispersion, hysteresis, friction losses and the viscosity of the medium. Moreover, attenuation increases with frequency, which can alter the reflected wave [41]. External noises, such as turbulence, vibrations and the noise due to the electronics used, also affect the received echo travelling in a medium. By only taking into account the air viscosity, the enlargement of the acoustic beam mainly depends on the displacement with respect to the source and the attenuation of the medium according to the Lambert–Beer law.

Obviously, depending on the specific application and frequency of the system, it is always desirable to improve the resolution, in order to reduce ambiguity during target positioning and tracking. Bimodal transducers can result in a worse performance in terms of the SNR at the input of the receiver. In some cases, multiple unimodal transducers are thus preferred, in order to achieve an electrical and mechanical decoupling. Previously developed ultrasonic transducers, based on the ferroelectric properties of PVDF, were investigated for robotic applications. The hemicylindric geometry has been theoretically and experimentally investigated in the range between 30 and 120 kHz [3,42]. The transducer was made with a strip of PVDF with a thickness of 28 μm, a width of 5 mm and a length which depended on the specific resonance frequency (f_r). The strip was metallized on both faces, with about 200 nm of aluminum and clamped on the short side in order to achieve a hemicylindrical geometry. The operating principle was based on the conversion of longitudinal motion into radial vibrations due to the clamped extremities (caused by the alternating voltage applied between the electrodes) allowing the generation of radial acoustic waves in the anterior (concave) and posterior (convex) sides [7,14]. The resonance frequency was inversely proportional to the bending radius and, therefore, could be easily manipulated by varying the curvature. Due to the very low-quality factor of the transducer (Q about 12), the signal is characterized by a broad spectrum. Deviation of the resonance frequency (~5%) can be observed with respect to the theoretical value due to assembly defects (not perfectly hemicylindrical, nonparallel electrode shapes), as well as parasitic resistances created during the realization of the external electrodes (e.g., silver paste, pressure contacts). Figure 1a reports the effective dimensions of few representative sensors and the related diameters, while Figure 1b shows the supporting structure used to maintain the geometry, the curved PVDF film, and the external contact.

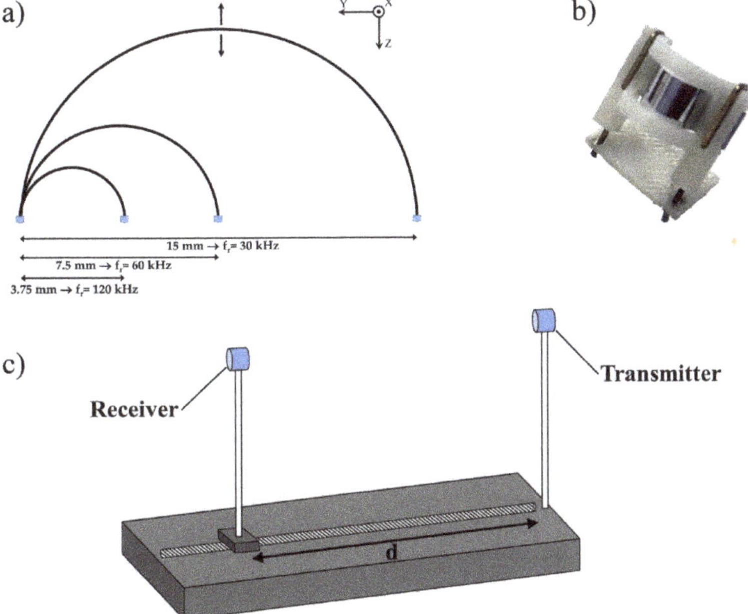

Figure 1. Schematic of hemicylindrical geometries resonating at 30, 60 and 120 kHz, respectively (**a**). A 60 kHz fabricated transducer (**b**) and Scheme of the experimental setup for the time of flight (TOF) evaluation between transmitter and receiver (**c**).

The experimental set-up was composed by two unimodal 60 kHz PVDF transducers, one transmitter and one receiver, facing each other at a variable distance d (Figure 1c). The transmitter was characterized by a sound pressure level (SPL) of 105 dB, considering a reference pressure of 20 μPa (0 dB) at 0.3 m. The receiver, instead, had a sensitivity of −80 dB, considering a reference sensitivity of 10 V/Pa (0 dB). Both unimodal transducers had a bandwidth of 5 kHz [25,43,44]. The acoustic beam was generated by driving the PVDF transmitter with a pulse of 10 sinusoidal cycles at 60 kHz, with a peak-to-peak voltage of 2V (Tektronix AFG3102), amplified by 36.5 dB through a power amplifier stage. The echo conditioning circuit was composed by a low noise amplifier, a band pass filter and a further amplifier stage. The PVDF receiver was shunted by a couple of diodes with the purpose of protecting the low-noise amplification stage from excessive amplitude voltage signals that the transmitting stage or other noise sources could capacitively induce. The ultrasonic beam was characterized according to the Institute of Electrical and Electronics Engineers (IEEE) international standard by means of intensity parameters. The spatial peak-temporal peak intensity (I_{sptp}), spatial peak time average (I_{spta}) and spatial peak pulse average (I_{sppa}) were determined over a plane 300 mm from the ultrasonic transmitter, using a wide-band system composed by a conditioning amplifier (Brüel and Kjaer NEXUS 2692-C) and a $\frac{1}{4}''$ free-field microphone, 4 to 100 kHz, 200 V polarization (Brüel and Kjaer, Type 4939). The ultrasonic signal was detected by the PVDF receiver, conditioned, and the was voltage recorded by a digital oscilloscope (Tektronix DPO 3054) [45].

2.2. Monitoring Routine

The transmitter was driven with a sinusoidal burst with a frequency f_r, allowing the generation of an acoustic signal, which was propagated toward the target (receiver), then transduced and conditioned, obtaining a voltage profile as shown in Figure 2. The cross-correlation gives a measure of waveform similarities while shifting one of them onto the other. Since the cross-correlation of white noise approaches to zero, the cross-correlation was inherently characterized by noise reduction. Moreover, in order to reduce the frequency and phase errors, the signal envelope was obtained before starting the signal processing [19]. Given two digital sequences $y_P(kT^S)$ and $y_E(kT_S)$ of the pulse and echo signals, respectively, where T_S is the sampling time, the cross-correlation is given by:

$$X_C = \sum_{-\infty}^{+\infty} y_P(kT_S) \cdot y_E(kT_S + nT_S) \tag{1}$$

The estimation of the time delay between the two sequences was evaluated trough the maximum of Xc. Let us now consider the signal as shown in Figure 2, used to represent the transmitted pulse (red shaded area) and the received echo (green shaded area). In the time domain the differences between the maximum of the echo signal (t_b) and the related pulse transmission time (t_a) represents the time elapsed between ultrasonic source transmission and echo reception.

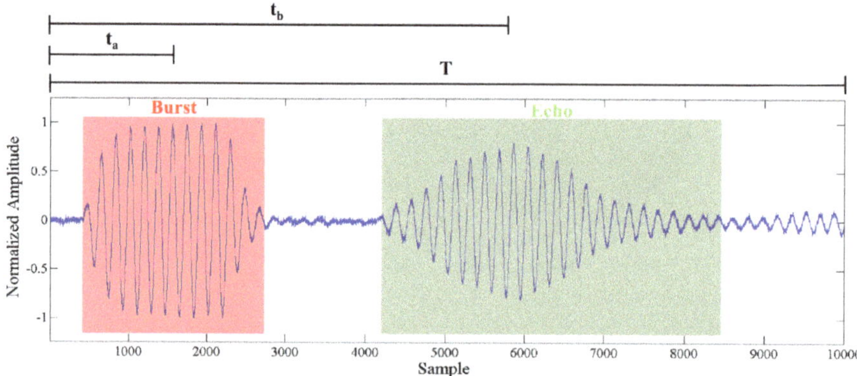

Figure 2. Pulse-echo signal (60 kHz) transmitted and received through a couple of hemicylindrical polyvinylidene fluoride (PVDF) transducers.

The distance was then computed by taking into account the sound velocity in air ($d = TOF \cdot v$). Even though variable (influence of temperature, humidity, etc.), the sound velocity in air can be modeled with good approximation by $v = 20.555 \cdot \sqrt{T}$, where T is the temperature in Kelvin, to take into account the environmental conditions [46,47]. Since the time reference is used in signal acquisition, an accurate pulse-echo acquisition is necessary. Synchronization can be inherently affected by frequency errors (i.e., nonconstant errors) and in case of multiple reference signals these errors can affect each other. A time shift can be observed also in the case of a single reference signal used to synchronize transmission and reception. These synchronization errors are due to different factors, such as local temperature random errors. This means that the TOF is affected by smaller variations happening continually (i.e., time shift of the pulse and echo maximum t_a and t_b). As shown in the flowchart (Figure 3), the processing technique starts with the acquisition of a pulse-echo signal at a given distance, named template signal, then the following steps were carried out: (i) selection of the pulse component $s_a(t)$) and echo component ($s_b(t)$, (ii) cross-correlation between two subsequent acquired signals and the pulse-echo, respectively, (iii) TOF evaluation and return to the acquisition of a new set of signals.

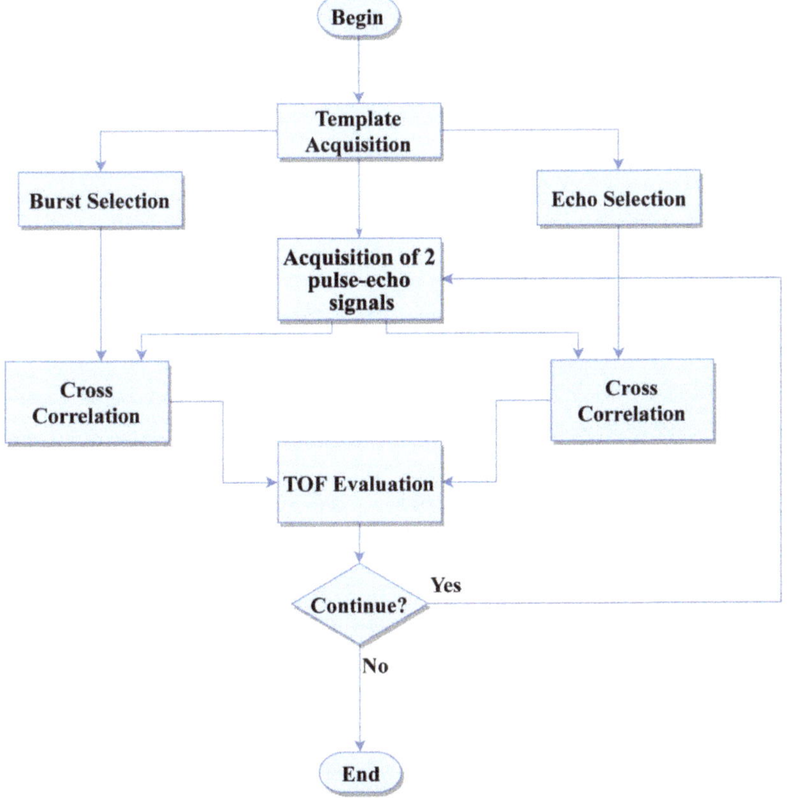

Figure 3. Transmission and reception model of the setup.

The use of a template signal allows for the referencing of all the cross-correlations to the same signal, which is expected to affect the accuracy of the TOF evaluation, especially on multiple cyclic transmissions/receptions. Moreover, in the proposed implementation, no envelope extraction was investigated. Considering two acquired pulse-echo signals, $s_1(t)$ and $s_2(t)$, shifted with respect to the template, similarly to what was done for the template signal (Figure 2), t_c, t_d, t_{c2}, and t_{d2} indicate the referenced time at pulse, and the maximum echo time of $s_1(t)$ and $s_2(t)$, respectively. The proposed TOF estimation through the modified cross-correlation technique according to the procedure previously described can be seen in Figure 4.

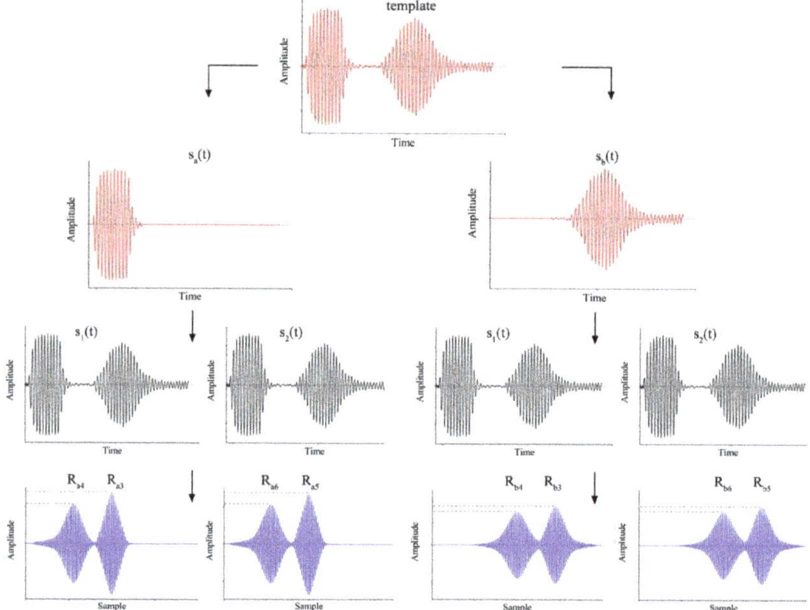

Figure 4. Ultrasound signal processing steps for TOF estimation through a modified cross-correlation-based technique: in red is reported the template signal, properly filtered in transmitter ($s_a(t)$) and receiver ($s_b(t)$); subsequently, two further acquired signals ($s_1(t)$ and $s_2(t)$ (in black)), were opportunely cross-correlated with the two templates, obtaining 4 cross-correlations (in blue); local cross-correlation maxima, related to the corresponding time shift for homologous (pulse–pulse) and nonhomologous (pulse-echo) signals, were obtained as reported in Table 1.

The cross-correlation between $s_a(t)$ and the template is in general characterized by two local maxima, the first, R_{a1}, related to the maximum overlap between homologous (pulse–pulse) signals, while the second, R_{a2}, related to the maximum overlap between nonhomologous (pulse-echo) signals (not shown in Figure 4). Similarly, the cross-correlation between $s_b(t)$ and the template evidenced other two local maxima, R_{b1} (pulse–pulse) and R_{b2} (pulse-echo). The same steps have been performed between the two template signals and $s_1(t)$, $s_2(t)$. According to the proposed technique, 4 cross-correlations were evaluated providing multiple maxima, each one related to a specific time shift. Moreover, two more maxima were related to the cross-correlation of the template signal with $s_a(t)$ and $s_b(t)$, which provides the calibrated initial position. A maxima evaluation of the pulse-echo and cross-correlation signals involves the selection of an appropriate Dirichlet window, with a time length L. The start and end of the window involves, firstly, the signal being rectified, binned (2 samples) and then set to a threshold (average value of the processed signal) with a window length overestimation of 10% (Figure 5). As each cross-correlation sample correlated with a specific time shift, the combination of information carried out by multiple cross-correlations can be used to retrieve the TOF related to the signals $s_1(t)$ and $s_2(t)$.

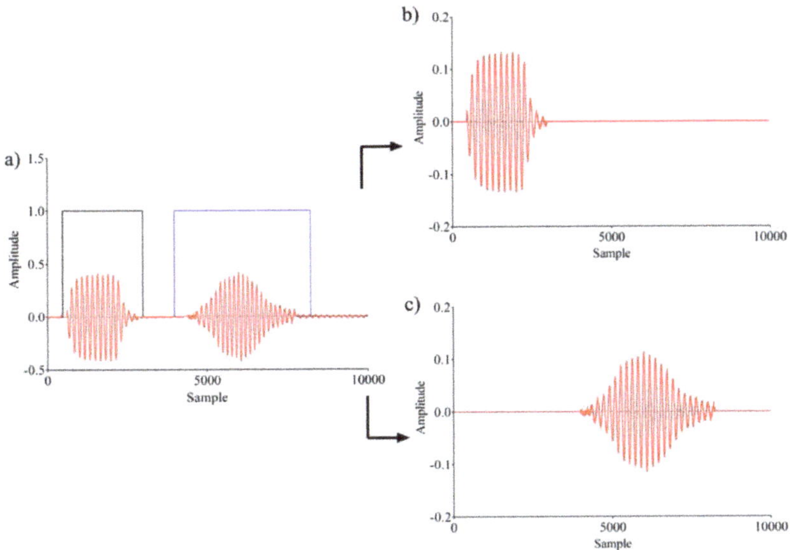

Figure 5. Representative signal windowing for maximum evaluation.

In Table 1, the local cross-correlation maximum is related to the specific time shift on which the proposed implementation is based. Therefore, the TOF evaluation is not affected by the choice of pulse–pulse or pulse-echo local maximum and, considering that the reference signal is the same, these times are expected to be more accurate than the times of flight evaluated through threshold and standard cross-correlation methods.

Table 1. Local cross-correlation maximum related to the corresponding time shift for homologous (pulse–pulse) and nonhomologous (pulse-echo) signals.

	Time	Pulse–Pulse Maxima	Pulse-Echo Maxima
R_{a1}	T		
R_{a2}	$(T - t_d) + t_a$		
R_{b1}	$(T - t_c) + t_b$		
R_{b2}	$T - d$		$R_{b1} - R_{a2} = TOF_0 + k$
R_{a3}	T	$R_{a1} - R_{b2} = TOF_0$	$k = t_b - t_a$
R_{b3}	$(T - t_{c1}) + t_a$	$R_{a3} - R_{b4} = TOF_1$ $d_1/v = (TOF_1 - TOF_0)$	$R_{b3} - R_{a4} = TOF_1 + k$ $d_1/v = (TOF_1 + k) - (TOF_0 + k) = TOF_1 - TOF_0$
R_{a4}	$(T - t_{d1}) + t_a$	$R_{a5} - R_{b6} = TOF_2$ $d_2/v = TOF_2 - TOF_1$	$R_{b5} - R_{a6} = TOF_2 + k$ $d_2 = (TOF_2 + k) - (TOF_1 + k) = TOF_2 - TOF_1$
R_{b4}	$T - d_1$		
R_{a5}	T		
R_{b5}	$(T - t_{c2}) + t_a$		
R_{a6}	$(T - t_{d2}) + t_a$		
R_{b6}	$T - d$		

T = pulse-echo acquisition time; d = temporal distance between the reference signal and the shifted signal $s_1(t)$; d_1 = temporal distance between the reference signal and the shifted signal $s_2(t)$; TOF= temporal distance between the shifted signals $s_1(t)$ and $s_2(t)$;

In this way, the distance between the transmitter and receiver can be evaluated by observing TOF increments with respect to the template signal (placed at a calibrated distance, related to TOF_0). As we can verify, TOF_1 and TOF_2 can be alternatively obtained by analyzing the homologous or nonhomologous components of the cross-correlation. The reliability of the three methods were compared by the standard error SE = $\sqrt{(\sigma^2/n)}$, where σ^2 is the sample variance and n is the sample size. Since a recursive evaluation is often required in positioning systems, investigations were performed by moving the receiver back and forth.

2.3. Experimental Validation

A set-up was fabricated in order to investigate the performance comparison between the threshold, standard and modified cross-correlation technique (Figure 1c). The system includes a threaded rod (M10 with a pitch of 1.5 mm), which is rotated by a 4-phase unipolar stepper motor (RS Components, Corby, UK) with a 7.5° step angle, 0.24 Nm holding torque and a positioning accuracy of 5%. The stepper motor has been driven by using a national instrument DAQ6015 board. A hemicylindrical ultrasonic transmitter was fixed solidly to the threaded rod, while the receiver had been placed at a reference position.

On the base of the number of steps and therefore the angular variation of the bar, the linear movement could be traced, apart from the errors due to the motor positioning and mechanical tolerances on the bar, which are assumed constant during the experimental evaluation. Considering the step angle and the pitch, the minimum longitudinal distance was evaluated by $d_L = (p \cdot \varphi)/360$ (i.e., 0.03 mm). The supports, instead, gave the right height and the right alignment to the two sensors, so that the obstacles in the immediate vicinity did not create multiple reflections and, therefore, an echo signal with the presence of unwanted components. The impedance analysis and frequency response of the PVDF transducer evidenced the characteristic electric resonance feature (Figure 6a) and the bandwidth (Figure 6b) of the hemicylindrical sensor [48–50].

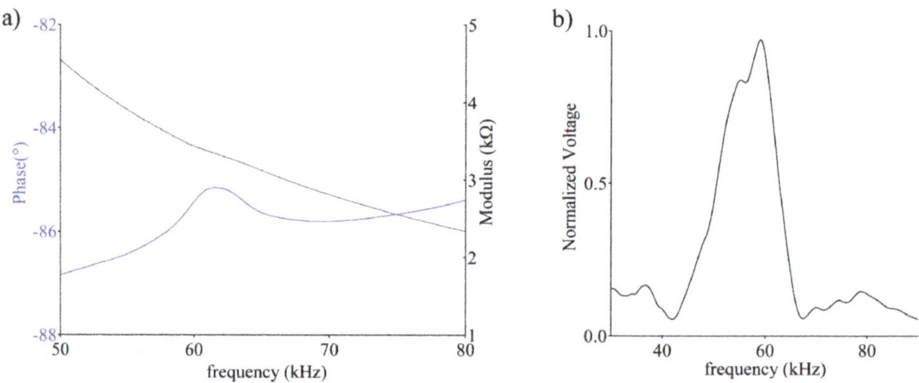

Figure 6. Impedance analysis (**a**) and frequency response (**b**) of a 60 kHz hemicylindrical PVDF transducer.

Starting from a predefined transmitter/receiver distance (set to 0.3 m), the stepper motor was driven to obtain a variable number of the turns from 1 (d_L = 1.25 mm corresponding to 48 motor steps) up to 5 (d_L = 6.25 mm corresponding to 240 motor steps) and the distance was maintained within 0.6 m. For each position, the signal acquisition was repeated four times for the statistical analysis. The effect of the pulse length was also investigated by changing the number of cycle N from 5 up to 15, corresponding to a pulse time duration of 83.3, 106.6 and 249.9 µs. The relationship between the actual distance and the relationship evaluated by the threshold, standard and modified cross-correlations were then compared.

3. Results

Three excitation pulse signals were used to drive the PVDF transmitter. The stability of the excitation source was of ±1 ppm ±1 µHz, 0 to 50 °C, with expected amplitude variations < 10 mV.

Therefore, the pulse was stable and controllable enough to be used as a reference signal for the cross-correlation method. The SNR was evaluated to be > 30 dB during all the acquisitions. As depicted in Figure 1c, the analyzed case is that of a transmitter facing a receiver with a separation distance controlled by the stepper motor.

The stepper motor was controlled by changing the turns and the TOF was subsequently evaluated with each method. Subsequently, the distance is computed taking into account the sound velocity in air by compensating the temperature fluctuation through a sensor, resulting in an uncertainty on the sound velocity of less than 0.05 m/s [46]. Figure 7a–c shows the comparison among threshold, standard and modified cross-correlation in the evaluation of TOF using a variable pulse length as previously reported, respectively. As expected, the standard and modified cross-correlation techniques performed better in terms of standard error (SE) and linearity with respect to the threshold technique (Figure 7d).

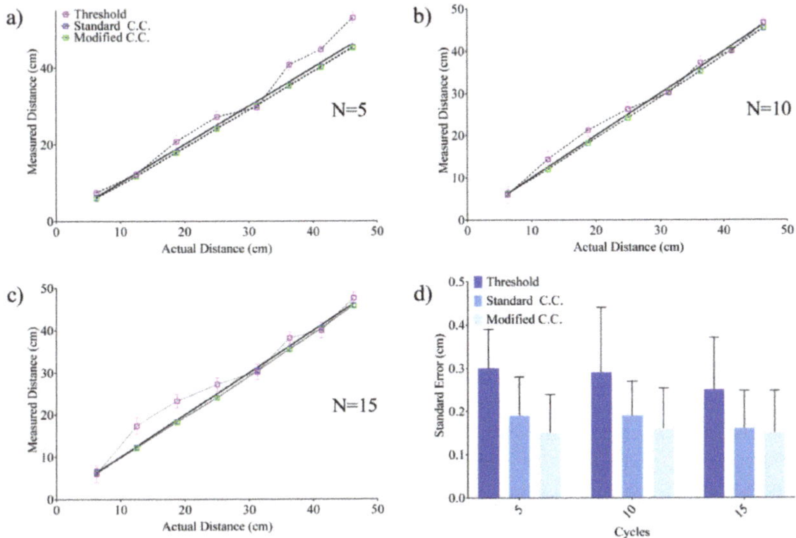

Figure 7. Comparison of distance evaluation used for a pulse length of (**a**) 5 sinusoidal cycles, (**b**) 10 sinusoidal cycles and (**c**) 15 sinusoidal cycles at 60 kHz. (**d**) Standard error in the distance evaluation.

The absolute mean errors reported in Figure 8a–c are representative of a target moving in a range of 40 cm, while the standard and modified cross-correlation techniques were used by varying the number of cycles N. Figure 8d reports the maximum error observed in the previously reported cases. In all cases, the results evidenced a nonlinear behavior, which however can be reduced by increasing the number of cycles (Figure 7d).

We additionally evaluated the computational time of both the standard and modified cross-correlations. In light of the results, the modified algorithm requires 70% of an additional computational load in the estimation of the TOF, which can be acceptable in most low frequency positioning systems.

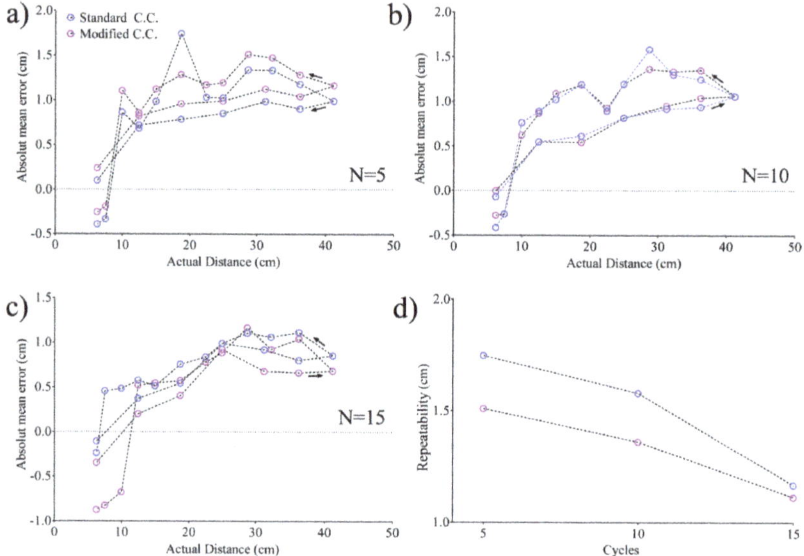

Figure 8. Repeatability evaluation through subsequent cycles of the TOF evaluation at increased/decreased distances using for a pulse length of (**a**) 5 sinusoidal cycles, (**b**) 10 sinusoidal cycles and (**c**) 15 sinusoidal cycles at 60 kHz. (**d**) Maximum repeatability error vs. pulse length.

4. Discussion

Based on the proposed technique, a template signal was evaluated as a reference signal for all the TOF evaluations in order to reduce errors due to synchronization that can be inherently affected by the range. The overall model is suitable in positioning systems working in a confined unstructured environment in which the distance between the transmitter and the target can be evaluated by observing TOF increments with respect to the calibrated position. In Figure 7d, it is clearly shown that a standard and modified cross-correlation exhibits a better performance than the threshold method. When increasing the number of cycles, no differences were highlighted between cross-correlation techniques, while the threshold method evidenced a deteriorated performance. Moreover, remarkable improvements with respect to the threshold technique are clearly observed with a reduction in SE in the order of 45%. Further improvements were also observed with respect to conventional cross-correlations which has been estimated in the order of 20%. This is mainly due to the use of a calibrated reference signal, which reduces the smaller variations that happen continually (i.e., time shift of the pulse and echo maximum). As previously highlighted, it is evident that this improvement is counterbalanced by a higher computational load. Moreover, no significant differences were observed by changing the pulse length in the range from 83.3 up to 249.9 µs, evidencing that it is possible to choose the pulse length in accordance with the requirements of the application without affecting the performances. Interestingly, the experimental results evidenced an enhancement in repeatability of about 10% by continuously changing the distance of the target back and forth, which means that it is possible to compensate for hysteresis-like behavior in the use of a recursive positioning system. Although the computational cost of the algorithm is higher than that of the compared techniques, it still guarantees the possibility of obtaining data in real-time for the specific application. In fact, an algorithm has been conceived for the monitoring and tracking of the driver, where a more accurate knowledge of driver dynamics can be used complementarily with other systems, providing shared information (e.g., the calibration of alcohol monitoring systems). The use of a single template signal for all the TOF evaluations can be advantageously applied in positioning systems based on multiple transmission/reception points, to reduce the time shift introduced by multiple reference signals. Moreover, the implementation of the combination of multiple data retrieved from a standard cross-correlation can reduce the time shift that can also be observed in the case of a single reference signal.

5. Conclusions

A study on a modified algorithm based on the cross-correlation technique for the evaluation of time of flight specifically designed for a recursive data evaluation was investigated. This proposed algorithm was implemented in MATLAB and a comparison with threshold and standard cross-correlation techniques was presented. The conventional resolution in SONAR is limited by the wavelength and, subsequently, different signal processing techniques, such as those based on cross-correlations. Of course, one of the ways to improve the overall performance of the system is to increase the ultrasound source frequency (i.e., a lower wavelength), and different SONAR systems were recently proposed in order to allow a frequency shift using wideband transducers. Obviously, ultrasonic attenuation in air dramatically increases as the frequency increases. The modified algorithm evidenced improvements with respect to both threshold and conventional cross-correlation techniques, with a reduction in the standard error of about 45% and 20%, respectively. On the other hand, an increase of 70% of computational load has been estimated in the evaluation of TOF. Nonintrusive on-board driver positioning can benefit the recursive nature of the algorithm and the electronic sensors investigated.

Author Contributions: Conceptualization, S.A.P., A.S.F., A.L.G.; methodology, A.S.F., A.L.G., S.A.P. D.C.C:, M.M.; software, M.G.B.; validation, S.A.P., M.G.B., M.M, formal analysis, S.A.P., M.G.B., D.C.C., M.M.; investigation, S.A.P., M.G.B.; resources, A.S.F., A.L.G.; data curation, S.A.P., M.G.B., writing—original draft preparation, S.A.P., M.G.B.; writing—review and editing, S.A.P., M.G,B., D.C.C., M.M., A.L.G., A.S.F.; visualization, S.A.P., M.G.B., D.C.C., M.M., A.L.G., A.S.F.; supervision, A.S.F., A.L.G.; project administration, A.S.F., A.L.G.; funding acquisition, A.S.F., A.L.G. All authors have read and agreed to the published version of the manuscript.

Funding: This research received no external funding.

Conflicts of Interest: The authors declare no conflict of interest.

References

1. Rocchi, A.; Santecchia, E.; Ciciulla, F.; Mengucci, P.; Barucca, G. Characterization and optimization of level measurement by an ultrasonic sensor system. *IEEE Sens. J.* **2019**, *19*, 3077–3084. [CrossRef]
2. Rocchia, A.; Santecchia, E.; Barucca, G.; Menguccia, P. Optimization of distances measurement by an ultrasonic sensor. *Mater. Today Proc.* **2019**, *19*, 33–39.
3. Malvasi, A.; Baldini, D. *Pick Up and Oocyte Management*; Springer: London, UK, 2020.
4. Santagati, G.E.; Melodia, T. Experimental Evaluation of Impulsive Ultrasonic Intra-Body Communications for Implantable Biomedical Devices. *IEEE Trans. Mob. Comput.* **2016**, *16*, 367–380. [CrossRef]
5. Bianco, M.G.; Pullano, S.A.; Citraro, R.; Russo, E.; De Sarro, G.; de Villers Sidani, E.; Fiorillo, A.S. Neural Modulation of the Primary Auditory Cortex by Intracortical Microstimulation with a Bio-Inspired Electronic System. *Bioengineering* **2020**, *7*, 23. [CrossRef]
6. Saad, M.; Bleakley, C.J.; Nigram, V.; Kettle, P. Ultrasonic hand gesture recognition for mobile devices. *J. Multimodal User Interfaces* **2018**, *12*, 31–39. [CrossRef]
7. Fiorillo, A.S. Design and Characterization of a PVDF Ultrasonic Range Sensor. *IEEE Trans. Ultrason. Ferroelectr. Freq. Control.* **1992**, *39*, 688–692. [CrossRef]
8. Pinggera, P.; Franke, U.; Mester, R. High-performance long range obstacle detection using stereo vision. In Proceedings of the 2015 IEEE/RSJ International Conference on Intelligent Robots and Systems (IROS), Hamburg, Germany, 28 September–2 October 2015.
9. Rabadan, J.; Guerra, V.; Rodríguez, R.; Rufo, J.; Luna-Rivera, M.; Perez-Jimenez, R. Hybrid Visible Light and Ultrasound-Based Sensor for Distance Estimation. *Sensors* **2017**, *17*, 330. [CrossRef]
10. Dahl, T.; Ealo, J.L.; Bang, H.J.; Holm, S.; Khuri-Yakub, P. Applications of airborne ultrasound in human-computer interaction. *Ultrasonics* **2014**, *54*, 1912–1921. [CrossRef]
11. Carotenuto, R.; Merenda, M.; Iero, D.; Della Corte, F. Mobile Synchronization Recovery for Ultrasonic Indoor Positioning. *Sensors* **2020**, *20*, 702. [CrossRef]

12. Yan, X.; Wu, Q.; Wang, X.; Sun, X. Semicool Temperature Compensation Algorithm Based on the Double Exponential Model in the Ultrasonic Positioning System. *IEEE Trans. Instrum. Meas.* **2020**, *69*, 995–1010. [CrossRef]
13. Toda, M.; Thompson, M. Detailed investigations of polymer/metal multilayer matching layer and backing absorber structures for wideband ultrasonic transducers. *IEEE Trans. Ultrason. Ferroelectr. Freq. Control.* **2012**, *59*, 231–242. [CrossRef] [PubMed]
14. Fiorillo, A.S.; Grimaldi, D.; Paolino, D.; Pullano, S.A. Low-frequency ultrasound in medicine: An in vivo evaluation. *IEEE Trans. Instrum. Meas.* **2012**, *61*, 1658–1663. [CrossRef]
15. Gurkan, K.; Akan, A. Simulation and Measurement of Air-Coupled Semi-Circular and Conical PVDF Sensors. *IEEE Sens. J.* **2016**, *16*, 983–988. [CrossRef]
16. Gohl, P.; Honegger, D.; Omari, S.; Achtelik, M.; Pollefeys, M.; Siegwart, R. Omnidirectional visual obstacle detection using embedded. In Proceedings of the 2015 IEEE/RSJ International Conference on Intelligent Robots and Systems (IROS), Hamburg, Germany, 28 September–2 October 2015.
17. Ionescu, R.; Carotenuto, R.; Urbani, F. 3D localization and tracking of objects using miniature microphones. *Wirel. Sens. Netw.* **2011**, *3*, 147. [CrossRef]
18. Chassagne, L.; Bruneau, O.; Bialek, A.; Falguière, C.; Broussard, E.; Barrois, O. Ultrasonic sensor triangulation for accurate 3D relative positioning of humanoid robot feet. *IEEE Sens. J.* **2015**, *15*, 2856–2865. [CrossRef]
19. Jia, L.; Xue, B.; Chen, S.; Wu, H.; Yang, X.; Zhai, J.; Zeng, Z. A High-Resolution Ultrasonic Ranging System Using Laser Sensing and a Cross-Correlation Method. *Appl. Sci.* **2019**, *9*, 1483. [CrossRef]
20. Villladangos, J.M.; Ureña, J.; García, J.J.; Mazo, M.; Hernández, Á.; Jiménez, A.; Ruíz, D.; Marziani, C.D. Measuring Time-of-Flight in an Ultrasonic LPS System Using Generalized Cross-Correlation. *Sensors* **2011**, *11*, 10326–10342. [CrossRef]
21. Toda, M. High sensitivity and wideband design for impedance matching layer between protection Metal and PZT. In Proceedings of the IEEE International Ultrasonics Symposium, Washington, DC, USA, 6–9 September 2017.
22. Howcroft, J.; Wallace, B.; Goubran, R.; Marshall, S.; Porter, M.M.; Knoefel, F. Trip-Based Measures of Naturalistic Driving: Considerations and Connections with Cognitive Status in Older Adult Drivers. *IEEE Trans. Instrum. Meas.* **2019**, *68*, 2451–2459. [CrossRef]
23. Lavergne, T.; Škvor, Z.; Husník, L.; Bruneau, M. On the modeling of an emitting cylindrical transducer with a piezoelectric polymer membrane. *Acta Acust. United Acust.* **2016**, *102*, 705–713. [CrossRef]
24. Park, J.; Je, Y.; Lee, H.; Moon, W. Design of an ultrasonic sensor for measuring distance and detecting obstacles. *Ultrasonics* **2010**, *50*, 340–346. [CrossRef]
25. Fiorillo, A.S.; Pullano, S.A.; Bianco, M.G.; Critello, C.D. Bioinspired US sensor for broadband applications. *Sens. Actuators A Phys.* **2019**, *294*, 148–153. [CrossRef]
26. Chen, J.; Zhao, J.; Lin, L.; Sun, X. Truncated Conical PVDF Film Transducer for Air Ultrasound. *IEEE Sens. J.* **2019**, *19*, 8618–8625. [CrossRef]
27. Corcoran, A.; Moss, C.F. Sensing in a noisy world: Lessons from an auditory specialist, the echolocating bat. *J. Exp. Biol.* **2017**, *220*, 4554–4566. [CrossRef]
28. Fiorillo, A.S.; D'Angelo, G. Echo signals processing with neural network in bat-like sonars based on PVDF. In Proceedings of the IEEE Ultrasonics Symposium, Munich, Germany, 8–11 October 2002.
29. Grimaldi, D. Time-of-Flight Measurement of Ultrasonic Pulse Echoes Using Wavelet Networks. *IEEE Trans. Instrum. Meas.* **2006**, *55*, 5–13. [CrossRef]
30. Angrisani, L.; Bechou, L.; Dallet, D.; Daponte, P.; Ousten, Y. Detection and location of defects in electronic devices by means of scanning ultrasonic microscopy and the wavelet transform. *Measurement* **2002**, *31*, 77–91. [CrossRef]
31. Jackson, J.C.; Summan, R.; Dobie, G.I.; Whiteley, S.M.; Gareth Pierce, S.; Hayward, G. Time-of-Flight Measurement Techniques for Airborne Ultrasonic Ranging. *IEEE Trans. Ultrason. Ferroelectr. Freq. Control.* **2013**, *60*, 343–355. [CrossRef]
32. Khyam, M.O.; Ge, S.S.; Li, X.; Pickering, M. Orthogonal Chirp-Based Ultrasonic Positioning. *Sensors* **2017**, *17*, 976. [CrossRef]
33. Zhang, H.; Wang, Y.; Zhang, X.; Wang, D.; Jin, B. Design and Performance Analysis of an Intrinsically Safe Ultrasonic Ranging Sensor. *Sensors* **2016**, *16*, 867. [CrossRef]
34. Suga, N. Cortical Computational Maps for Auditory Imaging. *Neural Netw.* **1990**, *3*, 3–21. [CrossRef]

35. Inoue, S.; Kimyou, M.; Kashimori, Y.; Hoshino, O.; Kamba, T. A neural model of medial geniculate body and auditory cortex detecting target distance independently of target velocity in echolocation. *Neurocomputing* **2000**, *32–33*, 833–841. [CrossRef]
36. Zafari, F.; Gkelias, A.; Leung, K.K. A Survey of Indoor Localization Systems and Technologies. *IEEE Commun. Surv. Tutor.* **2019**, *21*, 2568–2599. [CrossRef]
37. Pelenis, D.; Barauskas, D.; Vanagas, G.; Dzikaras, M.; Viržonis, D. CMUT-based biosensor with convolutional neural network signal processing. *Ultrasonics* **2019**, *99*, 105956. [CrossRef] [PubMed]
38. Kim, E.Y. Wheelchair Navigation System for Disabled and Elderly People. *Sensors* **2016**, *16*, 1806. [CrossRef] [PubMed]
39. Ealo, J.L.; Camacho, J.J.; Fritsch, C. Airborne ultrasonic phased arrays using ferroelectrets: A new fabrication approach. *IEEE Trans. Ultrason. Ferroelectr. Freq. Control.* **2009**, *56*, 848–858. [CrossRef] [PubMed]
40. Marioli, D.; Narduzzi, C.; Offelli, C.; Petri, D.; Sardini, E.; Taroni, A. Digital time of flight measurement for ultrasonic sensors. *IEEE Trans. Instrum. Meas. Technol.* **1992**, *41*, 93–97. [CrossRef]
41. Gurbatov, S.N.; Rudenko, O.V.; Saichev, A.I. Types of Acoustic Nonlinearities and Methods of Nonlinear Acoustic Diagnostics. *Waves Struct. Nonlinear Nondispersive Media* **2011**, 271–307. [CrossRef]
42. Toda, M.; Tosima, S. Theory of curved, clamped, piezoelectric film, air-borne transducers. *IEEE Trans. Ultrason. Ferroelectr. Freq. Control.* **2000**, *47*, 1421–1431. [CrossRef]
43. Fiorillo, A.S.; Pullano, S.A.; Critello, C.D. Spiral-Shaped Biologically-Inspired Ultrasonic Sensor. *Trans. Ultrason. Ferroelectr. Freq. Control.* **2019**, *67*, 635–642. [CrossRef]
44. Fiorillo, A.S.; Pullano, S.A.; Bianco, M.G.; Critello, C.D. Ultrasonic Transducers Shaped in Archimedean and Fibonacci Spiral: A Comparison. *Sensors* **2020**, *20*, 2800. [CrossRef]
45. Fiorillo, A.S.; Pullano, S.A.; Menniti, M.; Bianco, M.G.; Critello, C.D. Ultrasonic Transducer for Broadband Applications. *Lect. Notes Electr. Eng.* **2020**, 629. [CrossRef]
46. Pullano, S.A.; Islam, S.K.; Fiorillo, A.S. Pyroelectric Sensor for Temperature Monitoring of Biological Fluids in Microchannel Devices. *IEEE Sens. J.* **2014**, *14*, 2725–2730. [CrossRef]
47. Carullo, A.; Parvis, M. An ultrasonic sensor for distance measurement in automotive applications. *IEEE Sens. J.* **2001**, *1*, 143–147. [CrossRef]
48. Pullano, S.A.; Fiorillo, A.S.; La Gatta, A.; Lamonaca, F.; Carnì, D.L. Comprehensive System for the Evaluation of the Attention Level of a Driver. In Proceedings of the IEEE International Symposium on Medical Measurements and Applications, Benevento, Italy, 15–18 May 2016.
49. Wang, H.-Y.; Zhao, M.-M.; Beurier, G.; Wang, X.-G. Automobile Driver Posture Monitoring Systems: A Review. *China J. Highw. Transp.* **2019**, *2*, 1–18.
50. Pullano, S.A.; Fiorillo, A.S.; Vanello, N.; Landini, L. Obstacle Detection System based on Low Quality Factor Ultrasonic Transducers for Medical Devices. In Proceedings of the IEEE International Symposium on Medical Measurements and Applications, Benevento, Italy, 15–18 May 2016.

Article

Adaptive Residual Weighted K-Nearest Neighbor Fingerprint Positioning Algorithm Based on Visible Light Communication

Shiwu Xu [1,2], Chih-Cheng Chen [3,4], Yi Wu [1,*], Xufang Wang [1] and Fen Wei [1]

[1] Key Laboratory of OptoElectronic Science and Technology for Medicine of Ministry of Education, College of Photonic and Electronic Engineering, Fujian Normal University, Fuzhou 350007, China; shiwuxu0501@163.com (S.X.); fzwxf@fjnu.edu.cn (X.W.); fafu.weifen@gmail.com (F.W.)

[2] Concord University College, Fujian Normal University, Fuzhou 350117, China

[3] Department of Aeronautical Engineering, Chaoyang University of Technology, Taichung 413310, Taiwan; ccc@gm.cyut.edu.tw

[4] Information and Engineering College, Jimei University, Fujian 361021, China

*   Correspondence: wuyi@fjnu.edu.cn; Tel.: +86-0591-2286-8159

Received: 6 July 2020; Accepted: 6 August 2020; Published: 8 August 2020

Abstract: The weighted K-nearest neighbor (WKNN) algorithm is a commonly used fingerprint positioning, the difficulty of which lies in how to optimize the value of K to obtain the minimum positioning error. In this paper, we propose an adaptive residual weighted K-nearest neighbor (ARWKNN) fingerprint positioning algorithm based on visible light communication. Firstly, the target matches the fingerprints according to the received signal strength indication (RSSI) vector. Secondly, K is a dynamic value according to the matched RSSI residual. Simulation results show the ARWKNN algorithm presents a reduced average positioning error when compared with random forest (81.82%), extreme learning machine (83.93%), artificial neural network (86.06%), grid-independent least square (60.15%), self-adaptive WKNN (43.84%), WKNN (47.81%), and KNN (73.36%). These results were obtained when the signal-to-noise ratio was set to 20 dB, and Manhattan distance was used in a two-dimensional (2-D) space. The ARWKNN algorithm based on Clark distance and minimum maximum distance metrics produces the minimum average positioning error in 2-D and 3-D, respectively. Compared with self-adaptive WKNN (SAWKNN), WKNN and KNN algorithms, the ARWKNN algorithm achieves a significant reduction in the average positioning error while maintaining similar algorithm complexity.

Keywords: visible light communication; indoor positioning system; fingerprint positioning; weighted K-nearest neighbor; distance metric

1. Introduction

Positioning systems can be divided into outdoor positioning system (OPS) and indoor positioning system (IPS). The OPS usually uses global positioning system (GPS) to obtain the coordinates of the target. Since the GPS signal is not able to penetrate the wall and other obstacles, GPS cannot be applied in the indoor positioning scene [1]. As a supplement to OPS, IPS has attracted increasing attention among researchers. At present, there are two main research areas on IPS. One is based on radio frequency communication technology, such as radio frequency identification (RFID) [2], wireless sensor network (WSN) [3], ultra-wideband (UWB) [4], wireless fidelity (WiFi) [5], Bluetooth [6], etc. The other is based on visible light communication (VLC) [7]. IPS can be divided into range-based IPS and range-free IPS. The methods of range-based include time of arrival (TOA), angle of arrival (AOA), and received signal strength indication (RSSI), etc. [8,9]. The range-free IPS usually uses fingerprint matching to achieve positioning [10]. Compared with radio frequency communication technology, using a light-emitting diode (LED) to achieve indoor positioning has the following advantages: (1) LED communication uses the visible light spectrum, which can be applied to some areas where electromagnetic radiation is prohibited, such as operating rooms and gas stations; (2) generally, LED is uniformly distributed on the ceiling, there is mainly line-of-sight (LoS) communication between transceivers and receivers; (3) existing LED lighting devices can be used directly, and the receiver can use integrated photodiode (PD) devices [11,12]; (4) the signal-to-noise ratio (SNR) is usually very high due to lighting requirements.

Typical fingerprint-based localization algorithms usually use machine-learning algorithms [13], for example, random forest (RF) [14], K-nearest neighbor (KNN) [15], extreme learning machine (ELM) [16], artificial neural network (ANN) [17], etc. In [11], for indoor positioning based on VLC, three classical machine-learning algorithms, RF, ELM and KNN are adopted to train multiple classifiers based on received signal strength indication (RSSI) fingerprints, and a grid-independent least square (GI-LS) algorithm was proposed to combine the outputs of these classifiers. Experimental results show that compared with RF, KNN and ELM algorithms, the positioning error based on the GI-LS algorithm is lower. In machine learning algorithms, K-nearest neighbor (KNN) is one of the most widely used fingerprint positioning. The KNN fingerprint positioning algorithm [15] works in two stages. The first one is run offline, and it consists of generating a set of fingerprint points in the application area. In the second step, the target measures an RSSI vector of M LEDs, which is then matched with the K nearest fingerprints obtained previously offline. When K fingerprints have different weights, this method is called the weighted K-nearest neighbor (WKNN) fingerprint positioning algorithm. The WKNN localization algorithm is based on the shortest physical distance between fingerprints and the target position [12,18,19], which usually adopts two ranging methods: Euclidean distance [18] and Manhattan distance [19]. In Hu et al. [19], for indoor positioning based on WiFi, a self-adaptive WKNN (SAWKNN) algorithm with a dynamic K was proposed. Experimental results show that the positioning error based on the SAWKNN algorithm is lower than that of the WKNN algorithm. In most cases, M LEDs are laid out on the ceiling of the same horizontal plane. The traditional trilateration method and least linear multiplication method can only solve the two-dimensional (2-D) coordinates of targets [20], and the height of the target from the floor needs to be known in advance, which is not feasible in many applications. A Newton–Raphson method was proposed in Şahin et al. [21] and Mathias et al. [22] to estimate the PD location. For a non-convex optimization problem of 3-D positioning, it is easy for the least linear multiplication method and Newton–Raphson method to fall into the local optimal solution, resulting in large positioning error. Particle swarm optimization [23] and differential evolution algorithm [24] are adopted to perform 3-D visible light positioning, which will increase the complexity of the algorithm. In Van et al. [25], compared with trilateration method in the case of ambient light interference and without ambient light interference, simulation results show that the positioning accuracy of the WKNN algorithm is improved by 36% and 50%, respectively. In Alam et al. [12], experiment results show that the average positioning error of the fingerprints established by Lambertian regeneration model is close to that of the actual RSSI measurement fingerprints, which are 2.7 cm and 2.2 cm, respectively. Therefore, the WKNN positioning algorithm based on VLC does not need a large number of human resources to acquire the fingerprints. In Gligorić et al. [26], a visible light localization algorithm based on compressed sensing (CS) was proposed. The orthogonal matching pursuit (OMP) reconstruction algorithm [27] is used to determine the overlapping region, and the KNN algorithm is used to determine the coordinates of the target. In Zhang et al. [28], an visible light inversion positioning system based on CS and a 4-sparse 2-D fingerprint matching algorithm was proposed. When CS is used to realize fingerprint positioning, the measurement matrix needs to satisfy the restricted isometry property (RIP) attribute, and the orthogonal decomposition of the measurement matrix is needed [29,30], which will increase the complexity of the algorithm. The fingerprint positioning algorithm based on CS must satisfy O (K log (N)), where this is the value of the number of measurements M (i.e., the number of LEDs) [29]. When the neighboring fingerprints K

and the number of fingerprints N become larger, a high-density LED layout is required to satisfy the compression sensing reconstruction condition. However, an excessively dense LED layout not only wastes resources but also increases interference between LEDs. Although the WKNN algorithm is a commonly used fingerprint positioning, the difficulty lies in how to optimize the value of K to obtain the minimum positioning error. Compared with the traditional WKNN positioning algorithm, this paper makes the following contributions:

1. A novel adaptive residual weighted K-nearest neighbor (ARWKNN) fingerprint positioning algorithm is proposed, and K is a dynamic value according to the matched RSSI residual. As far as the authors know, most fingerprint positioning algorithms based on VLC only consider a fixed neighbor value, e.g., [12,25,28].
2. The impact of modulation bandwidth, transmit power, the signal-to-noise ratio, the maximum number of neighboring fingerprints, the sampling interval, the number of LEDs, the sampling ratio, and distance metric on positioning errors are analyzed in detail. The distribution of optimal K and the complexity of the algorithm are also analyzed in detail. The results can provide a useful reference for the design of the actual VLP system.
3. Simulation results show that the ARWKNN algorithm based on Clark distance (CLD) and minimum maximum distance (MMD) metrics produces the smallest average positioning error in 2-D and 3-D, respectively, as far as the authors know, this is the first work to report the impact of CLD and MMD metrics on the positioning error of the fingerprint positioning algorithm.
4. Simulation results show that when the SNR is 20 dB, in 2-D, compared with the fingerprint positioning algorithm based on RF [14], ELM [16], ANN [17], GI-LS [11], SAWKNN [19], WKNN [12] or KNN [15,25] algorithms, the average positioning error of the ARWKNN algorithm based on Manhattan distance can be reduced by 81.82%, 83.93%, 86.06%, 60.15%, 43.84%, 47.81%, and 73.36%, respectively. Compared with SAWKNN, WKNN and KNN algorithms, the ARWKNN algorithm can significantly reduce the average positioning error while maintaining similar algorithm complexity.

The rest of this paper is organized as follows: the ARWKNN algorithm is proposed in Section 2. Simulation results are shown and discussed in Section 3. Finally, Section 4 concludes this paper.

Notation: Matrices and vectors are in boldface. The field of real numbers is denoted by $\mathbb{R}$. $\|.\|_2$ is the 2 norm of the vector. $|\cdot|$ is the absolute value, and $\lceil\ \rceil$ denotes the rounding up operator. The transpose operation is denoted by $[.]^T$.

2. Design of the Adaptive Residual Weighted K-Nearest Neighbor (ARWKNN) Algorithm

2.1. System Model

The positioning model is shown in Figure 1a. If there is M_{total} LEDs in the room, the target checks and selects M LEDs that has the highest RSSI for positioning. For simplicity, we assume that the target appears in a 3-D space with M LEDs. The coordinates of M LEDs are $\beta_i = [x_{\text{LED-}i}, y_{\text{LED-}i}, z_{\text{LED-}i}]^T$, for $i = 1, 2, \ldots, M$. It is assumed that M LEDs are evenly distributed on the same horizontal plane, i.e., $z_{\text{LED-}i} = z_{\text{LED}}$, z_{LED} is the height from the floor to the LED. $\alpha_i \in \mathbb{R}^{3\times1}$ represents the angle of the ith LED. $\theta_j \in \mathbb{R}^{3\times1}$ and $\gamma_j \in \mathbb{R}^{3\times1}$ represent the coordinate and angle of the jth fingerprint point, respectively, for $j = 1, 2, \ldots, N$, N represents the number of fingerprint points. Suppose the target moves in an interval from h_{L} to h_{H} at the z-axis, h_{L} and h_{H} are the minimum and maximum vertical distance from the floor to the target, respectively.

We use S to denote the spacing of the fingerprints, as shown in Figure 1a. m, n and l are used to represent the collection directions of fingerprints in x-axis, y-axis and z-axis, respectively, the meanings of m, n, and l are shown in Table 1. To make it easier to understand, an example is given, as shown in Figure 1b. In Figure 1b, columns are arranged from left to right (in the positive direction of the x-axis), rows are arranged from bottom to top (in the positive direction of the y-axis), and dimensions are arranged from low to high (in the positive direction of the z-axis). The starting point of fingerprint collection is $\theta_{init} = [x_{init}, y_{init}, z_{init}]^T$, x_{init}, y_{init}, and z_{init} are given by:

$$\begin{cases} x_{init} = \min(x_{LED-i}) \\ y_{init} = \min(y_{LED-i}) \\ z_{init} = h_L \end{cases} \tag{1}$$

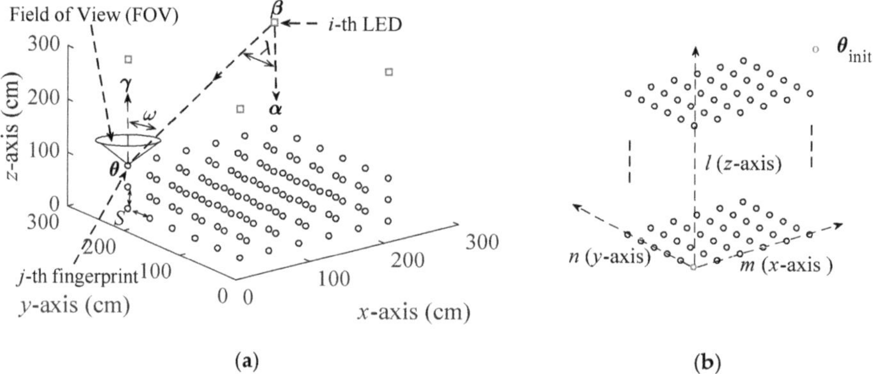

Figure 1. Fingerprint positioning based on visible light communication (VLC): (**a**) The positioning model; (**b**) The collection directions of fingeprints in x-axis, y-axis and z-axis.

Table 1. The meaning of the indices.

Indices	Meaning
m	The number of columns corresponding to the fingerprint point
n	The number of rows corresponding to the fingerprint point
l	The number of dimensions corresponding to the fingerprint point

Then, in the positioning space, the coordinates corresponding to the fingerprint points in the l dimension, i.e., the m column and the n row are

$$\begin{cases} x_{fin-m} = x_{init} + S(m-1), & m = 1, 2, \ldots, \left\lceil \frac{L_1}{S} + 1 \right\rceil \\ y_{fin-n} = y_{init} + S(n-1), & n = 1, 2, \ldots, \left\lceil \frac{L_2}{S} + 1 \right\rceil \\ z_{fin-l} = z_{init} + S(l-1), & l = 1, 2, \ldots, \left\lceil \frac{L_3}{S} + 1 \right\rceil \end{cases} \tag{2}$$

where $L_1 = \max(x_{LED-i}) - \min(x_{LED-i})$, $L_2 = \max(y_{LED-i}) - \min(y_{LED-i})$ and $L_3 = h_H - h_L$. Then the distance $d_{l,m,n-i}$ between each fingerprint point and the ith LED can be obtained as:

$$d_{l,m,n-i} = \sqrt{\left(x_{fin-m} - x_{LED-i}\right)^2 + \left(y_{fin-n} - y_{LED-i}\right)^2 + \left(z_{fin-l} - z_{LED-i}\right)^2} \tag{3}$$

2.2. Fingerprint Matrix Construction

We use $\Phi \in \mathbb{R}^{M \times N}$ to denote the measurement matrix of the fingerprints, which is given by:

$$\Phi = \begin{pmatrix} \phi_{1,1,1-1} & \phi_{1,1,2-1} & \cdots & \phi_{l,m,n-1} \\ \phi_{1,1,1-2} & \phi_{1,1,2-2} & \cdots & \phi_{l,m,n-2} \\ \vdots & \vdots & \ddots & \vdots \\ \phi_{1,1,1-M} & \phi_{1,1,2-M} & \cdots & \phi_{l,m,n-M} \end{pmatrix} \tag{4}$$

where N is given by:

$$N = \left\lceil \frac{L_1}{S} + 1 \right\rceil \left\lceil \frac{L_2}{S} + 1 \right\rceil \left\lceil \frac{L_3}{S} + 1 \right\rceil \tag{5}$$

And $\phi_{l,m,n-i}$ represents the RSSI, which is given by:

$$\phi_{l,m,n-i} = 10\log_{10}\left(P_{l,m,n-i}\right) \tag{6}$$

where $P_{l,m,n-i}$ represents the optical power value from the ith LED received by the fingerprint point in the l dimension, m column and n row within the positioning area.

2.3. Measurement Vector

Suppose the coordinates of targets in 3-D are $\mathbf{\Psi}_k = [x_{\text{target-}k}, y_{\text{target-}k}, z_{\text{target-}k}]^{\text{T}}$, for $k = 1, 2, \ldots,$ C, and C represents the number of targets. Thus, the receiving signal intensity vector $\mathbf{Y}_k$ of M LEDs collected by the kth target is given by:

$$\mathbf{Y}_k = [Y_{k,1}, Y_{k,2}, \ldots, Y_{k,M}]^{\text{T}} \tag{7}$$

where $Y_{k,i}$ is given by

$$Y_{k,i} = 10\log_{10}\left(P_{k,i}\right) \tag{8}$$

where $P_{k,i}$ represents the optical power value of the ith LED received by the kth target.

2.4. Measurement Model

In this paper, the measurement matrix $\mathbf{\Phi}$ and measurement vector $\mathbf{Y}_k$ are generated by the Lambertian radiation model. Because the LED is distributed on the ceiling, there is mainly LoS communication between the fingerprint point and the LED. Without loss of generality, this paper only considers the Lambertian radiation model of the LoS, which are widely adopted in papers such as [12,28,30–32], the received light power value of the fingerprint point is:

$$P_{\text{Re}} = P_{\text{Tr}}\frac{A_{PD}(b+1)T_s g}{2\pi d^2}(\cos(\lambda_i))^b \cos(\omega_i) \tag{9}$$

where P_{Re} represents the received light power value; P_{Tr} represents the transmit power of the LED; d is the distance between the transmitter and the receiver; T_s and g are the optical filter gain and optical concentrator gain, respectively; b is the Lambertian order; $\lambda_{1/2}$ is the half-power angles of the LED; A_{PD} is the effective area of the PD detection; The field of view (FOV) of PD is defined as ω_{FOV}, and $0 < \omega_i < \omega_{\text{FOV}}$. λ_i and ω_i are the radiation and incident angles, i.e., the transmitter's normal and receiver's normal, respectively, as shown in Figure 1a.

2.5. Channel Access Method

As LEDs transmit a unique identification (ID) code independently, however, signals sent from different LEDs will interfere with each other at the receiver. In order to receive the power from different LEDs, we also use time division multiplexing to achieve this goal [20,31,32], and in a real scenario, we can also use different modulation frequencies, such as Guo et al. [11] and Alam et al. [12]. M LEDs have synchronous frames [20,31], and different LEDs use different time slots to transmit signals within each frame cycle, when one LED transmits the ID code, other LEDs emit a constant light intensity (CLI) for illumination purposes only. The frame structure is shown in Figure 2. After photoelectric conversion, a high-pass filter can be used to filter out the power from other LEDs [20].

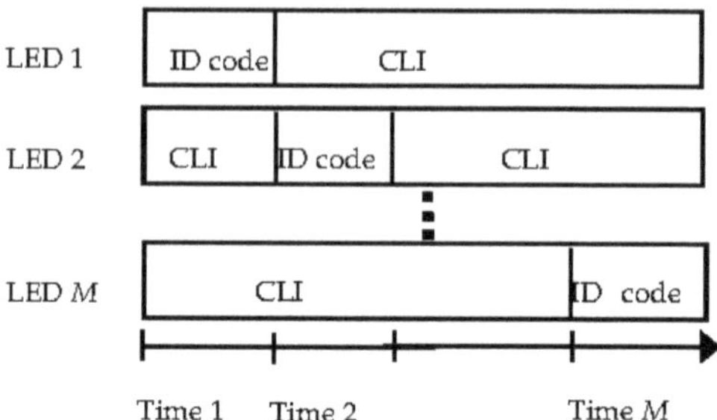

Figure 2. Frame structure of the positioning system for one period.

2.6. Setting of K

According to the principle of fingerprint positioning, the purpose of positioning is to find K fingerprint points that are close to the target. When in a different experimental environment, K generally takes different values, such as in Xue et al. [15], the optimal positioning accuracy is obtained when $K = 5$; in Alam et al. [12] and Zhang et al. [28], the optimal positioning accuracy is obtained when $K = 4$; in Van et al. [25], the optimal positioning accuracy is obtained when $K = 3$. One thing they all have in common is that K is a fixed value. In this paper, N fingerprint points are evenly distributed in the 2-D or 3-D space. In a specific time, there are K fingerprint points close to the same target, which is called the KNN fingerprint positioning algorithm. For example, when the target exactly matches the fingerprint point, as shown in Figure 3a, obviously, the optimal positioning accuracy is obtained when $K = 1$. When the target falls on a straight line formed by two fingerprint points, as shown in Figure 3b, i.e., $K = 2$. When the target is in a triangular area composed of three fingerprint points, as shown in Figure 3c, i.e., $K = 3$. If the 3-D fingerprints map is adopted, and the target is obviously located in a minimum cube composed of 8 fingerprint points with a high probability, i.e., $K = 8$, as shown in Figure 3d.

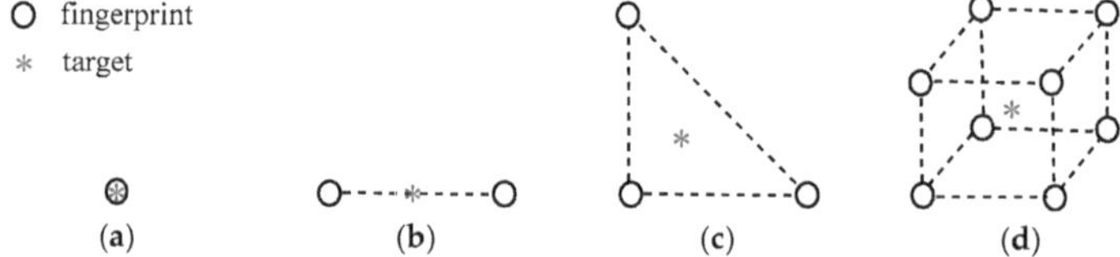

Figure 3. The relationship between target and fingerprints: (**a**) $K = 1$; (**b**) $K = 2$; (**c**) $K = 3$; (**d**) $K = 8$.

Figure 4 is the positioning error of five targets at different 3-D positions using WKNN algorithm, for K increases from 1 to 8. When $K = 4$, the positioning error of target 1 is minimal. When $K = 3$, the positioning error of target 2 is minimal. When $K = 8$, the positioning error of target 3 is minimal. When $K = 1$, the positioning error of target 4 is minimal. When $K = 6$, the positioning error of target 5 is minimal. It can also be seen from Figure 4 that the positioning error varies with the K value fluctuation, and there is no monotonous increasing or decreasing relationship. In a 2-D visible light localization, the average positioning error based on the WKNN algorithm can be minimized when $K = 3$ or $K = 4$, e.g., [12,25,28]. In the 3-D visible light localization, the average positioning error based on the WKNN algorithm can be minimized when $K = 8$, which will be discussed in Section 3 . The minimum mean positioning error does not mean that the positioning error of each target is the smallest, so the dynamic K value can effectively reduce the positioning error of different targets. To address this issue, this paper proposes an adaptive residual weighted K-nearest neighbor fingerprint positioning algorithm, which is called ARWKNN fingerprint positioning algorithm.

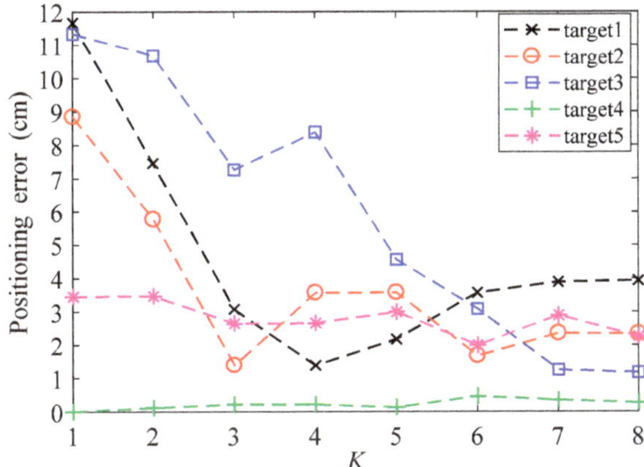

Figure 4. Positioning errors of 5 targets using weighted *K*-nearest neighbor (WKNN) algorithm.

2.7. ARWKNN Algorithm

The WKNN fingerprint positioning algorithm is based on the shortest RSSI physical distance between the fingerprint and the target position. The positioning error for the WKNN algorithm is affected by the weight of the fingerprint point and this weight is affected by the *K* value. If the optimal *K* value can be obtained, the positioning error can be reduced, so a novel ARWKNN algorithm is proposed in this paper. The pseudo-code of the ARWKNN algorithm is shown in Algorithm 1. In Algorithm 1, if we only consider Steps 1, 2 and 5, then it is the WKNN algorithm, and in Step 5, if the location of the target is estimated by averaging the coordinates of *K* fingerprints, then it is the KNN algorithm. By contrast with the KNN and WKNN algorithms, the ARWKNN algorithm also performs Step 3 and 4 in Algorithm 1. Because there is no prior information about the location of the target, that is, the value of $\mathbf{\Psi}_k$ is unknown, but we known the fingerprint matrix $\mathbf{\Phi}$ and the target RSSI measurement vector $\mathbf{Y}_k$, we can adaptively select the *K* value by matching the residual between the measured and calculated RSSI values. Therefore, the purpose of Steps 3 and 4 in algorithm 1 is to obtain the optimal *K* value, i.e., the *K* value corresponding to the smallest RSSI matching residual. In Algorithm 1, because the maximum number of neighboring fingerprint points K_{max} is much smaller than the total number of fingerprint points *N*, the ARWKNN algorithm has a large reduction in the average positioning error while maintaining similar algorithm complexity, which will be discussed in Section 3.4.

Algorithm 1. ARWKNN algorithm

Input: the maximum number of nearest neighbor fingerprints K_{max}, fingerprint matrix $\mathbf{\Phi}$, and the kth target measurement vector $\mathbf{Y}_k$.
Output: The coordinates of the kth target, i.e., $\mathbf{\Psi}_k$.
Step 1: Calculate the distance from the kth target to N fingerprint points.

$$dis_{l,m,n-k} = \left(\sum_{i=1}^{M} |\phi_{l,m,n-i} - Y_{k,i}|^r \right)^{1/r}$$

where $r = 1$ represents the Manhattan distance, $r = 2$ represents the Euclidean distance.
Step 2: Sort the distance values in ascending order, i.e.,

$$[\mathbf{X}, \mathbf{I}] = \text{sort } (\mathbf{dis}).$$

where $\mathbf{dis} = [dis_{1,1,1-k}, dis_{1,1,2-k}, \dots, dis_{l,m,n-k}]^T \in \mathbb{R}^{N \times 1}$, $\mathbf{X} \in \mathbb{R}^{N \times 1}$ represents the distance vector after sorting, and $\mathbf{I} \in \mathbb{R}^{N \times 1}$ represents the corresponding index set.
Step 3: Calculate the matched RSSI residuals.
$K = 1$,
while $K \leq K_{max}$ **do**
    for $ii = 1: K$
    $\mathbf{A}(:, ii) = \mathbf{\Phi}(:, \mathbf{I}(ii))$;
    end for
where $\mathbf{A} \in \mathbb{R}^{M \times K}$ represents finding the K column values corresponding to the fingerprint matrix $\mathbf{\Phi}$ according to the index set $\mathbf{I}$.
Calculate the kth target RSSI vector via K nearest neighbor fingerprints,

$$\widetilde{\mathbf{Y}}_k = \mathbf{AB},$$

where $\mathbf{B} = [B_1, B_2, \dots, B_K]^T \in \mathbb{R}^{K \times 1}$ and
$B_t = \frac{1}{\mathbf{X}(t)} / \sum_{tt=1}^{K} \frac{1}{\mathbf{X}(tt)}$, for $t = 1, 2, \dots, K$,
Calculate the matched RSSI residual between the measured and calculated RSSI values,

$$\mathbf{E}_{\text{residual}} = \mathbf{Y}_k - \widetilde{\mathbf{Y}}_k,$$

and calculate the sum of the absolute values of the residuals,

$$\mathbf{E}_{\text{sum}}(K) = \sum_{i=1}^{M} \left| \mathbf{E}_{\text{residual}}(i) \right|,$$

$K = K + 1$.
end while
Step 4: Output the K value, i.e.,

$$K = \text{argmin}(\mathbf{E}_{\text{sum}}); \text{ s.t. } 1 \leq K \leq K_{max}$$

Step 5: Calculate the coordinates of the kth target,

$$\mathbf{\Psi}_k = \frac{\sum_{t=1}^{K} \frac{1}{\mathbf{X}(t)} \mathbf{\theta}_{\mathbf{I}(t)}}{\sum_{t=1}^{K} \frac{1}{\mathbf{X}(t)}}$$

where $\mathbf{\theta}_{\mathbf{I}(t)}$ represents the coordinates of the corresponding fingerprint point found according to the index set $\mathbf{I}$.

---

3. Simulation Analysis

In this Section, the ARWKNN algorithm is compared with RF [14], ELM [16], ANN [17], GI-LS [11], SAWKNN [19], WKNN [12] or KNN [15,25] algorithms. The basic principle of the fingerprint positioning algorithm based on RF, ELM, ANN, and GI-LS machine learning is as follows [11,13]: Firstly, the positioning area is divided into several equal grid points according to the sampling interval S, RSSI measurements are obtained by placing the receiver at different grid points, and each grid point represents a category. Secondly, machine-learning algorithms are used to train the category to which each grid point belongs. Thirdly, the RSSI measurements obtained in the online phase are compared with the derived model to predict the location of the target.

3.1. Error Definition

Suppose the actual coordinates of targets are $\widetilde{\mathbf{\Psi}}_k \in \mathbb{R}^{3 \times 1}$, then the positioning error E_k is defined as:

$$E_k = \|\mathbf{\Psi}_k - \widetilde{\mathbf{\Psi}}_k\|_2 \tag{10}$$

and the average positioning error E_{APE} is defined as:

$$E_{APE} = \frac{1}{C} \sum_{k=1}^{C} E_k \tag{11}$$

3.2. Noise Model of Visible Light Communication (VLC)

In indoor VLC, the noise σ_{noise} includes shot noise σ_{shot} and thermal noise $\sigma_{thermal}$ [33], which are given by:

$$\sigma^2_{noise} = \sigma^2_{shot} + \sigma^2_{thermal} \tag{12}$$

$$\sigma^2_{shot} = 2qR_{PD}P_rB + 2qI_{bg}I_2B \tag{13}$$

$$\sigma^2_{thermal} = \frac{8\pi k_B T_K}{G_0}\eta A_{PD}I_2B^2 + \frac{16\pi^2 k_B T_K \Gamma}{g_m}\eta^2 A^2_{PD}I_3B^3 \tag{14}$$

where q is elementary charge, R_{PD} is the responsivity of the PD, B is the equivalent noise bandwidth, P_r indicates the received power from M LEDs, k_B is the Boltzmann's constant, T_K is the absolute temperature, G_0 is the open loop gain, η is the fixed capacitance of PD, I_{bg} is the background light current, Γ is the channel noise factor, g_m is the field effect transistor (FET) transconductance, I_2 and I_3 are the noise bandwidth factors.

According to the noise model, the signal-to-noise ratio (SNR) is given by [32]

$$SNR(dB) = 10\log_{10}\frac{\left(R_{PD}\frac{P_r}{A_{PD}}\right)^2}{\sigma^2_{noise}} \tag{15}$$

3.3. Simulation Parameters

Without loss of generality, we suppose $\alpha_i = [0, 0, -1]^T$ and $\gamma_j = [0, 0, 1]^T$, i.e., $\cos(\lambda_i) = \cos(\omega_i) = h_{l,m,n-i}/d_{l,m,n-i}$, $h_{l,m,n-i}$ is the z-axis distance from the fingerprint point to the ith LED in the l dimension, the m column and the n row, which are widely adopted in papers such as [12,20,28]. The parameter setting of the Lambertian radiation model is as follows: $T_s = g = 1$, $\lambda_{1/2} = \pi/3$, $\omega_{FOV} = \pi/2$, $A_{PD} = 1$ cm^2, $b = 1$, which follow from a typical LED setup. M LEDs are evenly distributed in a 3-D space with an area of 200 cm $\times$ 200 cm $\times$ 300 cm, min (x_{LED-i}) = min (y_{LED-i}) = 0 cm, max (x_{LED-i}) = max (y_{LED-i}) = 200 cm, and z_{LED} = 300 cm. $C = 200$, i.e., 200 targets randomly appear in the 3-D or 2-D positioning area. In the KNN and WKNN algorithms, K is a fixed value, that is, $K = K_{max}$. The parameter setting of the noise model is as follows [33]: T_K = 295 K, G_0 = 10, g_m = 30 mS, Γ = 1.5, I_2 = 0.562, I_3 = 0.0868, R_{PD} = 0.54

A/W, $\eta = 112$ pF/cm^2, $I_{bg} = 5100$ μA. Unless otherwise specified, $P_{Tr} = 6$ W, $M = 4$, $r = 1$ (i.e., Manhattan distance). In 3-D, $h_L = 20$ cm and $h_H = 100$ cm. In 2-D, $h_L = h_H = 100$ cm. The simulation tool is MATLAB R2017a.

For simplicity, unless otherwise specified, we only consider the 2-D case, and $S = 20$ cm. In order to obtain the optimal classification accuracy of ANN, ELM, and RF algorithms, and the optimal positioning accuracy of KNN, WKNN, and SAWKNN algorithms. The optimal parameters obtained through offline training and learning are as follows: In KNN, WKNN, ARWKNN and SAWKNN algorithms, $K_{max} = 4$. In the Section 3.4, we will also discuss the impact of different K_{max} values on the average positioning error. For the optimal number of hidden nodes and trees, the classification method is the same as that in Guo et al. [11], i.e., each grid point represents a category, and the cross-validation method is adopted based on experience adjustment. For the optimal number of hidden nodes, the cross-validation method has a range of 100 to 700 and a step size of 50. For the optimal number of trees, the cross-validation method has a range of 10 to 50 and a step size of 5. After comprehensive evaluation of the positioning accuracy and classification accuracy, the optimal number of hidden nodes and trees are selected to be 600 and 40, respectively. The impact of γ_{th} on the average positioning error is shown in Figure 5, it can be seen from the Figure 5 that minimum average positioning error is achieved when γ_{th} is within the range of [30%, 50%], so, the value of γ_{th} is selected to be 40%, which denotes the threshold of two RSSI difference values that can be considered similar [19].

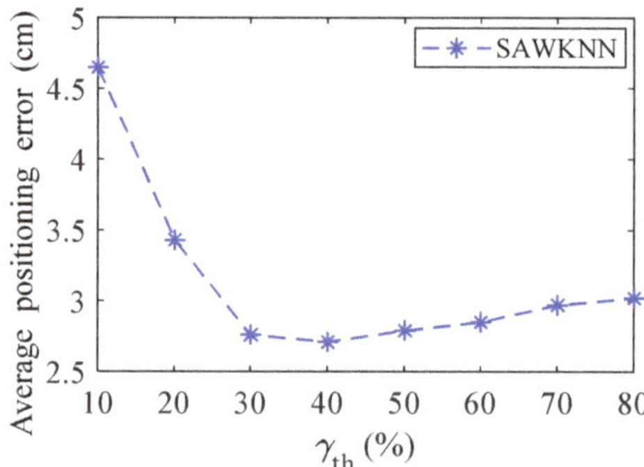

Figure 5. The impact of γ_{th} on the average positioning error.

3.4. Result Analysis

We only consider positioning in this paper, so $B = 640$ KHz will be able to label 3.4×10^{38} LEDs [34], which is far exceeds the actual needs. The SNR experimental results are shown in Table 2. If $B = 640$ KHz, typical SNR for indoor visible light communication ranges from 42.97 to 60.92 dB, and the average value reaches 52.45 dB. In addition to indoor positioning, LEDs can also provide high-speed data rate, If $B = 100$ MHz, the average SNR can also reach 28.86 dB.

Table 2. Typical signal-to-noise ratio (SNR) in indoor environment.

	Minimum	Maximum	Average
SNR ($B = 640$ KHz)	42.97 dB	60.92 dB	52.45 dB
SNR ($B = 100$ MHz)	19.72 dB	37.35 dB	28.86 dB

When $P_{tr} = 6$ W, the average positioning errors of eight algorithms are analyzed when B is within 50 MHz to 400 MHz, the results are shown in Figure 6. As the value of modulation bandwidth increases, the average positioning errors of eight algorithms increase. The higher the modulation bandwidth, the lower the SNR and the higher the average positioning errors. As only positioning is considered in this paper, a very high modulation bandwidth is not necessary. With a high-modulation bandwidth, it may be more suitable to modulate the transmission signal of the LED by modified orthogonal frequency division multiplexing (OFDM) to achieve indoor positioning [22,35,36], but this is beyond the scope of this paper. It can also be seen from Figure 6 that when B is within 50 MHz to 400 MHz, the average positioning error based on the ARWKNN algorithm is the smallest.

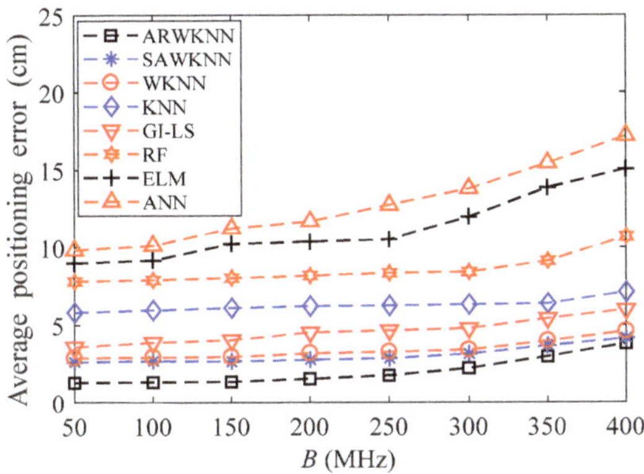

Figure 6. The impact of B on the average positioning error with $K_{max} = 4$.

When $B = 100$ MHz, the average positioning errors of eight algorithms are analyzed when P_{tr} is within 1 W to 6 W, the results are shown in Figure 7. As the P_{tr} increases, the average positioning errors of eight algorithms decrease. When $P_{tr} = 3$ W, the average positioning errors of eight algorithms are close to convergence. The higher the transmitting power, the higher the SNR and the smaller the average positioning errors. It can also be seen from Figure 7 that when P_{tr} is within 1 W to 6 W, the average positioning error based on the ARWKNN algorithm is the smallest.

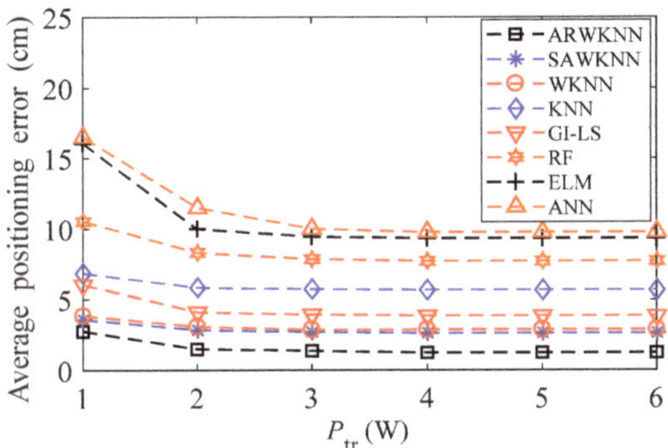

Figure 7. The impact of P_{tr} on the average positioning error with $K_{max} = 4$.

The average positioning errors of eight algorithms under different SNR are compared, simulation results are shown in Figure 8. As shown in Figure 8, when SNR = 10 dB, the average positioning errors of eight algorithms are large due to severe noise interference. As the SNR increases, the average positioning errors of eight algorithms decrease. When SNR = 20 dB, the average positioning errors of eight algorithms are close to convergence. Since fingerprint positioning based on RF, ELM and ANN algorithms can only determine the category of the target, compared with WKNN algorithm, the positioning error is larger. When the SNR is higher than 15, the average positioning error based on the ARWKNN algorithm is the smallest. Due to lighting requirements and LoS communication, within the typical SNR range of indoor visible light communication, the average positioning error based on the ARWKNN algorithm is significantly lower than that of RF, ELM, ANN, GI-LS, SAWKNN, WKNN and KNN algorithms. The average positioning error based on the SAWKNN algorithm is lower than that of the WKNN algorithm. The GI-LS algorithm uses the complementary advantages of KNN, RF, and ELM classifiers to weight the estimation results, the average positioning error based on the GI-LS algorithm is lower then that of KNN, RF, ELM and ANN algorithms, but higher then WKNN, ARWKNN and SAWKNN algorithms.

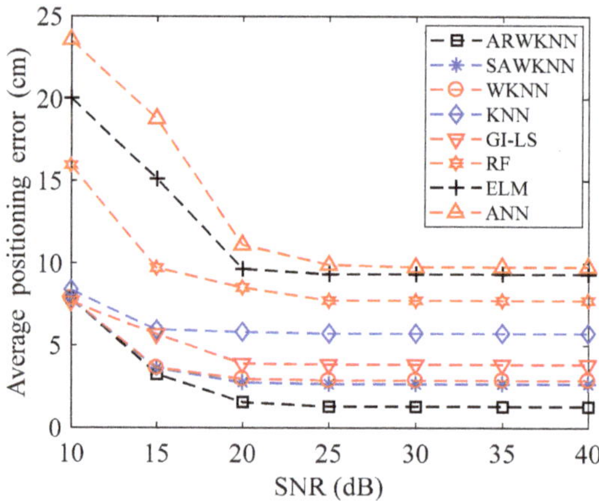

Figure 8. The impact of SNR on the average positioning error with $K_{max} = 4$.

When SNR = 20 dB, the average positioning errors based on ARWKNN, RF, ELM, ANN, GI-LS, SAWKNN, WKNN and KNN algorithms are shown in Table 3. It can be seen from Table 3 that compared with RF, ELM, ANN, GI-LS, SAWKNN, WKNN and KNN algorithms, the average positioning error based on the ARWKNN algorithm can be reduced by 81.82%, 83.93%, 86.06%, 60.15%, 43.84%, 47.81%, and 73.36%, respectively.

Table 3. Average positioning error for each algorithm with SNR = 20 dB.

Algorithm	Average Positioning Error
ARWKNN	1.55 cm
RF	8.53 cm
ELM	9.65 cm
ANN	11.12 cm
GI-LS	3.89 cm
SAWKNN	2.76 cm
WKNN	2.97 cm
KNN	5.82 cm

When SNR = 20 dB, the simulation results of cumulative distribution function (CDF) are shown in Figure 9. It can be seen from Figure 9 that the CDF of positioning errors based on the ARWKNN algorithm is significantly better than that of the RF, ELM, ANN, GI-LS, SAWKNN, WKNN and KNN algorithms. The KNN algorithm is one of the simplest of all machine learning algorithms. Compared with the RF, ELM, ANN and GI-LS algorithms, fingerprint positioning based on the ARWKNN algorithm, not only has lower complexity, but also has lower positioning error. Fingerprint positioning is based on machine-learning algorithms, which require a large amount of data for training and learning. If there are not enough training data, the positioning error will be large, and a large amount of training data will increase the complexity of the algorithm. Compared with the SAWKNN, WKNN, and KNN algorithms, the ARWKNN algorithm can significantly reduce the average positioning error while maintaining similar algorithm complexity, which will be discussed in the section of algorithm complexity analysis.

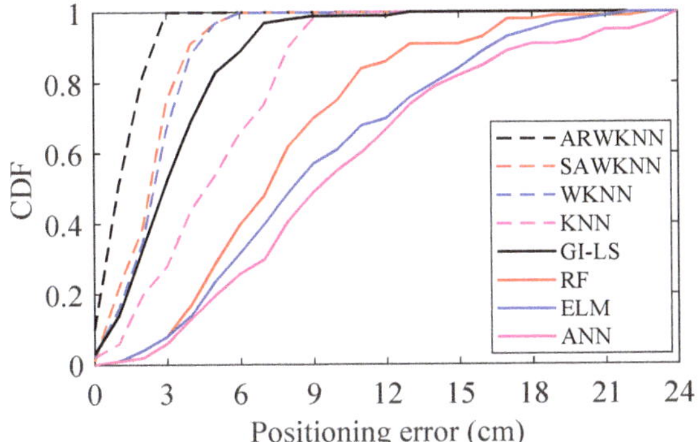

Figure 9. The cumulative distributions of positioning errors with $K_{max} = 4$.

When SNR = 20 dB, the average positioning errors of ARWKNN, SAWKNN, WKNN, and KNN algorithms are analyzed, in WKNN and KNN algorithms, K is a fixed value, that is, $K = K_{max}$. The simulation results of 2-D and 3-D are shown in Figures 10 and 11, respectively. As can be seen from Figure 10, when K_{max} is within 1 to 8, similar to the experimental results in most papers, in 2-D, the optimal K based on the WKNN algorithm is 3 or 4, which exactly conforms with the fact that the target will be located in a minimum triangle or square composed of 3 or 4 fingerprint points with a high probability. It can also be analyzed from Figure 10 that when K_{max} is greater than 3, the average positioning error based on the ARWKNN algorithm is significantly lower than that of the KNN, WKNN, and SAWKNN algorithms. From Figure 11, It can be seen that as the K_{max} increases from 1 to 12, the average positioning error based on the ARWKNN algorithm decreases. When $K_{max} = 8$, the average positioning error is not significantly reduced if the value of K_{max} continues to increase. Therefore, a reasonable value of K_{max} is taken as 8. From Figure 11, we can also see that when $K_{max} = 8$, the average positioning error based on the KNN and WKNN algorithms is the smallest, which exactly conforms that the target will be located in a minimum cube composed of 8 fingerprint points with a high probability. It can also be analyzed from Figure 11 that when K_{max} is greater than 6, the average positioning error based on the ARWKNN algorithm is significantly lower than that of the KNN, WKNN, and SAWKNN algorithms, and the advantages of the ARWKNN algorithm are more obvious as K_{max} increases.

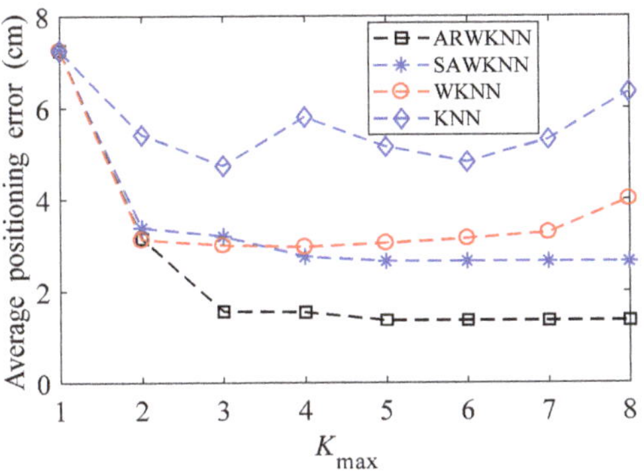

Figure 10. In 2-D, the impact of K_{max} on the average positioning error.

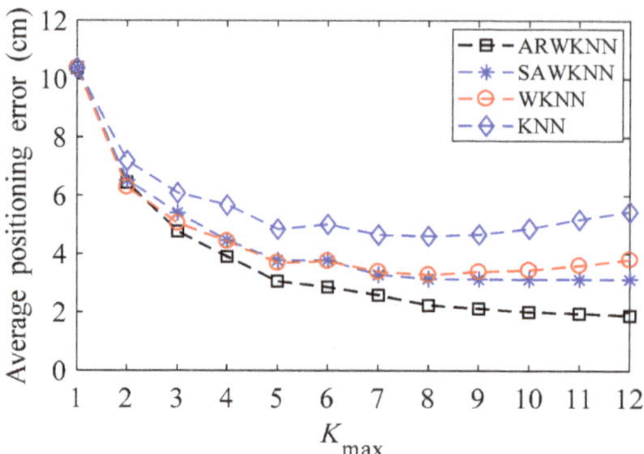

Figure 11. In 3-D, the impact of $K_{\max}$ on the average positioning error.

When SNR = 20 dB, the average positioning errors of ARWKNN, SAWKNN, WKNN, and KNN algorithms are analyzed with the variation of the fingerprint sampling point spacing S, the results of 2-D and 3-D are shown in Figures 12 and 13, respectively. It can be seen that as S decreases from 40 cm to 20 cm, whether in 2-D or 3-D, the average positioning error based on the ARWKNN algorithm is significantly lower than that of the KNN, WKNN, and SAWKNN algorithms, and the larger the S, the more obvious the advantage. As S decreases to 5 cm, the average positioning errors of four algorithms tend to be the same. The lower the value of S, the larger the number of fingerprint points N to be acquired, and the more complicated the algorithm becomes.

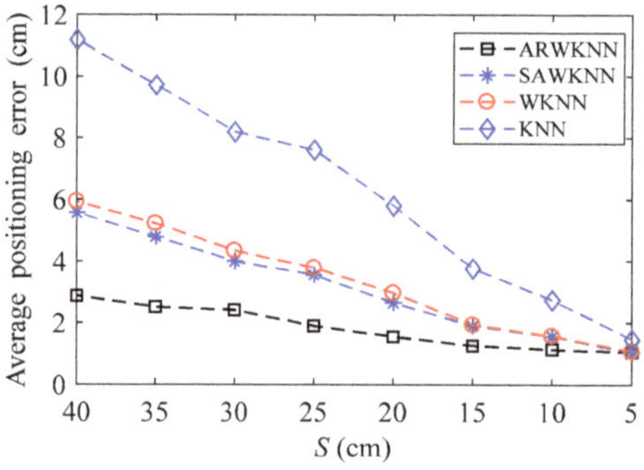

Figure 12. In 2-D, the impact of S on the average positioning error with $K_{\max} = 4$.

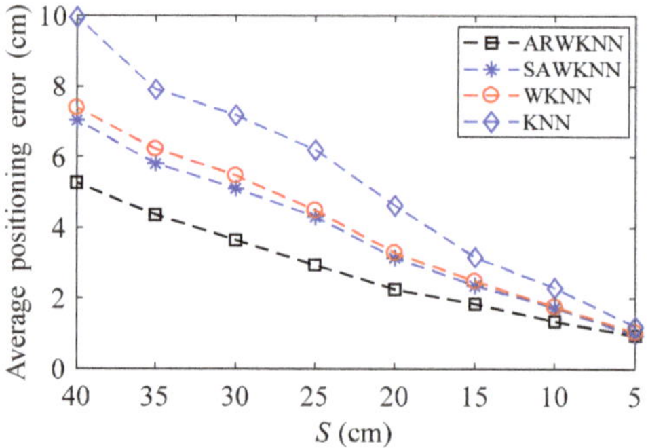

Figure 13. In 3-D, the impact of S on the average positioning error with $K_{\max} = 8$.

When SNR = 20 dB, the average positioning errors of ARWKNN, SAWKNN, WKNN, and KNN algorithms are analyzed when M is within 3 to 8, the results of 2-D and 3-D are shown in Figures 14 and 15, respectively. It can be seen that as M increases from 3 to 8, the average positioning errors based on the KNN, WKNN, SAWKNN and ARWKNN algorithms do not change much. Thus, only 4 LEDs are needed to achieve very low positioning error in this paper.

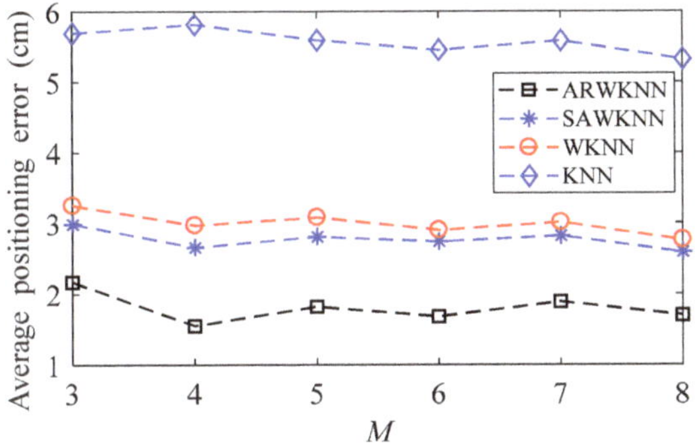

Figure 14. In 2-D, the impact of M on the average positioning error with $K_{\max} = 4$.

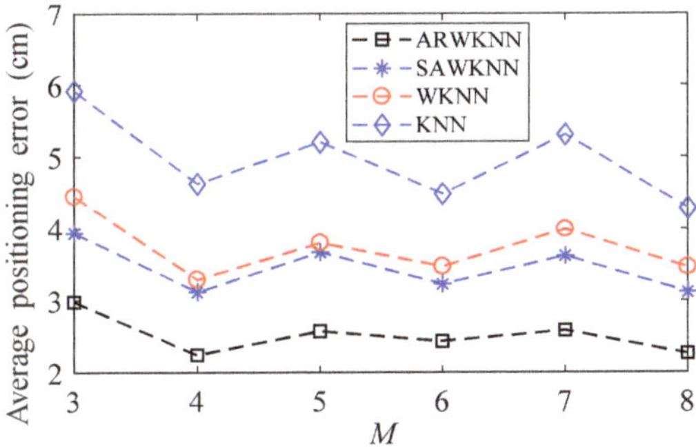

Figure 15. In 3-D, the impact of M on the average positioning error with $K_{\max} = 8$.

When SNR = 20 dB, in order to analyze the robustness of the algorithm, fingerprints adopt non-uniform distribution structure, i.e., the RSSI values in the fingerprint map are chosen randomly at different sampling ratios SR. The average positioning errors of the ARWKNN, SAWKNN, WKNN, and KNN algorithms are analyzed with the variation of the fingerprint sampling ratio SR, the results of 2-D and 3-D are shown in Figures 16 and 17, respectively. It can be seen that as SR increases from 50% to 100%, whether in 2-D or 3-D, the average positioning error based on the ARWKNN algorithm is significantly lower than that of the KNN, WKNN, and SAWKNN algorithms, and the larger the SR, the smaller the average positioning errors of the four algorithms. When $SR = 50\%$, the average positioning errors of the ARWKNN, SAWKNN, WKNN, and KNN algorithms are analyzed with the variation of the time, and the results of 2-D and 3-D are shown in Figures 18 and 19, respectively. It can be seen that as t increases from 1 to 50, whether in 2-D or 3-D, the average positioning error based on the ARWKNN algorithm is significantly lower than that of the KNN, WKNN, and SAWKNN algorithms. As can be seen from Figures 16–19, the ARWKNN algorithm has good robustness. When the fingerprint sampling rate is only 50%, lower positioning errors can still be achieved.

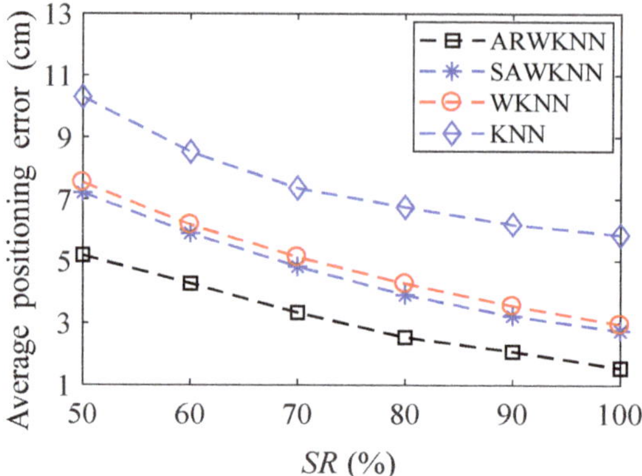

Figure 16. In 2-D, the impact of sampling ratio (*SR*) on the average positioning error with $K_{max} = 4$.

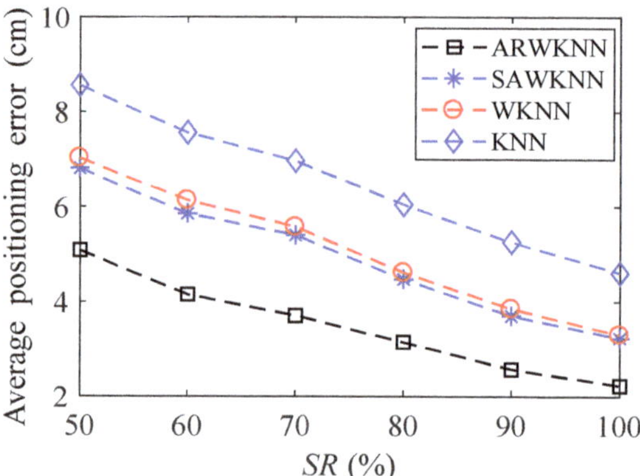

Figure 17. In 3-D, the impact of *SR* on the average positioning error with $K_{max} = 8$.

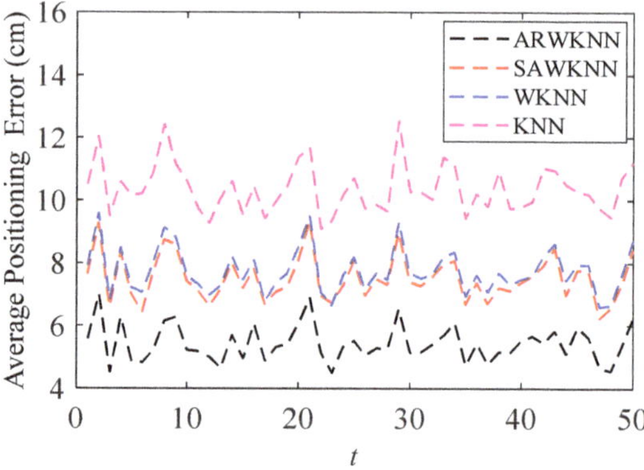

Figure 18. In 2-D, the impact of *t* on the average positioning error with $K_{max} = 4$.

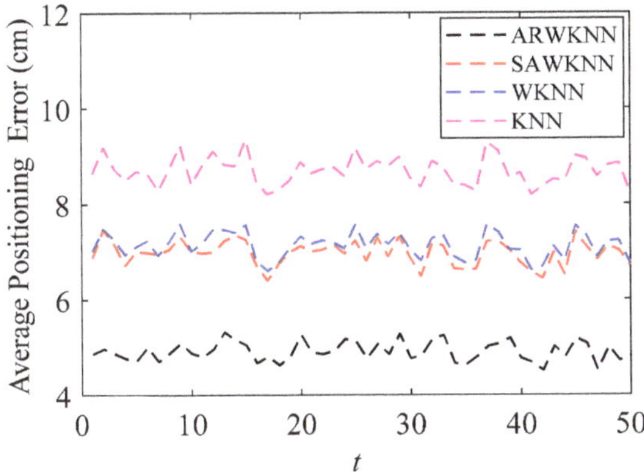

Figure 19. In 3-D, the impact of t on the average positioning error with $K_{max} = 8$.

The WKNN fingerprint positioning algorithm is based on the shortest RSSI physical distance between the fingerprint and the target position. It can be seen from Step 5 of the ARWKNN algorithm that the positioning error is affected by the weight of the fingerprint point and this weight is affected by the distance metric; therefore, it is necessary to analyze the impact of different distance metrics on the positioning error. In addition to Euclidean distance (ED) and Manhattan distance (MD), there are other distance metrics [12,37], such as:

Minimum maximum distance (MMD), which is defined as:

$$dis_{l,m,n-k} = 1 - \frac{\sum_{i=1}^{M} \left(\min\left(\left| \phi_{l,m,n-i} \right|, \left| Y_{k,i} \right| \right) \right)}{\sum_{i=1}^{M} \left(\max\left(\left| \phi_{l,m,n-i} \right|, \left| Y_{k,i} \right| \right) \right)} \tag{16}$$

Squared Euclidean distance (SED), which is defined as:

$$dis_{l,m,n-k} = \sum_{i=1}^{M} \left(\phi_{l,m,n-i} - Y_{k,i} \right)^2 \tag{17}$$

Chebyshev distance (CHD), which is defined as:

$$dis_{l,m,n-k} = \max_{i} \left| \phi_{l,m,n-i} - Y_{k,i} \right| \tag{18}$$

Squared-chord distance (SCD), which is defined as:

$$dis_{l,m,n-k} = \sum_{i=1}^{M} \left(\sqrt{\left| \phi_{l,m,n-i} \right|} - \sqrt{\left| Y_{k,i} \right|} \right)^2 \tag{19}$$

Wave hedges distance (WHD), which is defined as:

$$dis_{l,m,n-k} = \sum_{i=1}^{M} \left(1 - \frac{\min\left(\left| \phi_{l,m,n-i} \right|, \left| Y_{k,i} \right| \right)}{\max\left(\left| \phi_{l,m,n-i} \right|, \left| Y_{k,i} \right| \right)} \right) \tag{20}$$

Lorentzian distance (LD), which is defined as:

$$dis_{l,m,n-k} = \sum_{i=1}^{M} \ln\left(1 + \left|\phi_{l,m,n-i} - Y_{k,i}\right|\right) \tag{21}$$

Matusita distance (MTD), which is defined as:

$$dis_{l,m,n-k} = \sqrt{\sum_{i=1}^{M} \left(\sqrt{\left|\phi_{l,m,n-i}\right|} - \sqrt{\left|Y_{k,i}\right|}\right)^2} \tag{22}$$

Squared chi-squared distance (SCSD), which is defined as:

$$dis_{l,m,n-k} = \sum_{i=1}^{M} \frac{\left(\phi_{l,m,n-i} - Y_{k,i}\right)^2}{\left|\phi_{l,m,n-i} + Y_{k,i}\right|} \tag{23}$$

Canberra distance (CAD), which is defined as:

$$dis_{l,m,n-k} = \sum_{i=1}^{M} \frac{\left|\phi_{l,m,n-i} - Y_{k,i}\right|}{\left|\phi_{l,m,n-i} + Y_{k,i}\right|} \tag{24}$$

Clark distance (CLD), which is defined as:

$$dis_{l,m,n-k} = \sqrt{\sum_{i=1}^{M} \left(\frac{\left|\phi_{l,m,n-i} - Y_{k,i}\right|}{\left|\phi_{l,m,n-i} + Y_{k,i}\right|}\right)^2} \tag{25}$$

For different distance metrics, if the same γ_{th} value is used, the positioning error based on the SAWKNN algorithm will be greatly affected, so this section does not consider the SAWKNN algorithm. When SNR = 20 dB, we investigated 30 distance metrics and selected 12 distance metrics with the best performances, the results of which are shown in Tables 4 and 5. It can be seen from Tables 4 and 5 that when the KNN algorithm is used for positioning, ED and SED metrics produce the minimum average positioning error in 2-D and 3-D. In 2-D, the average positioning error based on the WKNN algorithm is similar to the experimental results in Alam et al. [12], we also get SCD and SCSD metrics produce the minimum average positioning error, but in 3-D, SED metric produces the minimum average positioning error. When the ARWKNN algorithm is used for positioning, the CLD metric produces the minimum average positioning error in 2-D and MMD metric produces the minimum average positioning error in 3-D. As far as the authors know, this is the first work to report the impact of CLD and MMD metrics on the positioning error of the fingerprint positioning algorithm. It can also be seen from Table 4 that the best values of the KNN, WKNN and ARWKNN algorithms are 4.84 cm, 2.03 cm and 1.45 cm, respectively. Compared with the KNN and WKNN algorithms, in 2-D, the minimum average positioning error of the ARWKNN algorithm can be reduced by 70.04%, and 28.57%, respectively. It can also be seen from Table 5 that the best values of the KNN, WKNN and ARWKNN algorithms are 4.46 cm, 3.05 cm and 2.18 cm, respectively. Compared with the KNN and WKNN algorithms, in 3-D, the minimum average positioning error of the ARWKNN algorithm can be reduced by 51.12%, and 28.52%, respectively. In 2-D or 3-D, the average positioning errors of the ARWKNN algorithm proposed in this paper are all smaller than that of the KNN and WKNN algorithms under 12 distance metrics.

Figure 20 shows the cumulative distributions of positioning errors for the ED and CLD metrics with various S values. As can be seen from Figure 20, in 2-D, compared with the ED metric, the CLD metric produces smaller positioning error. In addition, compared with the CLD metric, the positioning error of the ED metric increases faster when S becomes larger. Figure 21 shows the cumulative distributions of positioning errors for the ED and MMD metrics with various S values. As can be seen from Figure 21, in 3-D, compared with the ED metric, the MMD metric produces smaller positioning error. In addition, compared with the MMD metric, the positioning error of ED metric increases faster when S becomes larger. ED is a commonly used distance metric, however, as can be seen from Tables 4 and 5, in fact, the ED is not the most accurate metric for calculating weights when the WKNN and ARWKNN algorithms are used for positioning.

Table 4. In 2-D, the impact of distance metrics on the average positioning error with $K_{max} = 4$, best values for KNN, WKNN, and ARWKNN algorithms are highlighted in bold.

Distance Metrics	KNN	WKNN	ARWKNN
ED	**4.84 cm**	2.61 cm	1.51 cm
MD	5.82 cm	2.97 cm	1.55 cm
MMD	5.84 cm	2.97 cm	1.54 cm
SED	**4.84 cm**	2.16 cm	1.85 cm
CHD	5.22 cm	2.97 cm	1.61 cm
SCD	4.99 cm	**2.03 cm**	1.82 cm
WHD	5.79 cm	2.94 cm	1.54 cm
LD	6.17 cm	3.45 cm	1.81 cm
MTD	4.99 cm	2.61 cm	1.50 cm
SCSD	4.99 cm	**2.03 cm**	1.82 cm
CAD	5.97 cm	2.93 cm	1.53 cm
CLD	5.05 cm	2.62 cm	**1.45 cm**

Table 5. In 3-D, the impact of distance metrics on the average positioning error with $K_{max} = 8$, best values for KNN, WKNN, and ARWKNN algorithms are highlighted in bold.

Distance Metrics	KNN	WKNN	ARWKNN
ED	**4.46 cm**	3.30 cm	2.31 cm
MD	4.63 cm	3.29 cm	2.28 cm
MMD	4.69 cm	3.33 cm	**2.18 cm**
SED	**4.46 cm**	**3.05 cm**	2.42 cm
CHD	5.18 cm	4.12 cm	2.86 cm
SCD	4.53 cm	3.13 cm	2.58 cm
WHD	4.91 cm	3.53 cm	2.43 cm
LD	5.30 cm	4.16 cm	2.61 cm
MTD	4.53 cm	3.41 cm	2.45 cm
SCSD	4.53 cm	3.13 cm	2.58 cm
CAD	4.88 cm	3.51 cm	2.41 cm
CLD	4.64 cm	3.58 cm	2.60 cm

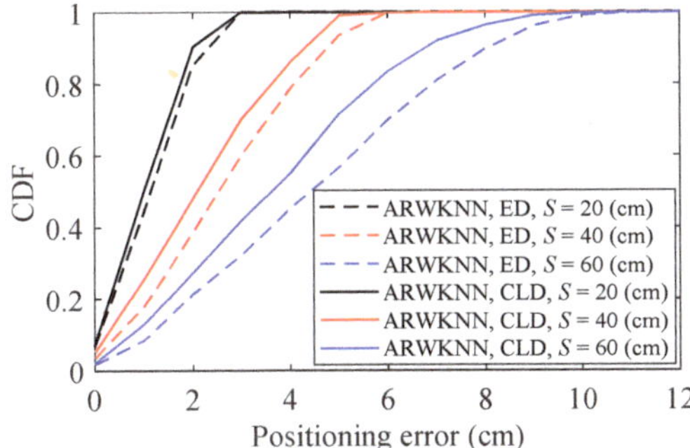

Figure 20. In 2-D, the cumulative distributions of positioning errors for the ED and CLD metrics with various S values. $K_{max} = 4$.

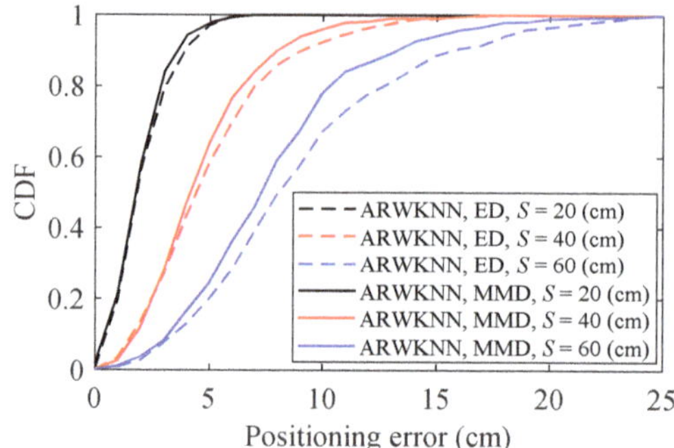

Figure 21. In 3-D, the cumulative distributions of positioning errors for the ED and CLD metrics with various S values. $K_{\max} = 8$.

SNR = 20 dB. To make the graph have a certain degree of discrimination, we only choose the ED, MMD, SED, SCD, and CLD metrics to analyze the cumulative distributions of optimal K. The results of 2-D and 3-D are shown in Figures 22 and 23, respectively. As can be seen from Figures 22 and 23, there are differences in the optimal K values for 200 targets, and there are also differences in cumulative distributions of the optimal K for five distance metrics. The optimal K cumulative distributions for ED, MMD and CLD are very close, and the optimal K cumulative distributions for SED and SCD are also very close.

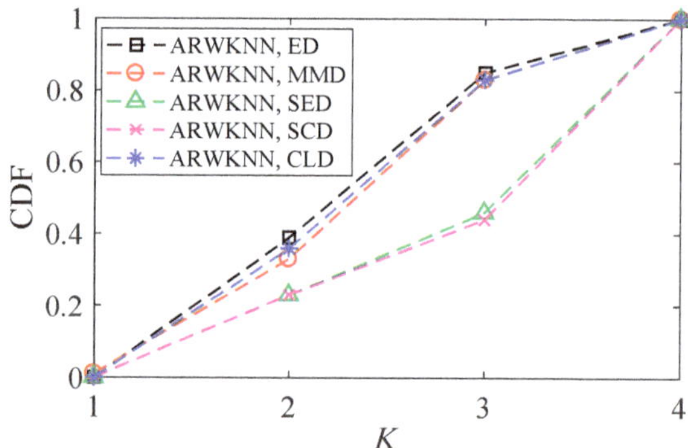

Figure 22. In 2-D, the cumulative distributions of the optimal K with $K_{\max} = 4$.

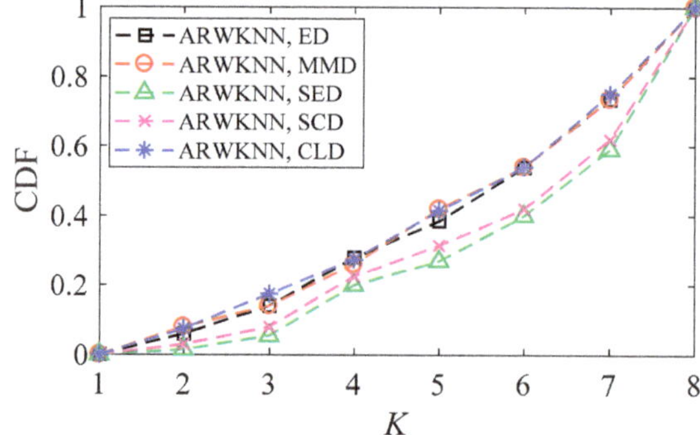

Figure 23. In 3-D, the cumulative distributions of the optimal K with $K_{\max} = 8$.

The complexity of the KNN and WKNN algorithms mainly depends on the size of N and the sorting operation of Step 2 in Algorithm 1. Compared with the KNN and WKNN algorithms, the ARWKNN algorithm also performs Step 3 loop function and Step 4 min function in Algorithm 1. The time complexity of Step 3 plus Step 4 depends on the size of K_{max}. Since K_{max} is much smaller than N, that is, the number of neighboring fingerprint points are much smaller than the total number of fingerprint points, the complexity of the ARWKNN algorithm is similar to the KNN and WKNN algorithms. In 3-D, when $K_{max} = 8$, the average computing time of 200 targets varying with S is analyzed, the result of which is shown in Table 6. It can be seen that when S is the same, the average calculation time of the KNN, WKNN, SAWKNN, and ARWKNN algorithms is almost the same. It can also be seen from Figure 13 that when S decreases, the average positioning errors of four algorithms decrease, but the complexity of the algorithm also increases. Therefore, according to the actual situation, the power consumption and positioning error of the algorithm can be compromised by selecting an appropriate S.

Table 6. Computational complexity analysis.

Algorithm	The Value of S	Average Computing Time
KNN	$S = 10$ cm	15.07 ms
WKNN	$S = 10$ cm	15.18 ms
SAWKNN	$S = 10$ cm	15.51 ms
ARWKNN	$S = 10$ cm	15.28 ms
KNN	$S = 20$ cm	8.62 ms
WKNN	$S = 20$ cm	8.68 ms
SAWKNN	$S = 20$ cm	8.95 ms
ARWKNN	$S = 20$ cm	8.91 ms

4. Conclusions

At present, the classical KNN and WKNN algorithms are mainly aimed at 2-D positioning, assuming that the height of the target from the floor is known, and it is not feasible to know the height of the target from the floor in advance. The least linear multiplication method and Newton–Raphson method are suitable for solving 2-D coordinates. Solving the 3-D coordinate is a non-convex optimization problem, which is easy to fall into a local optimal solution. In this paper, the shortcomings of the fingerprint positioning algorithm and the trilateration method are discussed, and an adaptive residual weighted K-nearest neighbor fingerprint positioning algorithm is proposed. Compared with the fingerprint positioning algorithm based on compressed sensing, the range-based WKNN algorithm can achieve high-precision positioning under the low-density LED layout. Compared with RF [14], ELM [16], ANN [17], and GI-LS [11] machine-learning algorithms, fingerprint positioning based on the ARWKNN algorithm not only has lower complexity, but also has lower positioning error. The impact of LEDs modulation bandwidth, LEDs transmit power, the signal-to-noise ratio, the maximum number of neighboring fingerprints, the sampling interval, the number of LEDs, the sampling ratio and distance metric on positioning errors are analyzed in detail. The distribution of optimal K and the complexity of the algorithm are also analyzed in detail. Simulation results show that the ARWKNN algorithm based on CLD and MMD metrics produces the smallest average positioning error in 2-D and 3-D, respectively. Compared with the SAWKNN [19], WKNN [12] and KNN [15,25] algorithms, the ARWKNN algorithm can significantly reduce the average positioning error while maintaining similar algorithm complexity.

Due to lighting requirements and LoS communication, the typical SNR of indoor visible light communication is relatively high, however, the RF, ELM, ANN, GI-LS, SAWKNN, WKNN, KNN, and ARWKNN algorithms have higher positioning error under low SNR conditions. Our next step is to design an efficient noise filtering algorithm to achieve higher positioning accuracy under low SNR conditions. LED communication can not only achieve high-precision positioning, but also achieve high rates. We will consider using modified OFDM to achieve high-precision positioning with high modulation bandwidth and provide a real scenario in the future.

Author Contributions: Conceptualization, S.X. and Y.W.; methodology, S.X. and Y.W.; software, S.X., X.W., and F.W.; validation, S.X., C.-C.C., and Y.W.; formal analysis, S.X. and Y.W.; investigation, S.X., C.-C.C., and Y.W.; resources, Y.W.; data curation, Y.W.; writing—original draft preparation, S.X.; writing—review and editing, Y.W., X.W., and F.W.; visualization, S.X., C.-C.C., Y.W., X.W., and F.W.; supervision, Y.W.; project administration, Y.W.; funding acquisition, Y.W. and X.W. All authors have read and agreed to the published version of the manuscript.

Funding: This research was funded by the National Natural Science Foundation of China, grant number U1805262, 61871131, 61701118 and 61901117, this research also was funded by Research Project of Fujian Provincial, grant number 2018H6007 and 2019J01267, this research also was funded by Special Fund for Marine Economic Development of Fujian Province, grant number ZHHY-2020-3, and this research also was funded by Fujian Provincial Engineering Technology Research Center of Photoelectric Sensing Application, grant number 2018003.

Conflicts of Interest: The authors declare no conflict of interest.

References

1. Jang, B.; Kim, H. Indoor positioning technologies without offline fingerprinting map: A survey. *IEEE Commun. Surv. Tutor.* **2019**, *21*, 508–525. [CrossRef]
2. Yao, C.Y.; Hsia, W.C. An indoor positioning system based on the dual-channel passive RFID technology. *IEEE Sens. J.* **2018**, *18*, 4654–4663. [CrossRef]
3. Gharghan, S.K.; Nordin, R.; Jawad, A.M.; Jawad, H.M.; Ismail, M. Adaptive neural fuzzy inference system for accurate localization of wireless sensor network in outdoor and indoor cycling applications. *IEEE Access* **2018**, *6*, 38475–38489. [CrossRef]
4. Abdulrahman, A.; Abdulmalik, A.S.; Mansour, A.; Ahmad, A.; Suheer, A.H.; Mai, A.A.A.; Hend, S.A.K. Ultra wideband indoor positioning technologies: Analysis and recent advances. *Sensors* **2016**, *16*, 707.
5. Zou, H.; Jin, M.; Jiang, H.; Xie, L.; Spanos, C.J. WinIPS: WiFi-based non-intrusive indoor positioning system with online radio map construction and adaptation. *IEEE Trans. Wirel. Commun.* **2017**, *16*, 8118–8130. [CrossRef]
6. Yadav, R.K.; Bhattarai, B.; Gang, H.S.; Pyun, J.Y. Trusted K nearest bayesian estimation for indoor positioning system. *IEEE Access* **2019**, *7*, 51484–51498. [CrossRef]
7. Trong-Hop, D.; Myungsik, Y. An in-depth survey of visible light communication based positioning systems. *Sensors* **2016**, *16*, 678.
8. Yuan, P.; Zhang, T.; Yang, N.; Xu, H.; Zhang, Q. Energy efficient network localisation using hybrid TOA/AOA measurements. *IET Commun.* **2019**, *13*, 963–971. [CrossRef]
9. Wang, Y.; Ho, K.C. Unified near-field and far-field localization for AOA and hybrid AOA-TDOA positionings. *IEEE Trans. Wirel. Commun.* **2018**, *17*, 1242–1254. [CrossRef]
10. Khalajmehrabadi, A.; Gatsis, N.; Akopian, D. Modern WLAN fingerprinting indoor positioning methods and deployment challenges. *IEEE Commun. Surv. Tutor.* **2017**, *19*, 1974–2002. [CrossRef]
11. Guo, X.; Shao, S.; Ansari, N.; Khreishah, A. Indoor localization using visible light via fusion of multiple classifiers. *IEEE Photon. J.* **2017**, *9*, 7803716. [CrossRef]
12. Alam, F.; Chew, M.T.; Wenge, T.; Gupta, G.S. An accurate visible light positioning system using regenerated fingerprint database based on calibrated propagation model. *IEEE Trans. Instrum. Meas.* **2019**, *68*, 2714–2723. [CrossRef]
13. Lovón-Melgarejo, J.; Castillo-Cara, M.; Huarcaya-Canal, O.; Orozco-Barbosa, L.; García-Varea, I. Comparative study of supervised learning and metaheuristic algorithms for the development of bluetooth-based indoor localization mechanisms. *IEEE Access* **2019**, *7*, 26123–26135. [CrossRef]
14. Breiman, L. Random forests. *Mach. Learn.* **2001**, *45*, 5–32. [CrossRef]
15. Xue, W.; Yu, K.; Hua, X.; Li, Q.; Qiu, W.; Zhou, B. APs' virtual positions-based reference point clustering and physical distance-based weighting for indoor Wi-Fi positioning. *IEEE Internet Things J.* **2018**, *5*, 3031–3042. [CrossRef]
16. Huang, G.B.; Zhou, H.; Ding, X.; Zhang, R. Extreme learning machine for regression and multiclass classification. *IEEE Trans. Syst. Man Cybern. B.* **2012**, *42*, 513–529. [CrossRef]

17. Zhang, S.; Du, P.; Chen, C.; Zhong, W.D.; Alphones, A. Robust 3D indoor VLP system based on ANN using hybrid RSS/PDOA. *IEEE Access* **2019**, *7*, 47769–47780. [CrossRef]
18. Fang, X.; Jiang, Z.; Nan, L.; Chen, L. Optimal weighted *K*-nearest neighbour algorithm for wireless sensor network fingerprint localisation in noisy environment. *IET Commun.* **2018**, *12*, 1171–1177. [CrossRef]
19. Hu, J.; Liu, D.; Yan, Z.; Liu, H. Experimental analysis on weight *K*-nearest neighbor indoor fingerprint positioning. *IEEE Internet Things J.* **2019**, *6*, 891–897. [CrossRef]
20. Gu, W.; Aminikashani, M.; Deng, P.; Kavehrad, M. Impact of multipath reflections on the performance of indoor visible light positioning systems. *IEEE/OSA J. Lightwave Technol.* **2016**, *34*, 2578–2587. [CrossRef]
21. Şahin, A.; Eroğlu, Y.S.; Güvenç, I.; Pala, N.; Yüksel, M. Hybrid 3D localization for visible light communication systems. *IEEE/OSA J. Lightwave Technol.* **2015**, *33*, 4589–4599. [CrossRef]
22. Mathias, L.C.; Melo, L.F.D.; Abrao, T. 3-D localization with multiple LEDs lamps in OFDM-VLC system. *IEEE Access* **2018**, *7*, 6249–6261. [CrossRef]
23. Zhou, B.; Lau, V.; Chen, Q.; Cao, Y. Simultaneous positioning and orientating for visible light communications: Algorithm design and performance analysis. *IEEE Trans. Veh. Technol.* **2018**, *67*, 11790–11804. [CrossRef]
24. Wu, Y.; Liu, X.; Guan, W.; Chen, B.; Chen, X.; Xie, C. High-speed 3D indoor localization system based on visible light communication using differential evolution algorithm. *Opt. Commun.* **2018**, *424*, 177–189. [CrossRef]
25. Van, M.T.; Tuan, N.V.; Son, T.T.; Le-Minh, H.; Burton, A. Weighted *k*-nearest neighbour model for indoor VLC positioning. *IET Commun.* **2017**, *11*, 864–871. [CrossRef]
26. Gligorić, K.; Ajmani, M.; Vukobratović, D.; Sinanović, S. Visible light communications-based indoor positioning via compressed sensing. *IEEE Commun. Lett.* **2018**, *22*, 1410–1413. [CrossRef]
27. Tropp, J.A.; Gilbert, A.C. Signal recovery from random measurements via orthogonal matching pursuit. *IEEE Trans. Inf. Theory* **2007**, *53*, 4655–4666. [CrossRef]
28. Zhang, R.; Zhong, W.D.; Qian, K.; Zhang, S.; Du, P. A reversed visible light multitarget localization system via sparse matrix reconstruction. *IEEE Internet Things J.* **2018**, *5*, 4223–4230. [CrossRef]
29. Feng, C.; Au, W.S.A.; Valaee, S.; Tan, Z. Received-signal-strength-based indoor positioning using compressive sensing. *IEEE Trans. Mob. Comput.* **2012**, *11*, 1983–1993. [CrossRef]
30. Keskin, M.F.; Sezer, A.D.; Gezici, S. Localization via visible light systems. *Proc. IEEE* **2018**, *106*, 1063–1088. [CrossRef]
31. Zhou, Z.; Kavehrad, M.; Deng, P. Indoor positioning algorithm using light-emitting diode visible light communications. *Opt. Eng.* **2012**, *51*, 085009. [CrossRef]
32. Yasir, M.; Ho, S.W.; Vellambi, B.N. Indoor positioning system using visible light and accelerometer. *IEEE/OSA J. Lightwave Technol.* **2014**, *32*, 3306–3316. [CrossRef]
33. Komine, T.; Nakagawa, M. Fundamental analysis for visible-light communication system using LED lights. *IEEE Trans. Consum. Electron.* **2004**, *50*, 100–107. [CrossRef]
34. Zhang, W.; Chowdhury, M.I.S.; Kavehrad, M. Asynchronous indoor positioning system based on visible light communications. *Opt. Eng.* **2014**, *53*, 045105. [CrossRef]
35. Li, H.; Wang, J.; Zhang, X.; Wu, R. Indoor visible light positioning combined with ellipse-based ACO-OFDM. *IET Commun.* **2018**, *12*, 2181–2187. [CrossRef]
36. Mossaad, M.S.A.; Hranilovic, S.; Lampe, L. Visible light communications using OFDM and multiple LEDs. *IEEE Trans. Commun.* **2015**, *63*, 4304–4313. [CrossRef]
37. Cha, S.H. Comprehensive survey on distance/similarity measures between probability density functions. *Int. J. Math. Models Methods Appl. Sci.* **2007**, *1*, 300–307.

Article

Simulating Signal Aberration and Ranging Error for Ultrasonic Indoor Positioning

Riccardo Carotenuto [1], Massimo Merenda [1,2,*], Demetrio Iero [1,2] and Francesco G. Della Corte [1,2]

[1] DIIES Department, University Mediterranea of Reggio Calabria, 89126 Reggio Calabria, Italy; r.carotenuto@unirc.it (R.C.); demetrio.iero@unirc.it (D.I.); francesco.dellacorte@unirc.it (F.G.D.C.)
[2] HWA srl, Spin-off University Mediterranea of Reggio Calabria, Via R. Campi II tr. 135, 89126 Reggio Calabria, Italy
* Correspondence: massimo.merenda@unirc.it; Tel.: +39-0965-1693-464

Received: 22 May 2020; Accepted: 20 June 2020; Published: 23 June 2020

Abstract: Increasing efforts toward the development of positioning techniques testify the growing interest for indoor position-based applications and services. Many applications require accurate indoor positioning or tracking of people and assets, and some market sectors are starting a rapid growth of products based on these technologies. Ultrasonic systems have already been demonstrating their effectiveness and to possess the desired positioning accuracy and refresh rates. In this work, it is shown that a typical signal used in ultrasonic positioning systems to estimate the range between the target and reference points—namely, the linear chirp—due to the effects of acoustic diffraction, in some cases, undergoes a shape aberration, depending on the shape and size of the transducer and on the angle under which the transducer is seen by the receiver. In the presence of such signal shape aberrations, even one of the most robust ranging techniques, which is based on cross-correlation, provides results affected by a much greater error than expected. Numerical simulations are carried out for a typical ultrasonic chirp, ultrasonic emitter, and range technique based on cross-correlation and for a typical office room, obtained using the academic acoustic simulation software Field II. Spatial distributions of the ranging error are provided, clearly showing the favorable low error regions. The work demonstrates that particular attention must be paid to the design of the acoustic section of the ultrasonic positioning systems, considering both the shape and size of the ultrasonic emitters and the shape of the acoustic signal used.

Keywords: acoustic diffraction; acoustic signal aberration; cross-correlation aberration; ultrasonic ranging

1. Introduction

Augmented reality (AR) and many other applications based on positioning are emerging technologies that need indoor positioning technology. Mall navigation, path finding in large hospitals or airports, the automatic guidance for unmanned cleaning and maintenance vehicles, surveillance systems, and others require positioning systems capable of operating inside buildings with high positioning accuracy [1–5]. Many accurate positioning systems use trilateration, a technique that has proven to work well indoors. The trilateration (multilateration) positioning technique requires three (many) range measurements between reference emitters and a sensor to be located. Distances and spatial positions with a high degree of precision at a relatively low cost can be provided by systems based on ultrasonic waves [6–10]. The most used ranging technique involves the estimation of the time of arrival (TOA) of a suitable ultrasonic signal. Typically, the TOA is estimated by finding in the received signal or in the postprocessed signal a specific that is easy to identify upon arrival. In some cases, the arrival time of the maximum peak of the envelope of an ultrasonic pulse is used. However, despite the simplicity of this technique, it is highly subject to acoustic disturbances, with errors in the order of wavelengths (centimeters) even in the presence of a high signal-to-noise ratio (SNR) [11–16].

Methods based on cross-correlation estimate the TOA with high accuracy and, in general, good acoustical noise immunity [13–16]. Digital cross-correlation techniques properly sample and analog-to-digital convert the received acoustical signal to obtain the array $C = S \star T$, where S is the numerical array of received signal samples, T is the digital reference signal, previously stored in the sensor processor memory as a numerical array of samples, and $\star$ denotes the cross-correlation operator. The best aligning of S and T in time is revealed by the peak of the cross-correlation. The inter-signals displacement, or lag τ, corresponding to the cross-correlation peak (i.e., τ_{MAX}), is proportional to the TOA [17,18].

The distance measurement accuracy of the order of the current space sampling, which is the distance covered by the ultrasound during the time sample interval and that can be made much smaller than the ultrasonic wavelength, is achieved by estimating the TOA through the cross-correlation peak. In addition, the cross-correlation peak is easily detectable when the signal T is a chirp. In practical systems, it is possible to achieve a range resolution up to the order of one-tenth of the wavelength used. For example, in [16], a range resolution of ± 1.2 mm was experimentally achieved using a 15–40-kHz chirp, with a wavelength range 22.86–8.57 mm, considering a sampling frequency of 1 MHz and, thus, space sampling of 0.34 mm with a sound speed of 343 m/s. When the signal and noise are uncorrelated, cross-correlation significantly increases the SNR. It is a known drawback that, in some cases, the cross correlation peak associated with the chirp that travels along the direct path (or line of sight, LOS) is lower than the signals coming from the reflection paths. In the presence of reflections, a number of signals from indirect paths combine to produce a peak higher than that of the direct path signal.

The acoustic field generated by acoustic transducers according to the shape and aperture of the transducer has been widely studied in the past considering impulsive or continuous sinusoidal wave signals. The closed-form solution of the generated acoustic field made it possible to derive simple approximate formulas to calculate the emission angle as a function of the wavelength, aperture, and distance from the emitter, the best known of which apply to circular apertures. In the far field, the emission cone semi-angle ϑ is approximately described by the well-known relationship [19]

$$\sin \vartheta = 1.22 \frac{\lambda}{D} \tag{1}$$

where λ is the emitted wavelength and D the diameter, or aperture, of the circular planar transducer.

Furthermore, it is well-known that, to obtain good results, the receiver must always operate within the emission cone of the emitter [20]. At present, no equivalent formulas are known in the case of any signals, such as chirp.

To design a positioning system based on ultrasonic signals, tools to evaluate the spatial coverage of each transducer in terms of quality (amplitude and level of deformation) of the received signal are needed. For any signals, there are no simplified formulas that give a reliable indication; therefore, the usage of numerical tools is mandatory.

The use tout court of finite element analysis (FEA), which is a very powerful general use tool, seems excessive for the design of positioning systems and definitively not practical from the computational point of view. In fact, for the systems under investigation, large spatial regions of several cubic meters and time windows of tens of milliseconds should be considered (see, e.g., [9,10,13]).

From these premises, it was, therefore, decided to use a powerful numerical tool, the academic Field II software [21], for the analysis of ultrasound positioning systems. Field II was developed and is currently very well-known for the simulation of complete ultrasonic imaging systems. A transducer for ultrasonic signals both for transmission and reception, and the formation process of images in the field of medical ultrasounds, are simulated. However, Field II has the numerical characteristics that make it a valid tool also in the field of ultrasonic positioning.

Among many other capabilities, which however fall outside the scope of this work, this tool is able to calculate the acoustic pressure field at any point in the space for transducers and signals of any shape, taking into account the attenuation properties of the propagation medium. In other words, Field II allows to modify in any way the transducer shape and size, and the signal applied to the transducer, to evaluate their effects in a trial and error design cycle, if necessary.

In this work, using Field II, the effectiveness of cross-correlation-based ranging techniques using a chirp signal when the diameter of the circular plane transducer used as ultrasonic emitter is changed is shown.

This work will show that, considering a chirp signal outside a certain emission cone generated by the transducer, the usual ranging technique introduces a significant error in calculating the emitter-receiver distance.

In perspective, the main advantages of the proposed approach are the possibility of examining the acoustic field over time and space at each point of the region of interest as a function of the aperture and of the type of signal emitted (e.g., of its bandwidth or shape) and the ability to easily test each algorithm dedicated to estimating the TOA in the various positions and operating situations.

This paper is structured as follows. Section 2 describes the proposed simulation setup, while Section 3 shows the simulation results, and Section 4, the discussion. Section 5 draws the paper's conclusions.

2. Field II and Simulation Setup

In this section, the operating principle of the Field II simulator is briefly outlined, and the simulation configuration is described in detail.

The acoustic field simulator Field II [22] employs the concept of spatial impulse responses [23–25]. The ultrasound field for both the pulsed and continuous wave cases is found using the linear systems theory. The spatial impulse response gives the emitted ultrasound field at a specific point in space as a function of time, when the transducer is excited by a Dirac delta function. In a second step, the field generated by any kind of excitation is found by convolving the spatial impulse response with the excitation function. Since the linear systems theory is used, any excitation can be considered. The impulse response is a function of the position relative to the transducer, hence the name spatial impulse response of the technique [26].

Briefly, the transducer surface is divided into small rectangles, introducing a transducer surface and field approximations that are as much smaller as the elements into which the transducer surface is divided are smaller. The approximation is reduced by using small rectangles, where the distance to the field point is large compared to the size of the rectangles. Typically, the element size is much smaller than the wavelength of the signal to be simulated. Each of the rectangular elements is considered a rectangular piston, of which the exact solution for the impulsive response is known [25]. A spherical wave is emitted by each of the small elements, and the impulsive responses due to each element are added together at each desired field point [26].

In the simulations that follow, the aim is to examine the acoustic field and the effectiveness of an established ranging technique based on the correlation in a typical $4 \times 4 \times 3$ m^3 room [27]. In particular, the simulation results will be examined on a grid of points belonging to a vertical section (see Section A of the room volume, Figure 1) and on a horizontal section halfway between the floor and the ceiling (see Section B of the room volume, Figure 1). The transducer is a circular planar and is placed in the center of the ceiling, in position $x = 0$, $y = 0$, and $z = 0$, and emits towards the floor of the room.

The transducer is immersed in the air, and a linearized air absorption (slope 39.3 dB/m·MHz, constant term -0.262 dB/m, i.e., about 0.917 dB/m @ 20 kHz and 1.703 dB/m @ 50 kHz) has been assumed around 40 kHz, corresponding to a transducer central frequency of 40 kHz at a temperature of 20 °C, a pressure of 1 atm, and a relative humidity of 55% [28,29].

The emitted and received signals at all points in the space depends on the shape and size of the emission surface (i.e., on the aperture D) of the transducer. In this work, a circular plane transducer was considered, with acoustic properties similar to those of the most commonly used transducers for positioning applications—for example, the typical piezoelectric transducer Murata MA40S4S (D = 9.9 mm) or Pro-wave 328ST/R160 (D = 13.1 mm) [30,31].

The emission disk is divided into a certain number of rectangles. In particular, square elements with sides 0.125 mm by 0.125 mm were used for all the simulations that follow. The transducer element size chosen in this work is a good compromise between the accuracy of the solution and computational resources engaged in the simulation. In Figure 2, for displaying purposes, in order to visualize the single elements, the dimension of the mathematical elements is 1 mm by 1 mm.

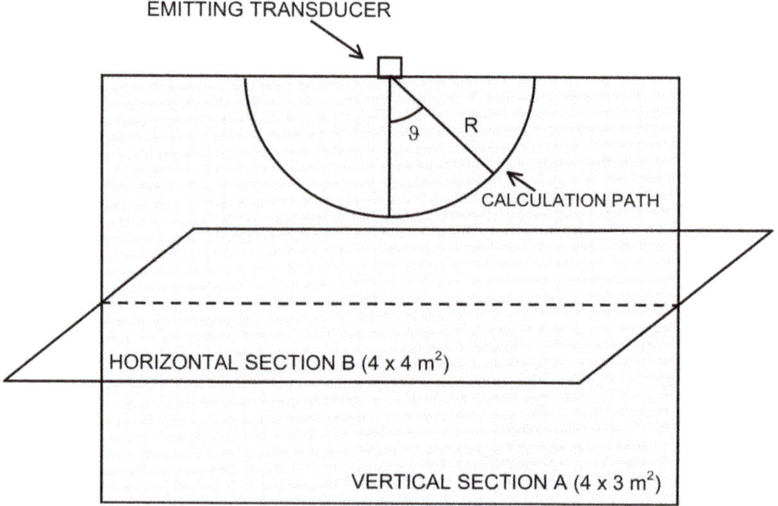

Figure 1. Simulation setup: the calculation path of the cross-correlation results shown in Figures 3–5; the vertical section A of the typical 4 m × 4 m × 3 m room along which the results displayed in Figure 6 are computed; the horizontal section B where the results displayed in Figure 7 are computed.

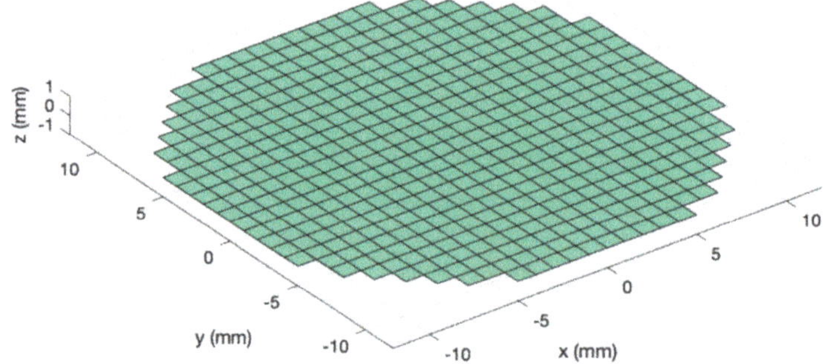

Figure 2. Example of a defined emitting transducer: circular and planar piston transducer with a diameter of D = 25 mm divided into square mathematical elements.

The signal used for the simulations is a linear chirp with a bandwidth of 30-50 kHz and 5.12-ms duration [10,27].

For simulation purposes, the signal was sampled at f_S = 1 MHz. The acoustic field was calculated in a set of points in the space for the duration of a time window compatible with the complete reception of the signal itself [26]. Once the simulation was completed, for each point of the space considered in the simulation, the behavior over time of the acoustic pressure generated by the complete excitation signal was obtained. This allowed any subsequent evaluation and postprocessing of the signal to be obtained. For example, the peak pressure and the total signal strength at each point can be calculated. Subsequently, an ideal receiver was assumed that linearly transduced the pressure signal into an electrical signal, downstream of a suitable sampling and numerical quantization, so that the cross-correlation vector C could be calculated.

3. Simulations Results

All the following simulations were performed for a set of acoustical apertures D = {25 mm, 20 mm, 15 mm, 13.1 mm, 8.5 mm, 6 mm}. The simulation includes acoustic diffractive phenomena, with the possibility of simulating transducers of every shape and every emitted signal. Finally, it is possible to test every ranging or positioning technique one intends to apply.

In Figure 3, it is possible to see the value of the pressure peak, the correlation peak, and the estimated distance when using the position of the correlation peak to evaluate the TOA at varying D within the previously defined set. Given the field symmetry, only the results for angles from 0° (on the axis) to 90° (laterally to the transducer) are shown on a path at a constant distance R = 1 m from the center of the transducer and with varying apertures of D.

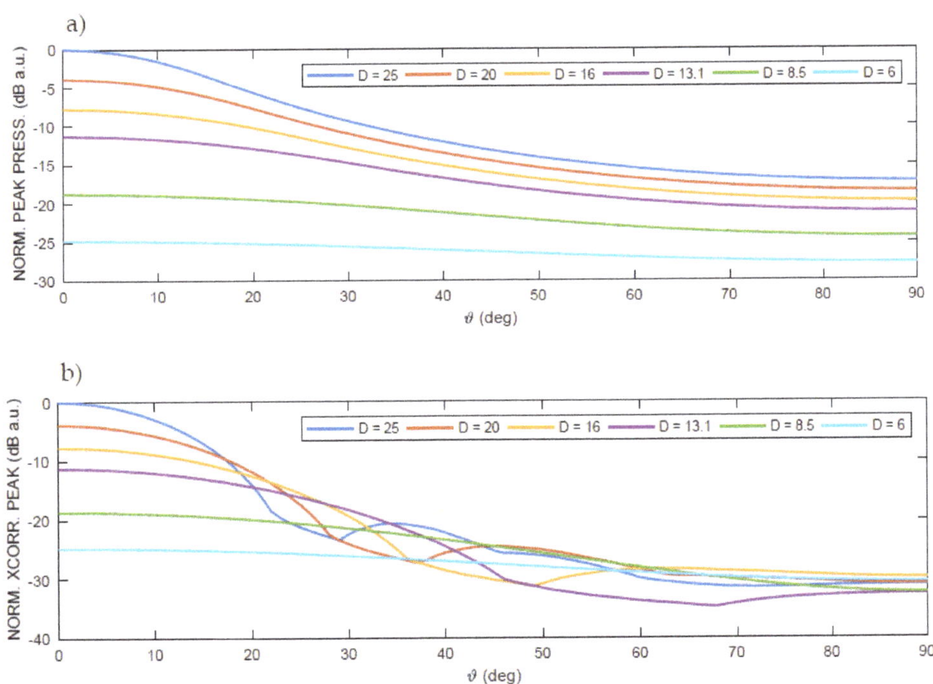

Figure 3. *Cont.*

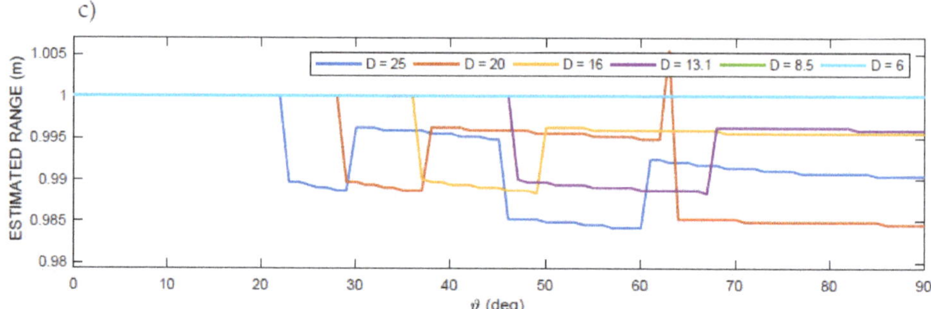

Figure 3. Numerical results at different transducer apertures D = {25, 20, 16, 13.1, 8.5, 6} mm along a semicircular path at distance R = 1 m from the emitting transducer, using a linear chirp with starting frequency f_L = 30 kHz and final frequency f_H = 50 kHz. A linearized air absorption around 40 kHz (slope 39.3 dB/m·MHz, constant term -0.262 dB/m, i.e., about 0.917 dB/m @ 20 kHz and 1.703 dB/m @ 50 kHz) has been assumed, considering the room temperature 20 °C and atmospheric pressure 1 atm: (**a**) acoustical pressure peak, displayed after normalization and dB conversion, (**b**) cross-correlation peak, displayed after normalization and dB conversion, and (**c**) estimated range from the position of the cross-correlation absolute peak. For the first four aperture diameters, the lag of the cross-correlation peak does not correspond to the correct time of arrival (TOA) (see also, Figure 4).

Finally, at the bottom of Figure 3, it is possible to see the results of the estimate of the distance R* using the usual technique based on the search for the maximum position of the cross-correlation peak (τ_{MAX}) [16,32], which, in favorable conditions, produces the correct estimate of the TOA and, from this, the estimate of the range R*, considering:

$$c_{air} = 331.5\sqrt{1 + \frac{T}{273.15}},\tag{2}$$

where c_{air} (m/s) is the speed of sound in the air, and T (°C) is the ambient temperature. In particular, the estimate of the range estimate R* is computed as follows:

$$R^* = \left(\frac{\tau_{MAX}}{f_S} - TOE\right)c_{air} - R_{cal}\tag{3}$$

where τ_{MAX} is the lag of the maximum peak, and R_{cal} is a calibration constant that takes into account all the fixed delays of the considered system. R_{cal} is independent of the range, and the time of emission of the ultrasonic signal (TOE) can be assumed known through some a priori operation.

The cross-correlation along the same path considered in Figure 3—that is, along a quarter of a circumference belonging to a plane passing through the emission axis of the transducer with a radius R = 1 m—is shown in Figure 4 as the grayscale amplitude of the cross-correlation with variable lag for each angle ϑ and for the six considered apertures.

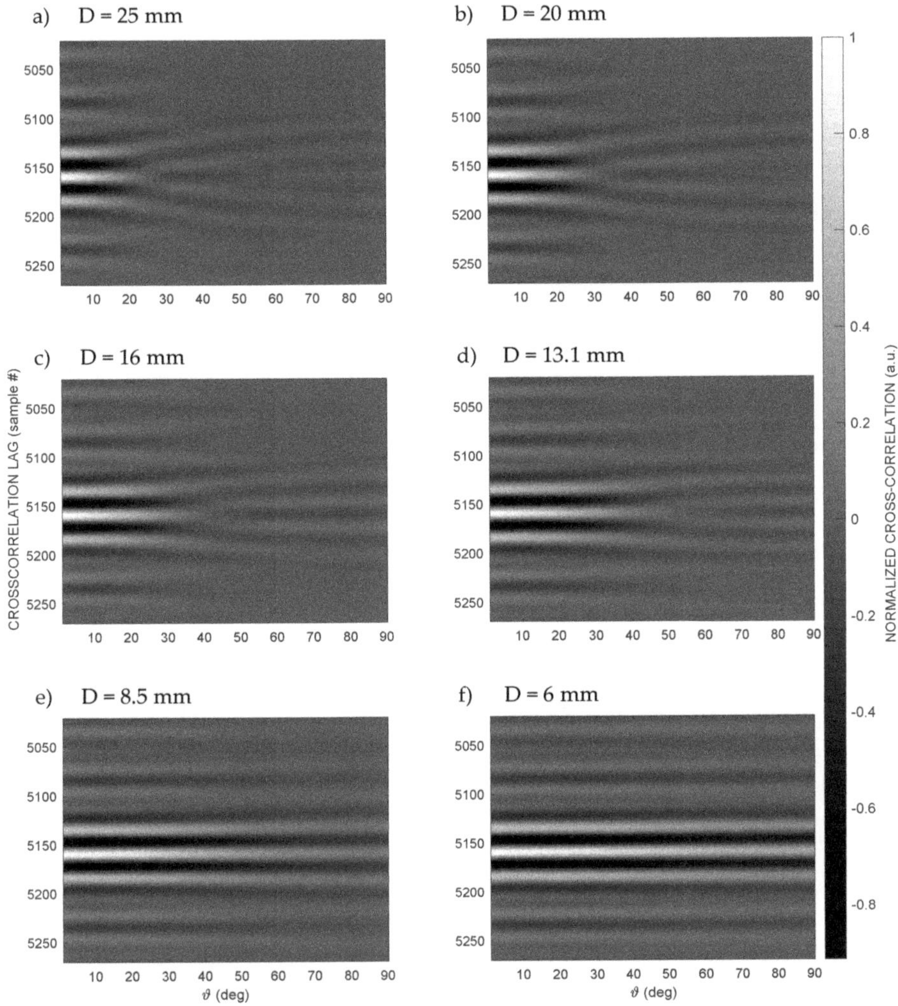

Figure 4. Cross-correlation values along a semicircular path at distance R = 1 m from the emitting transducer at different transducer apertures D = {25, 20, 16, 13.1, 8.5, 6} mm. From D = 25 mm down to D = 6 mm, it is possible to appreciate the progressive appearance of a single correlation peak, which makes the identification of the TOA univocal.

In Figure 5 are displayed the cross-correlations along a semicircular path at a distance R = 1 m from the emitting transducer for two different transducer apertures: D = 25 mm (Figure 5a) and D = 8.5 mm (Figure 5b). The cross-correlation values are normalized to their maximum value for each aperture.

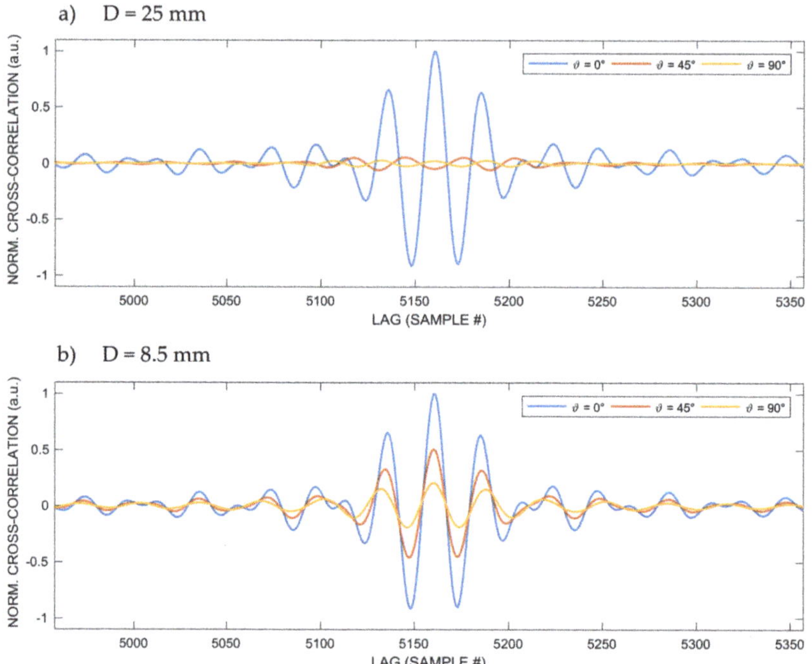

Figure 5. Cross-correlations along a semicircular path at distance R = 1 m from the emitting transducer for two different transducer apertures: (**a**) D = 25 mm and (**b**) D = 8.5 mm. For D = 25 mm, it is possible to see that the single unique peak of the cross-correlation for $\vartheta = 0°$ is no longer present at the 45° and 90° angles, while, for D = 8.5 mm, it is possible to appreciate the single correlation peak at all angles. The cross-correlation values are normalized to their maximum value for each aperture: for D = 8.5 mm, the amplitude relative reduction with respect to the increasing angle is much lower than for D = 25 mm, due to the much wider emission of the smaller aperture.

In order to observe in detail the extent and shape of the spatial regions within which it is possible to obtain the typical accuracy of the technique based on the cross-correlation, the ranging error on two rectangular grids of points (see Section A and B of the room volume, Figure 1) was evaluated as a function of D. The grid pitch is 5 cm in the x and z directions. For each point, the lag of the correlation peak (τ_{MAX}) and, from these, the estimates of the distance from the emitter through (3) were obtained. Finally, the competent ground-truth value at each point of the simulation grid was subtracted from the values just obtained, thus generating a grid of estimates of the ranging error.

Figure 6 shows the ranging error along a rectangular vertical section (Section of the room volume, Figure 1) of 3-m height and 4-m base passing through the center of the transducer, equal to the vertical section of the typical office room taken as a reference in some positioning works [27,33,34], for each aperture D of the set defined above. The grid pitch is 5 cm in the x and y directions.

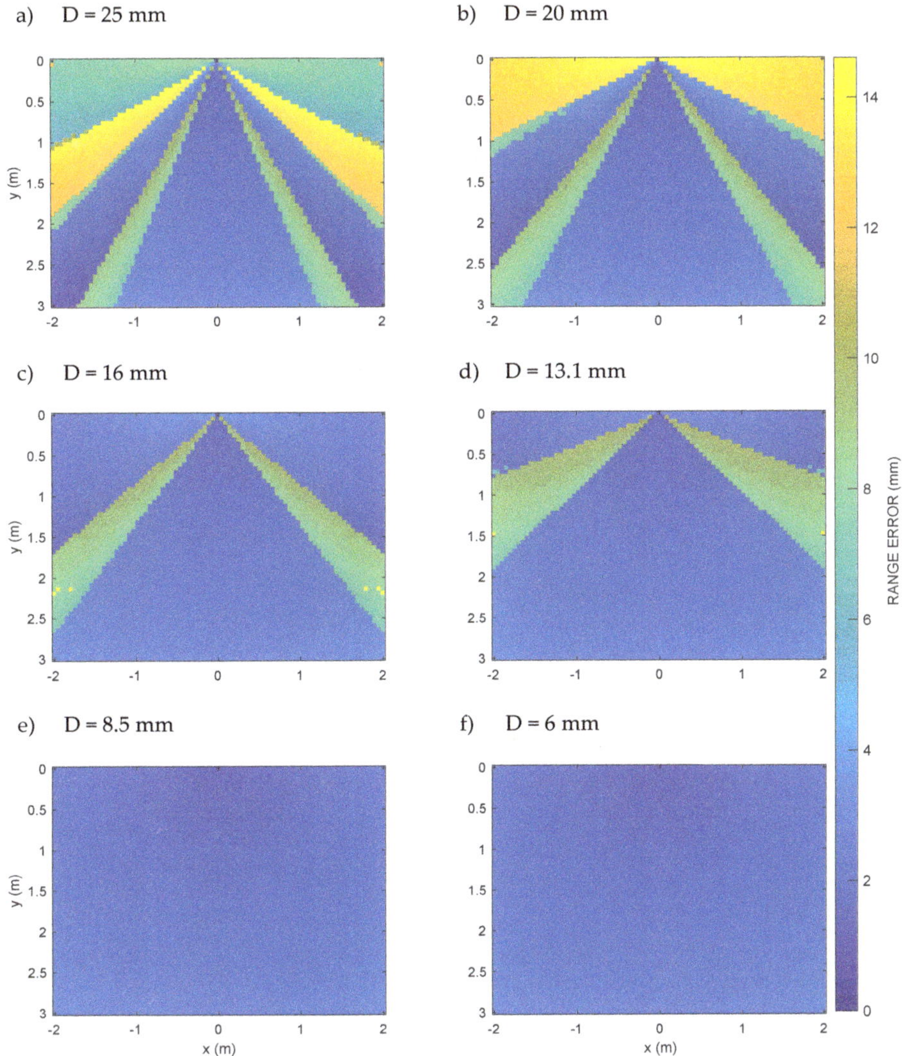

Figure 6. Computed range error at different transducer apertures D = {25, 20, 16, 13.1, 8.5, 6} mm in a dense grid of points (horizontal and vertical step = 0.05 m) belonging to the vertical 4 m × 3 m Section A (see Figure 1); it is possible to appreciate the progressive widening of the cone of the minimum ranging error going from D = 25 mm to D = 6 mm. For aperture values D = 8.5 mm and D = 6 mm, the low error region includes all the half-space in front of the transducer.

Finally, Figure 7 shows the behavior of the ranging error on a horizontal section of 4 m × 4 m at z = 1.5 m, or halfway between the ground and the ceiling (Figure 1b), for all the apertures considered.

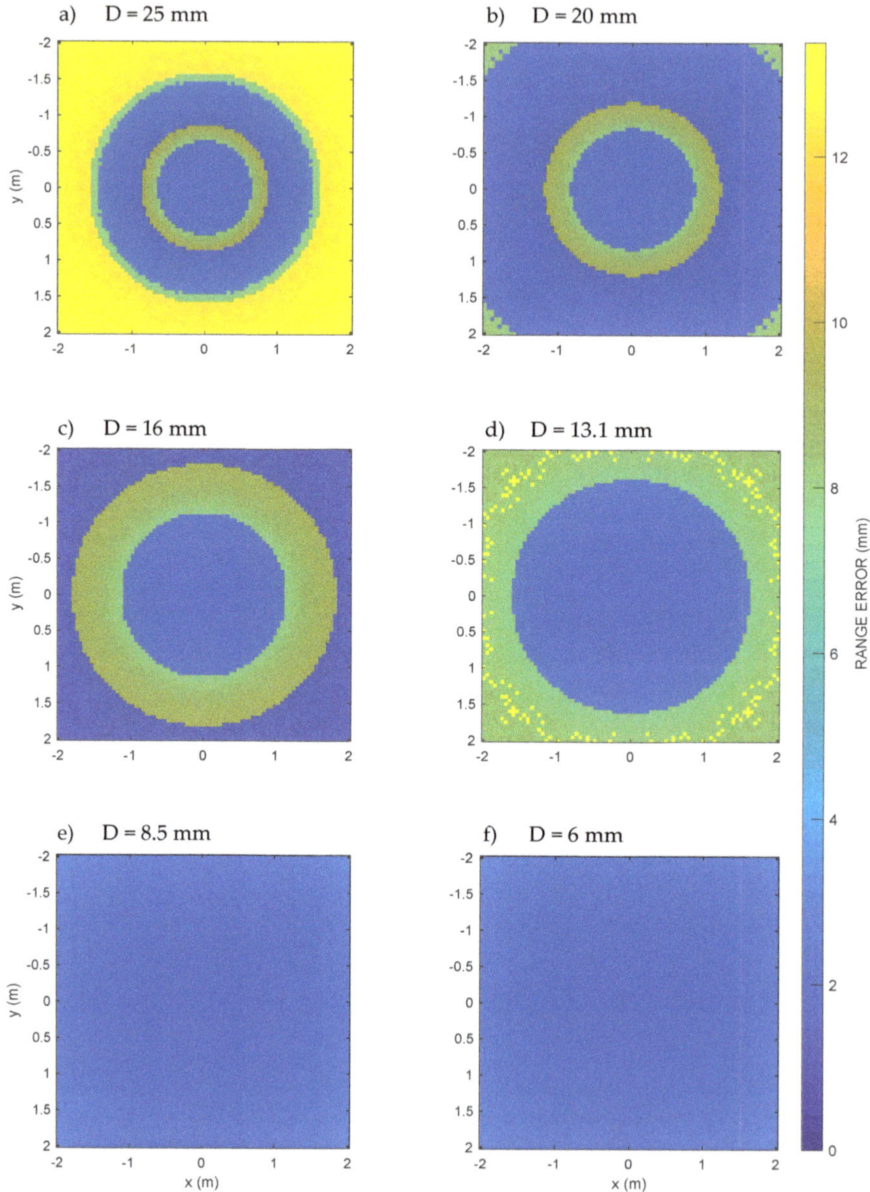

Figure 7. Computed range error at different transducer apertures D = {25, 20, 16, 13.1, 8.5, 6} mm in a dense grid of points (horizontal and vertical step = 0.05 m) belonging to the horizontal 4 m × 4 m Section B (see Figure 1); it is possible to appreciate the progressive widening of the disk of the minimum ranging error going from D = 25 mm to D = 6 mm. For aperture values D = 8.5 mm and D = 6 mm, the low error region includes all the half-space in front of the transducer.

For D from 25 mm to 13.1 mm, Figure 7a–d, it is possible to clearly recognize increasing low error circular areas, which are the circular sections of the low error cones already seen in Figure 6a–d. Using such apertures, therefore, it is not possible to cover the whole room for that height, and things go even worse for high heights, which, however, are certainly of interest for a three-dimensional internal positioning system. On the contrary, the volume of the room is completely covered by the last two apertures D = 8.5 mm and D = 6 mm, as can also be seen from the vertical sections of Figure 6e,f, which show ranging errors everywhere lower than about 3.09 mm. The ranging error is due to the numerical approximations and the sampling frequency chosen for the simulation.

4. Discussion

In Figure 3, as expected, the pressure peak drops smoothly and rapidly with the increasing angle ϑ. The behavior of the cross-correlation peak is different, abruptly varying at certain angles. This depends on the shape of the cross-correlation, as shown also in Figure 4.

In general, note that the pressure values are decreasing as the aperture decreases, since the surface power density in emission is kept constant, while the extension of the emitting surface decreases. Furthermore, since the cross-correlation is also proportional to the amplitude of the received signal, its peak value also decreases with the pressure signal. Note that, as the aperture D decreases, the angle up to which the correct estimate is obtained increases; for D = 25 mm, the largest ranging error is obtained, over 15 mm.

Additionally, consider that no noise has been added to show more clearly that the observed phenomenon is due only to the acoustic diffraction that exists regardless of the current SNR level.

In Figure 4, the grayscale amplitude of the cross-correlation with variable lag shows that, up to a certain angle ϑ, the shape of the cross-correlation remains regular, as expected, with only one clearly recognizable peak. In the regions included in the above angles, the correlation-based technique works very well, with errors in the order of the sampling rate of the signal. However, beyond a certain limit angle, the value of which increases as the transducer aperture D decreases, on the other hand, the cross-correlation deforms, with variations in the shape and multiplication of the peaks, with a trend similar to a bifurcation. For angles larger than this limit, it becomes unpractical to identify a peak corresponding to the TOA, simply because it no longer exists; in fact, the peaks of the cross-correlation beyond the limit angle no longer correspond to the correct lag proportional to the TOA, and, therefore, they produce incorrect estimates of the TOA and, as a consequence, of R*. This finally well explains the strange abrupt behavior of the range estimation of Figure 3c.

In Figure 5, it is possible to see that, for D = 25 mm, it is possible to appreciate that the single univocal peak of the cross-correlation for $\vartheta = 0°$ is not anymore present at angles 45° and 90°, while, for D = 8.5 mm, it is possible to identify the single correlation peak at all angles. Moreover, for D = 8.5 mm, the amplitude relative reduction with respect to the increasing angle is much lower than the reduction for D = 25 mm, due to the much wider emission of the smaller aperture.

In Figure 6a–d is displayed a zone with a shape similar to a triangle (similar to a cone in three dimensions), with the vertex corresponding to the center of the transducer, inside which the error is minimal, i.e., of the order of the error quantization due to the signal time sampling $1/f_S$. Therefore, let us define this low error triangular region (conical in three dimensions) as that of the correct operation of the ranging system and φ its angle at the vertex. In Table 1, it is possible to see φ and the maximum error as a function of the increasing aperture D values.

Table 1. Low error cone angle and range maximum error as a function of the aperture D.

Emitter aperture D (mm)	Low Error Cone Vertex Angle φ (°)	Range Maximum Error (mm)
6	180	3.3
8.5	180	3.3
13.1	93.51	13.3
16	73.26	13.7
20	55.84	14.2
25	44.62	14.6

Immediately outside this low error area, on the other hand, the abrupt appearance of a higher error, often greater than 11 mm, is observed, which is ultimately produced by the bifurcation of the cross-correlation values, as also shown in Figure 4.

Instead, in Figure 6e,f, no conical region is seen, but a fairly uniform error appears, which is about one order of magnitude lower than that seen in Figure 6a–d. This is the numerical error due to the numerical approximations and the sampling frequency chosen for the simulation, everywhere less than about 3.3 mm. The absence of the low error conical region is due to the fact that, for aperture values D = 8.5 mm and D = 6 mm, the low error region includes all the half-space in front of the transducer.

The isolated points of yellow color (relatively large ranging errors) in Figure 6c,d refer to positions where the peak detection error is large due to the similarity in the height of adjacent peaks of the cross-correlation (see also peaks of almost equal height for $\vartheta = 45°$ and $\vartheta = 90°$ in Figure 5a). The same phenomenon is also observed in Figure 7d.

In Figure 7a–d, for D from 25 mm to 13.1 mm, it is possible to clearly recognize increasing low error circular areas, which are the circular sections of the low error cones already seen in Figure 6a–d. Using such aperture values, therefore, it is not possible to cover the whole room for that height, and things go even worse for z higher than 1.5 m, which, however, is certainly of interest for a three-dimensional internal positioning system. On the contrary, the volume of the room is completely covered by the last two apertures D = 8.5 mm and D = 6 mm, as can also be seen from the vertical sections of Figure 6e,f, which show ranging errors everywhere lower than about 3.09 mm. The ranging error is mainly due to the numerical approximations and the sampling frequency chosen for the simulation.

As a significant result, for apertures D from 25 mm down to 13.1 mm, it is possible to clearly recognize the cone-shaped favorable zone. With these apertures, however, it is not possible to cover the whole room. In fact, the room can only be covered up to a height of less than 1 meter from the floor in the most favorable case. This unfortunately prevents, in many cases, from reaching the coverage required by three-dimensional indoor positioning systems. The room, on the other hand, is completely covered by the last two apertures, as can also be seen from the vertical Section A shown in Figure 6, where, in fact, the conical region is no longer recognized, since the low error area is now extended to the whole volume of the room.

The simulations presented demonstrate that, using Field II in the design phase, by varying the transducer aperture and the others parameters, it is therefore possible to check whether the acoustic coverage required by a specific application is reached, i.e., whether the region of interest for that application is within the region where the ranging error is sufficiently low or not.

5. Conclusions

In this paper, Field II, an acoustical simulation software well-established in the field of ultrasound medical imaging, has been applied to the simulation of the acoustic field in air produced by a circular transducer and to the evaluation of a ranging technique based on the measurement of this acoustic field.

The original contribution of this work is to show that Field II can be profitably applied to the problem of ranging with ultrasound in the air. As the first significant result, numerical simulations have shown that it is not enough to guarantee a certain acoustic pressure in a spatial region to reach a certain low level of error. In fact, depending on the angle at which the emitter is seen, the received chirp undergoes a significant aberration in shape compared to that emitted. Shape aberration also occurs to its cross-correlation, so the usual peak detection technique cannot detect the true TOA, regardless of the signal level or SNR.

Field II allows us to observe ranging errors greater than expected in the presence of signal shape aberrations, regardless of the SNR. This means that particular care must be taken in the acoustic design of an ultrasound positioning system and that the use of a numerical simulator such as Field II is necessary in the design phase. With such a tool, in fact, it is possible to evaluate effectively both the acoustic coverage and the accuracy of the ranging technique used.

In particular, it was possible to observe the total coverage of a typical 4 m × 4 m × 3 m room by using a circular aperture of diameter D = 8.5 mm or less, a 30–50-kHz linear chirp signal, and cross-correlation-based peak detection. In this case, the maximum ranging error obtained across the entire volume was about 3.3 mm. Instead, for larger D, outside the favorable regions shown by the numerical simulations, the ranging error increases up to 14.6 mm.

Many applications and services based on ultrasonic positioning systems can benefit from the presented simulation tool.

Author Contributions: Conceptualization, R.C.; data curation, R.C., M.M., and D.I.; formal analysis, R.C.; investigation, R.C., M.M., D.I., and F.G.D.C.; methodology, R.C.; software, R.C., M.M., and D.I.; supervision, R.C.; writing—original draft, R.C.; and writing—review and editing, R.C., M.M., D.I., and F.G.D.C. All authors have read and agreed to the published version of the manuscript.

Funding: This research received no external funding.

Acknowledgments: PAC Calabria 2014-2020, Asse Prioritario 12, Azione 10.5.12 is gratefully acknowledged by one of the authors (D.I.).

Conflicts of Interest: The authors declare no conflicts of interest.

References

1. Zafari, F.; Gkelias, A.; Leung, K.K. A survey of indoor localization systems and technologies. *IEEE Commun. Surv. Tutor.* **2019**, *21*, 2568–2599. [CrossRef]
2. Zhang, D.; Xia, F.; Yang, Z.; Yao, L. Localization Technologies for Indoor Human Tracking. In Proceedings of the 5th International Conference on Future Information Technology, Busan, Korea, 21–23 May 2010; pp. 1–6.
3. Przybyla, R.J.; Tang, H.Y.; Shelton, S.E.; Horsley, D.A.; Boser, B.E. 3D ultrasonic gesture recognition. In Proceedings of the 2014 IEEE International Solid-State Circuits Conference Digest of Technical Papers, San Francisco, CA, USA, 9–13 February 2014; Volume 57, pp. 210–211.
4. Paredes, A.; Alvarez, F.J.; Aguilera, T.; Villadangos, J.M. 3D indoor positioning of UAVs with spread spectrum ultrasound and time-of-flight cameras. *Sensors* **2018**, *18*, 89. [CrossRef]
5. Fedele, R.; Praticò, F.; Carotenuto, R.; Della Corte, F.G. Instrumented infrastructures for damage detection and management. In Proceedings of the 2017 5th IEEE International Conference on Models and Technologies for Intelligent Transportation Systems (MT-ITSO), Naples, Italy, 26–28 June 2017. [CrossRef]
6. Seco, F.; Jimenez, A.R.; Prieto, C.; Roa, J.; Koutsou, K. A survey of mathematical methods for indoor localization. In Proceedings of the 2009 IEEE International Symposium on Intelligent Signal Processing, Budapest, Hungary, 26–28 August 2009. [CrossRef]
7. Carotenuto, R.; Tripodi, P. Touchless 3D gestural interface using coded ultrasounds. In Proceedings of the 2012 IEEE International Ultrasonics Symposium, Dresden, Germany, 7–10 October 2012; pp. 146–149.
8. Filonenko, V.; Cullen, C.; Carswell, J. Indoor Positioning for Smartphones Using Asynchronous Ultrasound Trilateration. *ISPRS Int. J. Geo-Inf.* **2013**, *2*, 598–620. [CrossRef]
9. Ureña, J.; Hernández, Á.; García, J.J.; Villadangos, J.M. Acoustic Local Positioning with Encoded Emission Beacons. *Proc. IEEE* **2018**, *106*, 1042–1062. [CrossRef]
10. Carotenuto, R.; Merenda, M.; Iero, D.; della Corte, F.G. An indoor ultrasonic system for autonomous 3D positioning. *IEEE Trans. Instrum. Meas.* **2019**, *68*, 2507–2518. [CrossRef]
11. Sabatini, A.M.; Rocchi, A. Sampled baseband correlators for in-air ultrasonic rangefinders. *IEEE Trans. Ind. Electron.* **1998**, *45*, 341–350. [CrossRef]
12. Aldawi, F.J.; Longstaff, A.P.; Fletcher, S.; Mather, P.; Myers, A. A High Accuracy Ultrasound Distance Measurement System Using Binary Frequency Shift-Keyed Signal and Phase Detection. Available online: http://eprints.hud.ac.uk/id/eprint/3789/ (accessed on 28 April 2020).
13. Saad, M.M.; Bleakley, C.J.; Dobson, S. Robust high-accuracy ultrasonic range measurement system. *IEEE Trans. Instrum. Meas.* **2011**, *60*, 3334–3341. [CrossRef]
14. De Marziani, C.; Ureña, J.; Hernández, A.; García, J.J.; Alvarez, F.J.; Jiménez, A.; Pérez, M.C.; Villadangos, J.M.; Aparicio, J.; Alcoleas, R. Simultaneous round-trip time-of-flight measurements with encoded acoustic signals. *IEEE Sens. J.* **2012**, *12*, 2931–2940. [CrossRef]

15. Carotenuto, R. A range estimation system using coded ultrasound. *Sens. Actuators A-Phys.* **2016**, *238*, 104–111. [CrossRef]
16. Carotenuto, R.; Merenda, M.; Iero, D.; della Corte, F.G. Ranging RFID tags with ultrasound. *IEEE Sens. J.* **2018**, *18*, 2967–2975. [CrossRef]
17. Pérez, M.C.; Serrano, R.S.; Ureña, J.; Hernández, A.; de Marziani, C.; Alvarez, F.J. Correlator implementation for orthogonal CSS used in an ultrasonic LPS. *IEEE Sens. J.* **2012**, *12*, 2807–2816.
18. Jackson, J.C.; Summan, R.; Dobie, G.I.; Whiteley, S.M.; Pierce, S.G.; Hayward, G. Time-of-flight measurement techniques for airborne ultrasonic ranging. *IEEE Trans. Ultrason. Ferroelect. Freq. Control* **2013**, *60*, 343–355. [CrossRef] [PubMed]
19. Kinsler, L.; Frey, A.; Coppens, A.; Sanders, J. *Fundamentals of Acoustics*; John Wiley & Sons, Inc.: New York, NY, USA, 2000; ISBN 0-471-84789-5.
20. Gualda, D.; Ureña, J.; García, J.J.; Jiménez, A.; Pérez, M.C.; Ciudad, F. Coverage analysis of an ultrasonic local positioning system according to the angle of inclination of the beacons structure. In Proceedings of the IEEE IPIN, Pisa, Italy, 30 September–3 October 2019; pp. 1–7.
21. Jensen, J.A. Field: A program for simulating ultrasound systems. *Med. Biol. Eng. Comp.* **1996**, *34*, 351–352.
22. Field II Home Page. Available online: https://field-ii.dk/ (accessed on 28 April 2020).
23. Tupholme, G.E. Generation of acoustic pulses by baffled plane pistons. *Mathematika* **1969**, *16*, 209–224. [CrossRef]
24. Stepanishen, P.R. The time-dependent force and radiation impedance on a piston in a rigid infinite planar baffle. *J. Acoust. Soc. Am.* **1971**, *49*, 841–849. [CrossRef]
25. Stepanishen, P.R. Transient radiation from pistons in an infinite planar baffle. *J. Acoust. Soc. Am.* **1971**, *49*, 1629–1638. [CrossRef]
26. Jensen, J.A.; Svendsen, N.B. Calculation of pressure fields from arbitrarily shaped, apodized, and excited ultrasound transducers. *IEEE Trans. Ultrason. Ferroelectr. Freq. Control* **1992**, *39*, 262–267. [CrossRef]
27. Carotenuto, R.; Merenda, M.; Iero, D.; della Corte, F.G. Mobile Synchronization Recovery for Ultrasonic Indoor Positioning. *Sensors* **2020**, *20*, 702. [CrossRef] [PubMed]
28. Bass, H.E.; Sutherland, L.C.; Zuckerwar, A.J.; Blackstocks, D.T.; Hester, D.M. Atmospheric absorption of sound: Further developments. *J. Acoust. Soc. Am.* **1995**, *97*, 680–683. [CrossRef]
29. García, E.; Holm, S.; García, J.J.; Ureña, J. Link Budget for Low Bandwidth and Coded Ultrasonic Indoor Location Systems. In Proceedings of the Indoor Positioning and Indoor Navigation, Guimaraes, Portugal, 21–23 September 2011.
30. Specification of Ultrasonic Transducer Type MA40S4S. Available online: https://www.murata.com/products/productdata/8797589340190/MASPOPSE.pdf?1450668610000 (accessed on 28 April 2020).
31. Air Ultrasonic Ceramic Transducers: 328ST/R160. Available online: http://www.prowave.com.tw/english/products/ut/open-type/328s160c.htm (accessed on 28 April 2020).
32. Ens, A.; Reindl, L.M.; Bordoy, J.; Wendeberg, J.; Schindelhauer, C. Unsynchronized ultrasound system for TDOA localization. In Proceedings of the in International Conference on Indoor Positioning and Indoor Navigation (IPIN)), Busan, Korea, 27–30 October 2014; pp. 601–610.
33. Della Corte, F.G.; Merenda, M.; Bellizzi, G.G.; Isernia, T.; Carotenuto, R. Temperature Effects on the Efficiency of Dickson Charge Pumps for Radio Frequency Energy Harvesting. *IEEE Access* **2018**, *6*, 65729–65736. [CrossRef]
34. Medina, C.; Segura, J.C.; de la Torre, Á. Ultrasound Indoor Positioning System Based on a Low-Power Wireless Sensor Network Providing Sub-Centimeter Accuracy. *Sensors* **2013**, *13*, 3501–3526. [CrossRef] [PubMed]

MDPI

St. Alban-Anlage 66

4052 Basel

Switzerland

Tel. +41 61 683 77 34

Fax +41 61 302 89 18

www.mdpi.com

Sensors Editorial Office

E-mail: sensors@mdpi.com

www.mdpi.com/journal/sensors